S0-BXD-940

# Stratospheric Ozone Depletion/ UV-B Radiation in the Biosphere

# NATO ASI Series

## Advanced Science Institutes Series

*A series presenting the results of activities sponsored by the NATO Science Committee, which aims at the dissemination of advanced scientific and technological knowledge, with a view to strengthening links between scientific communities.*

The Series is published by an international board of publishers in conjunction with the NATO Scientific Affairs Division

| | |
|---|---|
| A Life Sciences<br>B Physics | Plenum Publishing Corporation<br>London and New York |
| C Mathematical and Physical Sciences<br>D Behavioural and Social Sciences<br>E Applied Sciences | Kluwer Academic Publishers<br>Dordrecht, Boston and London |
| F Computer and Systems Sciences<br>G Ecological Sciences<br>H Cell Biology<br>I Global Environmental Change | Springer-Verlag<br>Berlin Heidelberg New York<br>London Paris Tokyo Hong Kong<br>Barcelona Budapest |

## NATO-PCO DATABASE

The electronic index to the NATO ASI Series provides full bibliographical references (with keywords and/or abstracts) to more than 30000 contributions from international scientists published in all sections of the NATO ASI Series. Access to the NATO-PCO DATABASE compiled by the NATO Publication Coordination Office is possible in two ways:

- via online FILE 128 (NATO-PCO DATABASE) hosted by ESRIN, Via Galileo Galilei, I-00044 Frascati, Italy.
- via CD-ROM "NATO Science & Technology Disk" with user-friendly retrieval software in English, French and German (© WTV GmbH and DATAWARE Technologies Inc. 1992).

The CD-ROM can be ordered through any member of the Board of Publishers or through NATO-PCO, Overijse, Belgium.

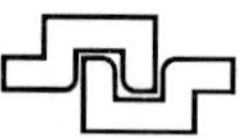

Series I: Global Environmental Change, Vol. 18

# Stratospheric Ozone Depletion/ UV-B Radiation in the Biosphere

Edited by

R. Hilton Biggs
Margaret E. B. Joyner

Horticultural Science Faculty
University of Florida
P. O. Box 110 692
Gainesville, FL 32611-0692
USA

Springer-Verlag
Berlin Heidelberg New York London Paris Tokyo
Hong Kong Barcelona Budapest
Published in cooperation with NATO Scientific Affairs Division

Proceedings of the NATO Advanced Research Workshop on Stratospheric Ozone Depletion/UV-B Radiation in the Biosphere, held in Gainesville, Florida, USA, June 14–18, 1993

ISBN 3-540-57810-2 Springer-Verlag Berlin Heidelberg New York
ISBN 0-387-57810-2 Springer-Verlag New York Berlin Heidelberg

CIP data applied for


Printed in Germany

Typesetting: Camera ready by authors
31/3130 - 5 4 3 2 1 0 - Printed on acid-free paper

To

Norton D. Strommen

scientist, humanist

# CONTENTS

# PREFACE

On June 14-18, 1993, we held a NATO Advanced Research Workshop on Stratospheric Ozone Depletion/UV-B Radiation in the Biosphere at Gainesville patterned after The Gordon Research format. Other agencies supporting the Workshop were NIGEC, EPA, and USDA.

The purpose of the Workshop was to assess the latest data available in the basic scientific disciplines associated with the problem of changes in stratospheric ozone and how they relate to changes in UV-B irradiance and how both relate to biology.

There are many layers of amplifiers, modulators, enhancers, stabilizers, and receptors, ranging from a change in ozone in the stratosphere to effects on biology, that can result in measurable changes. The major emphasis was on an in-depth review of the scientific databases to assist in the assessment of meaningful areas of research that need a better database for predictive models at all scientific levels from specific research models to global climate change models. Knowing what is presently available allows us to ask such questions as:

What are the interactive factors between the various research elements;
In the area of critical measurements and instrumentations, what measurements are needed to both validate ozone depletion and monitor ultraviolet flux changes in the biosphere;
How, when, and where will an integrated improved database be achieved, what levels and depth of data should we try to achieve, and, finally, how and by whom can the database be used.

These issues were addressed at the various working sessions and are incorporated as core concepts in these proceedings. In essence, each scientific presentation was well edited by the best process--scientific peer discussions.

To accomplish the tasks set for the workshop, there was an attendance by 107 researchers and observers from thirteen countries. The NATO countries represented were Great Britain, Germany, France, Sweden, the Netherlands, Italy, Portugal, Greece, Canada, and the United States. Other countries represented included New Zealand, the Philippines, and Argentina.

In the estimation of the director and the organizing committee and from testimonials by participants, the workshop accomplished and even exceeded the task intended. Groundwork was laid and cooperation was apparent between scientists working in both related and disparate areas and progress was made to further an international monitoring network, both from space and from ground level. Identified were databases that already exist and the ways to access these data, particularly the data that apply to health effects and efforts are being made to provide a better data base for biological assessments, including crops. These are the proceedings from that workshop:

In some sessions, tutorial lectures provided the cornerstone for discussion and the data provided to the participants stimulated discussion on the best-fit. These are incorporated in the workshop summaries.

It is hoped by the Director and the organizing committee that these Proceedings will add to the body of knowledge in this vital area.

WORKSHOP ORGANIZING COMMITTEE AND ADVISORS

R. Hilton Biggs, Co-Director
Joseph B. Knox, Co-Director
C. Bruce Baker
William F. Barnard
Jerry Elwood
Jim Gibson
Robert Saunders
Norton D. Strommen
Jan C. van der Leun
Robert Worrest

## Acknowledgments

The idea for a comprehensive multidisciplinary workshop grew out of a previous workshop oriented toward agricultural problems associated with an increase in UV-B radiation in the biosphere. It took the support and dedication of Drs. Joe Knox and Norton Stromnmen to bring it to fruition. We were committed to the idea that Dr. Stromnmen, the representative from NATO, stated so well: "This special NATO program fosters the development of strong interdisciplinary collaboration to investigate in depth the interactions and feedback between different components of the global earth system." To him and Dr. Knox, I give my personal thanks, for we accomplished our task. To the workshop participants, the verification of a job well done has been the subsequent workshop-stimulated activities that are international in scope, covering many of the problem areas addressed in the workshop. To you, also, thanks.

My special thanks go to Margaret Joyner, the Workshop Organizer and Proceedings Co-Editor, for her thoroughness and good-natured persistence, and to Beverley Booker who kept us on target financially and Judy Kite who adroitly handled the post-workshop details.

Finally, I would like to express profound thanks and appreciation to my wife, Dr. Susan V. Kossuth, for her dedication and help, and to my daughters, Robin, Dixie, and Mary, for silent forbearance and encouragement. This was not easy, since I transited to Emeritus status prior to the Conference.

Dr. R. Hilton Biggs

*Professor Emeritus, Plant Physiology and Biochemistry*
*Institute of Food and Agricultural Sciences*
*University of Florida*
*Gainesville, FL 32611*

November 1993

# STRATOSPHERIC OZONE DEPLETION AND CHANGES IN UV-B RADIATION

# TROPOSPHERIC CHEMISTRY CHANGES DUE TO INCREASED UV-B RADIATION

S. Madronich and C. Granier

National Center for Atmospheric Research, Boulder, Colorado 80307-3000

Key words: modeling, nitrous oxides, ozone depletion, photodissociation reactions, stratospheric ozone depletion, sulfur dioxide, tropospheric chemistry, UV-B radiation

## ABSTRACT

UV radiation is a principal driving force of tropospheric chemistry. Increases in tropospheric UV, calculated to result from stratospheric $O_3$ depletion, are likely to increase the reactivity of the troposphere, and thus change the concentrations of key species such as OH, $CH_4$, $O_3$, and $H_2O_2$. The geographical distributions of other species, such as $SO_2$ and $NO_x$, are also likely to be altered. Further study is needed for quantitative assessment of these indirect chemical effects of $O_3$ depletion.

## INTRODUCTION

Ultraviolet-B radiation (280-320 nm, or 3.88-4.43 eV) is sufficiently energetic to break the bonds of the key atmospheric trace gases ozone ($O_3$), nitrogen dioxide ($NO_2$), hydrogen peroxide ($H_2O_2$), formaldehyde ($CH_2O$), nitric acid ($HNO_3$), and so on. The resulting molecular fragments, O, NO, H, OH, HCO, and eventually $HO_2$ and organic peroxy radicals among others, are highly reactive and, together with their precursors, control the oxidizing (or self-cleaning) capacity of the lower atmosphere.

Stratospheric $O_3$ depletion is now apparently well under way (Gleason et al., 1993, and references therein), so that more UV-B radiation is penetrating to the troposphere and a general increase in chemical reactivity may be expected. However, the magnitude of these effects is difficult to calculate due to the highly non-linear nature of atmospheric chemistry, and the still poor understanding of the complex feedbacks which couple the chemistry of the troposphere to the stratosphere above and the biosphere below.

## INCREASES IN PHOTODISSOCIATION RATE COEFFICIENTS

Photodissociation reactions are of the general form

$$AB + h\nu \rightarrow A + B \qquad J,\ s^{-1}$$

with the reaction rate coefficient, or J value, is given by the expression (Madronich, 1987):

$$J = * F(\lambda)\ \sigma(\lambda)\ \phi(\lambda)\ d\lambda$$

where $\lambda$ is the wavelength (nm), $F(\lambda)$ is the spectral actinic flux (quanta $cm^{-2}\ s^{-1}\ nm^{-1}$), $\sigma(\lambda)$ is the molecular absorption cross section ($cm^2$ $molec^{-1}$), and $\sigma(\lambda)$ is the photodissociation quantum yield (molec $quanta^{-1}$). The spectral actinic flux in the UV-B region is a strong function of the total ozone column.

Table 1 shows some of the more important UV-driven reactions, and the sensitivity to stratospheric $O_3$ depletion computed for one set of conditions. The sensitivities for different conditions may vary some-

NATO ASI Series, Vol. I 18
Stratospheric Ozone Depletion/
UV-B Radiation in the Biosphere
Edited by R. H. Biggs and M. E. B. Joyner

Table 1: Key tropospheric photodissociation reactions and their sensitivity to stratospheric ozone depletion.

| | Reaction | s(J;Dobs)[a] |
|---|---|---|
| R1. | $O_3 + h\nu \rightarrow O_2 + O(^1D)$ | 1.83 |
| R2. | $HNO_3 + h\nu \rightarrow NO_2 + OH$ | 1.08 |
| R3. | $CH_3ONO_2 + h\nu \rightarrow CH_3O + NO_2$ | 0.93 |
| R4. | $CH_2O + h\nu \rightarrow H + HCO$ | 0.45 |
| R5. | $H_2O_2 + h\nu \rightarrow OH + OH$ | 0.36 |
| R6. | $CH_3OOH + h\nu \rightarrow CH_3O + OH$ | 0.30 |
| R7. | $CH_2O + h\nu \rightarrow H_2 + CO$ | 0.15 |
| R8. | $NO_3 + h\nu \rightarrow NO_2 + O(^3P)$ | 0.06 |
| R9. | $NO_2 + h\nu \rightarrow NO + O(^3P)$ | 0.02 |
| R10. | $O_3 + h\nu \rightarrow O_2 + O(^3P)$ | 0.01 |
| R11. | $HONO + h\nu \rightarrow NO + OH$ | 0.00 |

[a] The sensitivity, s(J;Dobs), of a photodissociation rate coefficient (J) is approximately the % increase in J for a 1% reduction in the total ozone column (Dobs); see Madronich and Granier (1992) for an exact definition. Calculated values are for 20°N, May 15, 3.4 km, 280 DU, diurnal average.

what. Clearly, the most sensitive J value is for the reaction

$$O_3 + h\nu \rightarrow O(^1D) + O_2 \qquad \text{(R1)}$$

The increase in $J_1$ calculated from the 1979-1992 TOMS $O_3$ column data record (Gleason et al., 1993) is shown in Figure 1. The overall global $J_1$ increase has been about 5.0 ±0.6% over the 14 year data record, or about 0.36 ±0.04% per year. The changes have been slightly higher in the southern hemisphere (0.40 ±0.05% per year) than in the northern hemisphere (0.32 ±0.05% per year).

## TROPOSPHERIC SELF-CLEANING CAPACITY

Translating changes in $J_1$ to changes in atmospheric chemical composition is not straightforward because tropospheric chemistry is highly non-linear. Of special interest are the concentrations of powerful oxidizers such as $O_3$, $H_2O_2$, and especially OH radicals, since these are primary agent for the removal of various gases from the troposphere. The complexity of the chemical response to UV changes may be appreciated through a highly simplified but illustrative sequence of reactions. Tropospheric $O_3$ comes partly from downward transport from the stratosphere, and, in the presence of nitrogen oxides ($NO_x = NO + NO_2$), from *in situ* photochemical production by the reaction sequence

$$NO_2 + h\nu\ (\lambda < 400\text{ nm}) \rightarrow NO + O(^3P) \qquad \text{(R9)}$$

$$O(^3P) + O_2 \rightarrow O_3 \qquad \text{(R12)}$$

The UV photolysis of $O_3$,

$$O_3 + h\nu\ (\lambda < 320\text{ nm}) \rightarrow O(^1D) + O_2 \qquad \text{(R1)}$$

is followed by quenching of the excited $O(^1D)$ atoms and their recombination with $O_2$ (R12) to give no net $O_3$ change. But some 1-10% of the time (depending on the relative humidity), an alternate fate of the $O(^1D)$ atoms is the reaction

$$O(^1D) + H_2O \rightarrow OH + OH. \qquad \text{(R13)}$$

which is therefore simultaneously a net loss of $O_{32}$ as

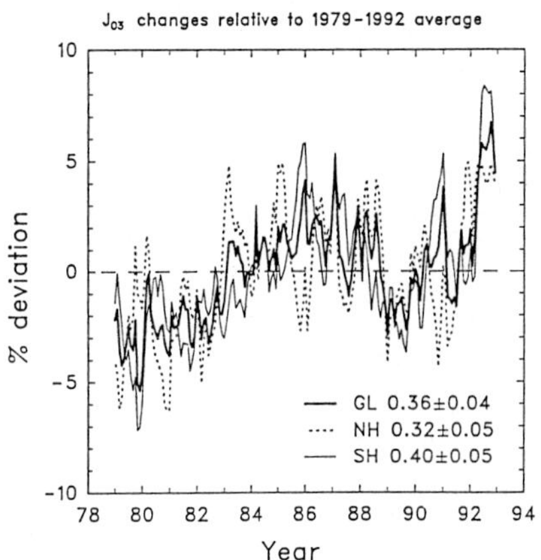


Figure 1. Change in mid-tropospheric rate coefficients for the photolysis reaction $O_3$ + hν g $O(^1D) + O_2$ from 1979 through 1992, computed from changes in the ozone column over the same period. Values are given as monthly deviations from the corresponding 1979-1992 averages. Thick solid curve is the area-weighted global average, thin solid line and dotted line are respectively the northern hemisphere and southern hemisphere area-weighted averages. Values next to legend are linear trends, and their corresponding uncertainties, expressed as percent per year relative to the 1979 intercept.

well as the primary production of two OH radicals. These OH radicals are rapidly converted to $HO_2$ radicals by

$$OH + O_3 \rightarrow HO_2 + O_2 \qquad (R14)$$

$$OH + CO + O_2 \rightarrow HO_2 + CO_2 \qquad (R15)$$

and by analogous reactions with $CH_4$, $SO_2$, and numerous other atmospheric trace gases. Rapid cycling between OH and $HO_2$ is maintained by

$$HO_2 + O_3 \rightarrow OH + 2O_2 \qquad (R16)$$

$$HO_2 + NO \rightarrow OH + NO_2. \qquad (R17)$$

Collectively, the $HO_x$ radicals ($HO_x$=OH+$HO_2$) are removed from this cycle by termination reactions

$$HO_2 + HO_2 \rightarrow H_2O_2 + O_2 \qquad (R18)$$

$$HO_2 + OH \rightarrow H_2O + O_2 \qquad (R19)$$

$$OH + NO_2 \rightarrow HNO_3 \qquad (R20)$$

where the soluble products may be removed from the gas phase by cloud water and precipitation. Alternatively, they may photolyze (R2 and R5 in Table 1) to regenerate the radicals. Finally, it should be noted that R16 represents an additional loss of $O_3$, while R17 (followed by R12) is a net production of ozone.

The effect of stratospheric $O_3$ depletion on such tropospheric reaction sequences is complex. A smaller stratospheric $O_3$ reservoir could result in lower $O_3$ input from the stratosphere to the troposphere, though changes in the character of cross-tropopause transport could also occur. Here, we consider only changes in tropospheric UV, which can increase both production and destruction of tropospheric $O_3$ and therefore $HO_x$. The sign and magnitude of the net effect is a complex function of the concentrations of various compounds (esp. $O_3$, $H_2O$, $H_2O_2$, $HNO_3$, and $NO_x$), and may well be different for different chemical regimes (e.g., polluted vs. pristine). Accurate estimates of the tropospheric response to UV changes are still a matter of active research, and the results appear to be sensitive to model formulation. Liu and Trainer (1988) used a simple box model (chemistry only, neglecting any atmospheric transport) to derive the changes in $O_3$ and OH as a function of prescribed $NO_x$ concentrations. Figure 2 shows that increased UV results in less $O_3$ for $NO_x$ levels less than about 100 ppt, and more $O_3$ at higher $NO_x$ levels due to the increased importance of reaction R17. Measured trends in surface $O_3$ at the south pole, where $NO_x$ is consistently low, do indeed suggest a reduction of $O_3$ due to UV increases associated with the $O_3$ hole (Schnell et al., 1991). Figure 3 shows that OH con-

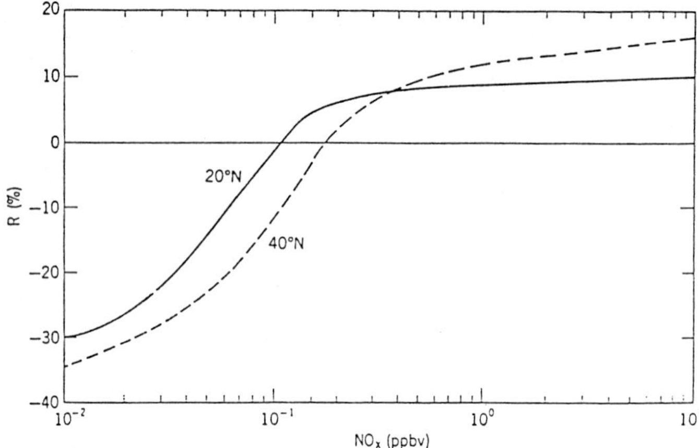


Figure 2. Predicted change in surface ozone concentration resulting from UV increases associated with a 20% reduction in column ozone, for summer conditions at indicated latitudes. From Liu and Tranier (1988).

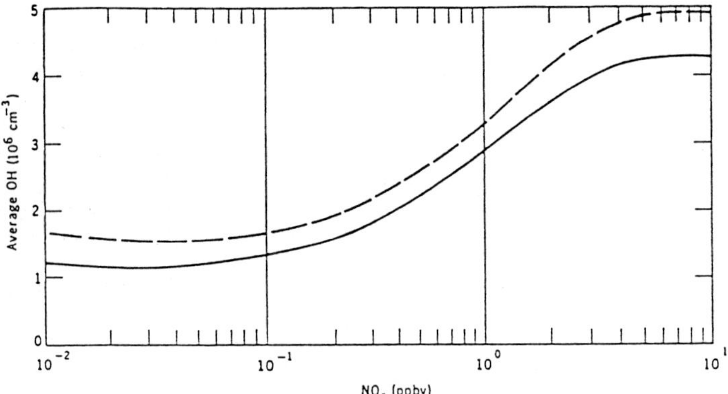


Figure 3. Surface OH concentration at 40°N, summer (solid line). Dashed line shows OH increase resulting from UV increases associated with a 20% reduction in column ozone. From Liu and Tranier (1988).

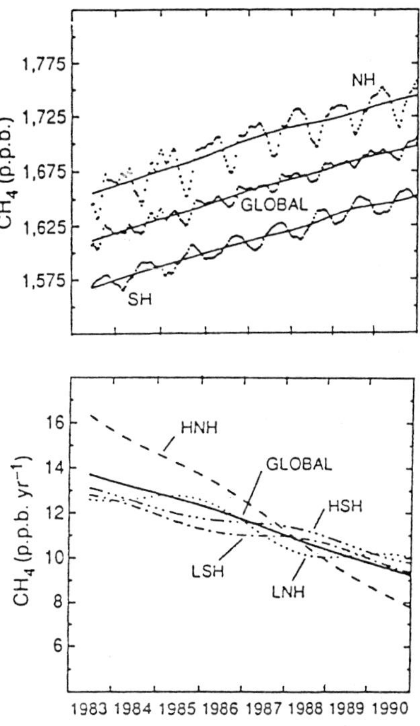


Figure 4. Upper panel gives measured atmospheric methane concentrations over 1983-1990. Lower panel gives smoothed trends for high and low latitudes of each hemisphere. From Steele et al. (1991).

centrations increase at all $NO_x$ levels, but the sensitivity of OH change to $O_3$ column change, s(OH;Dobs), or more directly to $J_1$ changes s(OH;$J_1$ ), ranges from about 0.9 at low $NO_x$ to 0.3 at high $NO_x$ consistent with the UV-sensitive reaction R1 being a more important source of OH than reaction R17 when $NO_x$ is low. Different results were obtained by Thompson et al. (1990) using a one-dimensonal model for different chemical regimes. Their results show a decrease in tropospheric $O_3$ concentrations for all regions considered, and imply values of s(OH;$J_1$ ) of 0.7 for urban mid-latitude regions and about 0.3 for marine low latitudes. Whether these differences arise from the inclusion of vertical transport or other model parameterizations remains unclear at this point. Recent preliminary results with a two-dimensional (vertical and latitudinal transport) model and a couple of three dimensional models (I.S.A. Isaksen, C. Granier, J. Lelieveld and J.P. Crutzen, per. comm., 1993) show globally and annually averaged values of s(OH;$J_1$ ) in the range 0.5-1.0. Thus it may be concluded that models currently predict that UV increases will increase the OH concentration, but the exact amount of the increase remains to be quantified.

## CHANGES IN $CH_4$ LIFETIMES

The importance of understanding the OH response to UV-B changes is illustrated by the oxidation of methane ($CH_4$),

$$OH + CH_4 \rightarrow CH_3 + H_2O \qquad (R21)$$

Sources of $CH_4$ are currently estimated at approximately 500 Tg $yr^{-1}$ while the major sink is thought to be reaction 21 with a lifetime of about 10 years. Pre-industrial atmospheric $CH_4$ concentrations were near 600 ppbv, but increases (most likely due to human activities) have brought the current value approaching 1,800 ppbv. Recent measurements of the rate of $CH_4$ increases (Figure 4) show that the increases have slowed in the past decade from about 14 ppbv $yr^{-1}$ to about 9 ppbv $yr^{-1}$. It has been proposed (Madronich and Granier, 1992) that the slowing of the $CH_4$ trend may be partly due to the increased OH which would be expected to result from tropospheric UV increases. Increases in OH concentrations will shorten the lifetime of $CH_4$ (and therefore its steady state value) by nearly the same proportion. The time-dependent response of $CH_4$ concentrations depends on the detail scenario for OH increase. Figure 5 shows the result of integrating forward in time the rate equation for $CH_4$,

$$d[CH_4]/dt = F_{CH4} - k_{21} [OH] [CH_4]$$

assuming that [OH] begins to increase at a rate of 0.36% per year in 1979 (i.e., at the same rate as the $J_1$ increase, and assuming for illustrative purposes that

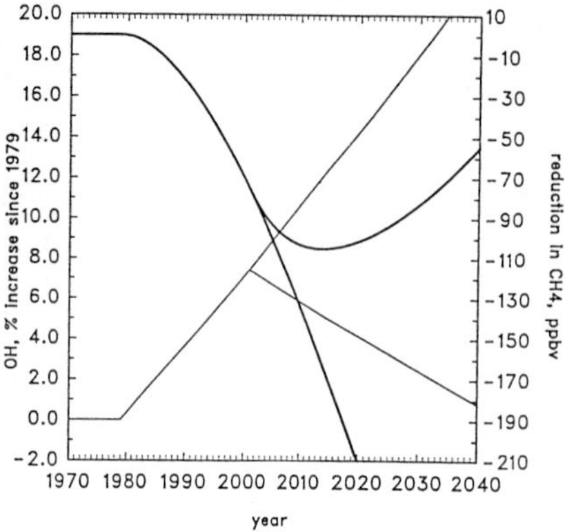


Figure 5. Reductions in instantaneous methane concentrations (thick curves, right scale) corresponding to UV-induced OH changes (thin lines, left scale). OH increases over 1979-2000 are estimated at 0.36 % $yr^{-1}$. Beginning in year 2000, two different scenarios are considered: OH is either allowed to increase at the same rate, or reduced at a rate of -0.18 % $yr^{-1}$.

$s(OH;J_1)$ is unity; for different values of $s(OH:J_1)$, the $CH_4$ reductions are approximately scaled in proportion). By 1993, the $CH_4$ is reduced by about 35 ppbv from the value it would have achieved without the OH increases, which may be compared with a measured atmospheric increase of about 150 ppbv over the same time (Khalil et al. 1993, Steele et al. 1992). Thus, it appears that a significant fraction of the slowing of the trend may be due to UV increases, and more detailed studies are under way to better quantify these effects.

## OTHER ATMOSPHERIC EFFECTS

Urban and regional scale air quality is another issue likely to be affected by increased UV levels. As has already been mentioned, most models predict increased photochemical smog formation (especially $O_3$) for highly polluted locations. Gery et al. (1987) have estimated that significantly more stringent regulation of non-methane hydrocarbon emissions will be required in urban U.S. locations, to maintain photochemical oxidants below air quality standards if UV levels are increased. De Leeuw and Van Rheineck Leyssius (1991) have similarly estimated that, given constant emissions, air quality violations will be much more frequent in European urban areas if UV levels are allowed to increase. Models also predict a substantial increase in $H_2O_2$ concentrations over larger geographical scales (Gery et al . 1987), which may have some impact on the geographical distribution of acid precipitation because $H_2O_2$ is a main agent in oxidizing S(IV) to S(VI) in cloud water. However, no quantitative estimates of these effects are yet available.

On the global scale, the UV-induced reductions of tropospheric $O_3$ and $CH_4$ (both of which are effective greenhouse gases) provide yet another way in which stratospheric $O_3$ depletion can affect the physical climate. Furthermore, stratospheric $O_3$ itself can be altered by UV-induced changes in tropospheric reactivity, since tropospheric gases may eventually find their way up to the stratosphere. Transport of $CH_4$ to the stratosphere provides a good example of this feedback, although even the sign of this effect is still poorly understood: stratospheric $CH_4$ is an effective sink for $O_3$-destroying Cl atoms, but is also a good source of water vapor, raising the possibility of enhanced formation of stratospheric particulates which are now known to reduce $O_3$ through heterogeneous processing. Likewise, the atmospheric residence time of CFC alternative compounds (HCFCs), and therefore the likelihood of their eventual transport to the stratosphere, is directly dependent on tropospheric OH levels.

The direct effects of UV increases on tropospheric chemistry are being investigated at several institutions, and it is likely that most major discrepancies which still exist between models will be resolved in the near future. There are, however, additional potential indirect effects which will require much further study before their importance can be assessed. Among the most difficult are those involving the biosphere-atmosphere exchange rates for various gases including $CO_2$, CO, COS, hydrocarbons, dimethyl sulfide, and organohalogens. Both terrestrial and aquatic ecosystems are likely to be affected by UV changes. For example, increased penetration of UV radiation into the ocean is expected to affect both living organisms (e.g., Smith et al. 1992) and non-living organic and

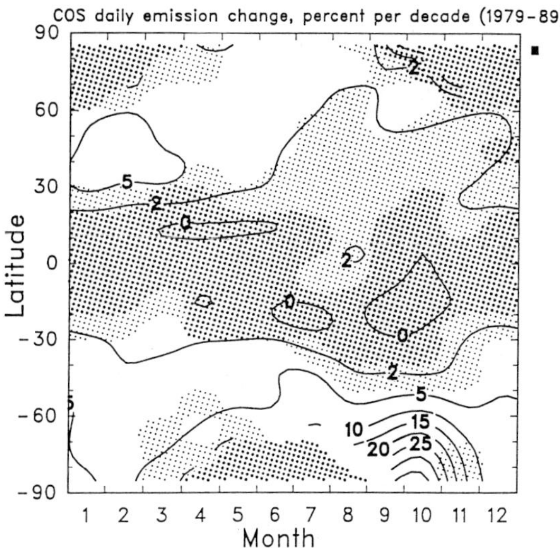


Figure 6. Trend in oceanic OCS emissions, due to increased UV radiation from $O_3$ depletion over 1979–1989. Contours are in % per decade, relative to the 1979 intercept values. Heavy shading indicates regions where trends differ from zero by less than one stadard devistion (1s), light shading by less than 2 s . Based on OCS production action spectrum of Zepp and Andreae (1990).

inorganic photochemical processing. An illustration of the magnitude of such potential effects is given in Figure 6, where the change in COS emissions has been estimated for the $O_3$ change between 1979 and 1989. COS is believed to be the primary source of sulfur atoms to the stratosphere, and of the resulting natural stratospheric sulfate aerosol layer. Current theories of stratospheric $O_3$ chemistry suggest that such aerosols may play a role in the destruction of $O_3$ , so that UV-induced increases in COS emissions may constitute a yet unquantified positive feedback on stratospheric $O_3$ depletion.

## CONCLUSIONS

Indisputable evidence for major changes in the chemical composition of the atmosphere has raised concerns about the quality of life in future decades. Global change scenarios of particular interest are those driven by human activities, above and beyond the natural trends and fluctuations that have brought the atmosphere to its present state. But the atmosphere is very complex, and numerous stratosphere-troposphere-biosphere linkages can be envisioned when the time scales of interest spans several decades. UV-induced changes are now among the best understood, but here too, major uncertainties remain.

For only a few of these uncertainties, resolution may be at hand through the use of more sophisticated computer models based on current understanding of the main processes. But most of the fundamental questions still arise from our lack of knowledge of basic physical, chemical, and biological processes, and of their response to those environmental parameters which are modified through other parts of the interacting atmosphere-surface system. While some coupling aspects can be reasonably addressed through improvements in numerical modeling, identification and understanding of the basic processes must still rely on well-supported experimental and intellectual scientific efforts.

**REFERENCES**

De Leeuw, F.A.A., and H.J. Van Rheineck Leyssius. 1991. Sensitivity of oxidant concentrations on changes in U.V. radiation and temperature. Atmos. Env., 25A:1025-1032.

Gery, M.W., R.D. Edmond, and G.Z. Whitten. 1987. Tropospheric Ultraviolet Radiation. U.S. Environmental Protection Agency report EPA/600/3-87/097.

Gleason J., P.K. Bhartia, J.R. Herman, R. McPeters, P. Newman, R.S. Stolarski, L. Flynn, G. Labow, D. Larki, C. Seftor, C. Wellemeyer, W.D. Komhyr, A.J. Miller, and W. Planet. 1993. Record low global ozone in 1992. Science 260:523-526.

Liu, S.C., and M. Tranier. 1988. Response of the tropospheric ozone and odd hydrogen radicals to column ozone change. J. Atmos. Chem. 6:221-233.

Khalil, M.A.K., R.A. Rasmussen, and F. Moraes. 1993. Atmospheric methane at Cape Meares: Analysis of a high-resolution data base and its environmental implications. J. Geophys. Res. 98:14753-14770.

Madronich, S. 1987. Photodissociation in the atmosphere 1. Actinic flux and the effects of ground reflections and clouds. J. Geophys. Res. 92:9740-9752.

Madronich, S., and C. Granier. 1992. Impact of recent total ozone changes on tropospheric ozone photodissociation, hydroxyl radicals, and methane trends. Geophys. Res. Lett. 19:465-467.

Schnell, R.C., S.C. Liu, S.J. Oltmans, R.S. Stone, D.J. Hoffman, E.G. Dutton, T. Deshler, W.T. Sturges, J.W. Harder, S.D. Sewell, M. Trainer, and J.M. Harris. 1991. Decreases of summer tropospheric ozone concentrations in Antarctica. Nature 351:726-729.

Smith, R.C., B.B. Prezelin, K.S. Baker, R.R. Bidigare, N.P. Boucher, T. Coley, D. Karentz, S. MacIntyre, H.A. Matlick, D. Menzies, M. Ondrusek, Z. Wan, and K.J. Waters. 1992. Ozone depletion: Ultraviolet radiation and phytoplankton biology in Antarctic waters. Science 255:952-959.

Steele, L.P., E.J. Dlugokencky, P.M. Lang, P.P. Tans, R.C. Martin, and K. A. Masarie. 1992. Slowing down of the global accumulation of atmospheric methane during the 1980s. Nature 358:313-316.

Thompson, A.M., M.A. Huntley, and R.W. Stewart. 1990. Perturbations to tropospheric oxidants, 1985-2035. J. Geophys. Res. 95:9829-9844.

Zepp, R.G. and M.O. Andreae. 1990. Photosensitized formation of carbonyl sulfide in sea water. *In* N.V. Blough and R.G. Zepp (eds.), Effects of Solar Ultraviolet Radiation on Biogeochemical Dynamics in Aquatic Environments. Woods Hole Oceanogr. Inst. Tech. Rep., WHOI-90-09, p.180.

# THE EFFECT OF CLOUDS ON ULTRAVIOLET RADIATION

W.F.J. Evans

Trent University, Peterborough, Ontario K9J 7B8 Canada

**ABSTRACT**

An observational program of the effects of clouds on ultraviolet radiation has been conducted in Southern ontario for the last three years. Generally when clouds are thick and especially when precipitation is occurring, the UV-B is strongly attenuated. When there are light clouds such as cirrus, the UV-B field can actually be enhanced by up to 25 percent. Rapid variations with enhancements have also been observed with light cumulus clouds.

Measurements with spatial variations showing a cloud edge enhancement effect are shown. This cloud edge enhancement effect can be explained by Mie scattering. It is proposed that jet contrails can actually increase the UV-B exposure levels near large airports.

NATO ASI Series, Vol. I 18
Stratospheric Ozone Depletion/
UV-B Radiation in the Biosphere
Edited by R. H. Biggs and M. E. B. Joyner

# EXPECTED CHANGES IN IRRADIANCES, PARTICULARLY UV-B, IN THE BIOSPHERE

E.C. Weatherhead

International Agency for Research on Cancer
150 Cours Albert Thomas
69372 Lyon Cedex 8, France

## ABSTRACT

Expected changes in UV radiation reaching the biosphere are fundamentally tied in with expected changes in stratospheric ozone. Predictions of changes in stratospheric ozone have a high degree of uncertainty associated with them. Present chemical models predict the decrease of a few percent over the next decade. Resultant trends in UV will be difficult to detect with present and currently proposed monitoring efforts. Changes in UV levels, however, are not limited to linear trend observations. Change in variance and episodic activity may also occur and may be detected by monitoring instruments.

Concern over possible changes in UV reaching the biosphere have brought about monitoring efforts of two types: efforts to monitor ozone concentrations and efforts to monitor absolute UV radiation directly. Both approaches are continued for the future and offer separate information about UV reaching the biosphere. There are benefits and limitations to both types of monitoring. Satellite monitoring of ozone allows for uniform assessment of world-wide changes in maximal UV doses. Radiative transfer models to estimate UV dosage based on ozone are reliable, however, only in clear sky conditions. Surface UV measurements allow direct measurement of biologically-active doses although devising instrumentation for UV monitoring is more difficult and offers more limitations than does ozone monitoring.

One of the most severe limitations to UV monitoring is that only a small number of geographic locations can be monitored. Aerosols, gaseous pollutants, and clouds affect the amount of UV reaching the biosphere. The effects of gaseous pollutants may alter observed UV trends by several percent per decade. Local changes in clouds can alter observed trends at a single location even more. Trends in the amounts of tropospheric pollutants, aerosols, and clouds can introduce trends in UV measurement that cannot be extrapolated to a broad geographic region, nor can they be extrapolated for predictions on future UV levels.

General conclusions cannot be drawn from UV measurements from a single location. Assessment of global changes in UV must be make from the aggregation of many monitoring sites. The use of co-located measurements of tropospheric aerosols, gases, and clouds will help develop radiative transfer modeling so that estimates may be made on world-wide changes in UV radiation.

NATO ASI Series, Vol. I 18
Stratospheric Ozone Depletion/
UV-B Radiation in the Biosphere
Edited by R. H. Biggs and M. E. B. Joyner

# UV RADIATION AS RELATED TO THE GREENHOUSE EFFECT

L.H. Allen, Jr.

U.S. Department of Agriculture, Agricultural Research Service
University of Florida, Gainesville, Florida 32611

Key words: aerosol, CFC, greenhouse effect, HCFC, HFC, ozone, refrigerant, UV radiation

**ABSTRACT**

The Greenhouse Effect is the term applied to the warming of the earth's surface that results from the fact that atmospheric gases have a greater transparency for incoming solar radiation than for outgoing longwave thermal radiation emitted by the earth's surface. Ultraviolet (UV) radiation contributes little directly to this whole global radiation balance. First, UV radiation is a small fraction of extraterrestrial solar radiation, and second, much of it is absorbed in the stratosphere in comparison with other radiation components. However, ozone ($O_3$) and chlorofluorocarbons (CFCs) are greenhouse effect gases. Ozone is important because it absorbs radiation in the middle of the "atmospheric window," within the 8 to 12 micrometer wavelength bandwidth where many of the other standard greenhouse-effect gases (water vapor, $CO_2$, $CH_4$, $N_2O$) have little effect. Tropospheric and stratospheric ozone play a role in the greenhouse effect. CFCs are important because 1) they have been recently introduced into the earth's atmosphere, 2) they have a strong warming potential for each increment added to the atmosphere, 3) they will continue to be released to the atmosphere from emissions from closed cell foams and fugitive refrigerant gases. CFC substitutes may have their own greenhouse effect impacts.

## INTRODUCTION

The extraterrestrial irradiance from the sun is about 1373 W $m^{-2}$ (Fröhlich, 1977). The effective incident radiation at the top of the atmosphere over a 24-hour cycle is one-fourth of this amount, about 340 W $m^{-2}$. Of this amount, about 103 W $m^{-2}$ is reflected back to space (assuming a planetary albedo of 0.3) and about 240 W $m^{-2}$ is absorbed at the earth's surface or in the atmosphere (Duxbury et al., 1993). This same amount of energy, 240 W $m^{-2}$, must be reradiated to maintain the earth in thermal balance. Most of the direct reradiation of energy to space from the surface of the earth would have to pass through the atmospheric window (the 8 to 12 µm bandwidth with little atmospheric absorption except by ozone). Figure 1 shows the emission spectrum of a perfect radiator at 300°K, which is representative or the temperature of the earths surface. The greenhouse effect gases of the atmosphere, $H_2O$ vapor, $CO_2$, $CH_4$, $N_2O$, $O_3$, and now the chlorofluorocarbons or CFCs, absorb part of the emitted thermal radiation and delay the loss of radiant energy from the earth's surface, and thus raise the temperature of the earth's surface. Clouds affect the earth's radiation balance in two ways. First, during daylight hours, clouds reflect solar radiation back to space. Secondly, day or night, clouds intercept outgoing long wavelength radiation and add to the greenhouse effect. The overall process of trapping of energy near the earth's surface by atmospheric gases and cloud cover is known as the greenhouse effect.

The present-day average global surface temperature is about 15°C which is about 33°C warmer than the effective 254°K temperature for emitting 240 W $m^{-2}$ of energy back to space. (For computation of effective surface temperature for reradiation of energy back to space, see p. 290, Wallace and Hobbs, 1977.)

NATO ASI Series, Vol. I 18
Stratospheric Ozone Depletion/
UV-B Radiation in the Biosphere
Edited by R. H. Biggs and M. E. B. Joyner

## CONTRIBUTION OF UV-B RADIATION TO THE EARTH'S RADIATION BUDGET

The solar constant is about 1373 W $m^{-2}$ (Fröhlich, 1977). The UV portion of the extraterrestrial solar energy is small compared to the total. The UV-A extraterrestrial irradiance (320 to 400 nm) is about 77.3 to 87.8 W $m^{-2}$ based on three sets of spectral data summarized by Pierce and Allen (1977). The UV-B extraterrestrial irradiance (280 to 320 nm is about 20.7 to 22.4 W $m^{-2}$ based on data by Pierce and Allen (1977) and Thekaekara (1974) as summarized in Heath and Thekaekara (1977). Thus, UV-A represents about 5.6% to 6.4% of the solar irradiance at the top of the atmosphere, and UV-B represents about 1.5 to 1.7% of total solar irradiance. The UV-C irradiance (less than 280 nm) is only about 10 W $m^{-2}$, or about 0.7% of extraterrestrial solar irradiance. Wavelengths below about 250 nm are absorbed by $O_2$ in the stratosphere, and ozone absorbs strongly below about 310 nm. Above this wavelength, the atmosphere is relatively transparent to beyond 1200 nm except for a strong $O_2$ absorption peak at 760 nm (Figure 1).

The percentages of solar UV radiation (290 to 400 nm) outside the atmosphere at zero air mass thickness, and at 1, 2, 3, and 5 air mass thicknesses (solar elevation angles of 90°, 30°, 19.47°, and 11.53°, respectively) are given in Table 1, based on data from the Handbook of Geophysics (1960). The decrease of solar radiation with decreasing solar elevation angle is much more pronounced at the shorter wavelengths as shown in Table 2.

The solar UV spectrum of wavelengths below 400 nm will contribute about 4% or less to the overall energy received from the sun at the earth's surface. Most of the energy at wavelengths of 320 nm or less will be absorbed by $O_3$ or $O_2$ in the stratosphere, and will contribute to elevated temperatures at the heights where the radiation is absorbed. Decreases in stratospheric ozone of 10% could contribute to only about 20% more biologically-effective UV-B radiation at the earth's surface. Since the total extraterrestrial UV-B spectrum (280 to 320 nm range) is about 22 W $m^{-2}$ out of 1373 W $m^{-2}$, and such a small percentage arrives at the earth's surface (only about 3% of 300 nm radiation at a 90° solar elevation angle, then only

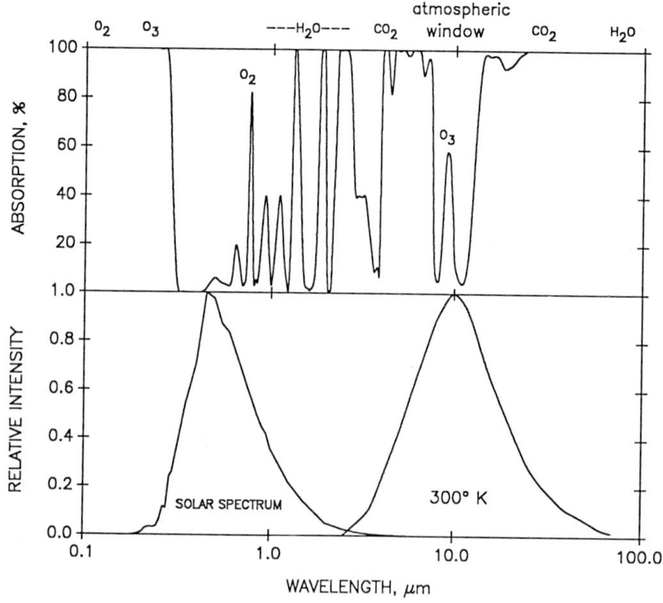


Figure 1. Absorption of radiation by the earth's atmosphere across the spectrum of 0.1 to 100.0 μm (from Uchijima, 1963) with absorption ranges of $O_2$, $O_3$, $H_2O$, and $CO_2$ indicated, and the relative intensity of the solar spectrum (from Thekaekara et al., 1969 and Thekaekara, 1974) in comparison with a perfect radiator at 300°K.

Table 1. Irradiance normal to sun's rays outside the atmosphere and at sea level, calculated for different air masses and wavelength bandwidths.[a]

| Bandwidth (μm) | Air mass 0 | 1 | 2 (Watts $m^{-2}$) | 3 | 5[b] |
|---|---|---|---|---|---|
| 0.29 - 0.40 | 94.6 (7.2)[c] | 40.1 (4.3) | 19.8 (2.7) | 10.0 (1.6) | 2.7 (0.6) |
| 0.40 - 0.70 | 540.0 (40.8) | 419.7 (45.2) | 327.8 (44.3) | 258.6 (42.6) | 163.7 (38.1) |
| 0.70 - 1.1 | 365.4 (27.6) | 309.2 (33.3) | 267.5 (36.2) | 233.4 (38.4) | 181.5 (42.2) |
| 1.1 - 1.5 | 162.5 (12.3) | 95.3 (10.3) | 70.7 (9.6) | 57.0 (9.4) | 40.7 (9.5) |
| 1.5 - 1.9 | 72.8 (5.5) | 50.8 (5.5) | 45.1 (6.1) | 41.0 (6.7) | 35.2 (8.2) |
| 1.9 - 10.0 | 86.8 (6.6) | 12.8 (1.4) | 9.2 (1.2) | 7.5 (1.2) | 5.8 (1.4) |
| Total | 1322.1 | 927.9 | 739.8 | 607.5 | 429.6 |
| Total normal to earth % outside the atmosphere | | 927.9 (70.2) | 369.9 (28.0) | 202.5 (15.2) | 85.9 (6.5) |

[a]Handbook of Geophysics (1960), The MacMillian Co., New York. Chapter 16, p. 16-18.
[b]Solar elevation angles for air mass thicknesses of 1, 2, 3, and 5, are 90°, 30°, 19°28', and 11°32', respectively.
[c]Values in parenthesis are the percentages with respect to the column total irradiance.

about 0.66 W $m^{-2}$ arrives at the earth's surface as solar UV-B energy. A 20% increase represents a change of about 0.12 W $m^{-2}$. Therefore, the maximum effect of a loss of 10% ozone represents a small change in irradiance at the earth's surface for only a small time of day when the sun approaches the zenith. The effect of a small incremental loss of ozone on the input of solar energy to the the earth's surface-troposphere system is likely to be greater than the estimate above because solar irradiance increases with increasing wavelength across the UV-B spectrum, while the absorptivity of ozone decreases across this same bandwidth. However, changes in UV radiation due to stratospheric ozone depletion should have a small effect on the radiation balance of the earth. This effect will be discussed later in consideration of the terrestrial radiation changes as well as the solar radiation changes.

## IMPACT OF CHLOROFLUOROCARBONS ON THE GREENHOUSE EFFECT

A serious question is the effect of chlorofluorocarbons (CFCs) on potential global warming. Before discussing this effect, we will examine the status of CFCs based on studies sponsored by the Alternative Fluorocarbons Environmental Acceptability Study (AFEAS).

The major CFCs based on production reported to AFEAS for the period 1970-1991 are CFC-11, CFC-12, CFC-113, HCFC-22, CFC-114, and CFC-115. Table 3 lists the chemical name and gives the chemical formula of these compounds.

The CFCs are used for many purposes, including refrigeration and air conditioning systems, blowing agent for closed-cell foam and open-cell foam, as an aerosol propellant, and in cleaning and drying. Also,

Table 2. Comparison at two wavelengths of total UV irradiance received normal to the sun's rays outside the earth's atmosphere (zero air mass) and at sea level normal to the earth's surface for various solar elevation angles and corresponding air mass thicknesses.[a]

| | Air mass thickness | | | | |
|---|---|---|---|---|---|
| | 0 | 1 | 1.15 | 2.92 | 5.76 |
| | Solar elevation angle | | | | |
| Wavelength | | 90° | 60° | 20° | 10° |
| --- nm --- | mW $m^{-2}$ per nm | | | | |
| 330 | 1027 | 624 (61)[b] | 505 (49) | 117 (11) | 41.2 (4) |
| 300 | 514 | 16.3 (3.2) | 8.4 (1.6) | 0.05 (0.01) | 0.01 (0.002) |

[a]Calculated from UV irradiance data of Thekaekara et al. (1969) and Thekaekara (1974); using solar radiation transmission functions of Shettle and Green (1974) for 0.28 atm-cm thickness of ozone.
[b]Values in parenthesis are percentages of the energy flux density with respect to zero air mass thickness (extraterrestrial radiation).

Table 3. Major CFCs based on production reported to AFEAS for the period 1970-1991, and potential alternative HCFCs and HFCs. (Adapted from PAFT, 1992, and several other sources.)

| Commercial name | Chemical name | Chemical formula | Boiling point (°C) |
|---|---|---|---|
| CFC-11 | trichlorotrifluoromethane | $CCl_3F$ | 23.8 |
| CFC-12 | dichlorodifluoromethane | $CCl_2F_2$ | -29.8 |
| CFC-113 | 1,1,2-trichlorotrifluoroethane | $CCl_2FCClF_2$ | 47.6 |
| CFC-114 | 1,2-dichlorotetrafluoroethane | $CClF_2CClF_2$ | 3.6 |
| CFC-115 | chloropentafluoroethane | $CClF_2CF_3$ | 39.0 |
| HCFC-22 | chlorodifluoromethane | $CHClF_2$ | -40.8 |
| HCFC-123 | 1,1-dichloro-2,2,2-trifluoroethane | $CHCl_2CF_3$ | 27.8 |
| HCFC-124 | 1-chloro-1,2,2,2-tetrafluoroethane | $CHClFCF_3$ | -12.1 |
| HCFC-141b | 1,1-dichloro-1-fluoroethane | $CH_3CCl_2F$ | 32.0 |
| CCFC-142b | 1-chloro-1,1-difluoroethane | $CH_3CClF_2$ | -9.8 |
| HCFC-225ca | dichloropentafluoropropane (isomer) | $CHCl_2CF_2CF_3$ | 50.0 |
| HCFC-225cb | dichloropentafluoropropane (isomer) | $CHClFCF_2ClF_2$ | 56.0 |
| HFC-32 | difluoromethane | $CH_2F_2$ | -51.9 |
| HFC-125 | pentafluoroethane | $CHF_2CF_3$ | -48.3 |
| HFC-134a | 1,1,1,2-tetrafluoroethane | $CH_2FCF_3$ | -26.5 |
| HFC-152a | 1,1-difluoroethane | $CH_3CHF_2$ | -24.0 |

Table 4. Total Sales of CFCs by Category for 1976-91 from Reporting Companies. (Adapted from AFEAS, 1992.)

Metric Tons x $10^3$

| | ----------Long Lifetime--------- | | | ---------Short Lifetime--------- | | | | |
|---|---|---|---|---|---|---|---|---|
| | Refrigeration | Blowing Agent Closed Cell Foam | | Blowing Agent Open Cell Foam | | Aerosol Propellant | All Other Uses | Total Sales |
| CFC-11 | 419 | | 1,779 | 854 | | 1,566 | 278 | 4,895 |
| CFC-12 | 2,877 | | 451 | 193 | | 1,906 | 319 | 5,746 |
| CFC-113 | - | 23 | - | - | 2,037 | - | < 1 | 2,060 |
| CFC-114 | - | 23 | - | - | 150 | - | < 1 | 173 |
| CFC-115 | - | 138 | - | - | < 1 | - | < 1 | 139 |
| HCFC-22 | - | 1,883 | - | - | 159 | - | 0 | 2,042 |
| Total | | 7,593 | | | 6,865 | | 597 | 15,055 |

they may be used as heat transfer fluids, in extended foams, in closed-cell thermoset foams, and other minor uses.

The total sales of CFCs by commercial name and category of use from 1976-1991 is shown in Table 4. The production and sales of CFCs peaked about 1987 to 1988. Total sales in the peak years was 1307 x $10^3$ metric tons and was 842 x $10^3$ metric tons in 1991. Thus, overall CFC sales appeared to decrease to about 64% of the peak years uses by 1991. Most of the reduction in sales was represented by CFC-11, CFC-12, and CFC-113. HCFC-22 sales increased somewhat. However, the non-reported production in 1991 was a larger fraction of world production than in previous years, so the actual sales in 1991 may have been higher than 64% of peak year production.

Table 5 shows the total sales of CFCs by category for 1991 and the percentage of 1991 sales to peak year sales. This table shows that short lifetime annual sales in 1991 had decreased to only 40% of the peak years value. The sales of CFC-11 and CFC-12 for aerosol propellant uses had decreased to 10% and 17%, respectively, of their 1987 usage. Furthermore, the sales of CFC-11 and CFC-12 for aerosol propellant uses had decreased to 6% and 8% of the 1976 usage.

Most of the CFC production that has been cumulated over time has been released to the atmosphere. Table 6 shows that about 90% of the total cumulative production has been released to the atmosphere. Much of this material still resides in the atmosphere since the stratosphere is the only real sink.

The rate of increase of CFC-11 and CFC-12 was somewhat linear from 1977 to 1989 (Elkins et al., 1993). This rate of increase seems to be in proportion to the rate of production and release (AFEAS, 1992). Since 1989 there has been a trend toward leveling of the rate of increase of these two CFCs. If this trend continues, we can expect the concentration of these CFCs to reach a maximum before the year 2000, and begin to decline (Elkins et al., 1993).

The ozone depletion potential of the five major CFCs is high (Table 7). There are obvious reasons for conversion of the CFC usages to HCFCs or HFCs based on the potential for decrease of ozone. However, there should be other factors taken into consideration, namely the infrared absorption of CFCs (and their potential substitutes, the HCFCs and the HFCs), the impacts of these gases on radiative forcing of climate, and the impacts of scenarios of the use of eachon greenhouse warming potential (GWP).

Table 5. Weight of the Total Sales of CFC Materials by Category for 1991 from Reporting Companies, and the Ratio of 1991 Sales to Peak Year Sales Expressed as a Percentage. (Adapted from AFAES, 1992.)

| | Metric Tons x $10^3$ for 1991 | | | | | | | |
|---|---|---|---|---|---|---|---|---|
| | -------Long Lifetime------- | | | ------Short Lifetime------- | | | | |
| | Refrigeration | | Blowing Agent Closed Cell Foam | Blowing Agent Open Cell Foam | | Aerosol Propellant | All Other Uses | Total Sales |
| CFC-11 | 20 | | 148 | 28 | | 11 | 7 | 213 |
| CFC-12 | 175 | | 19 | 1 | | 19 | 10 | 225 |
| CFC-113 | - | 1 | - | - | 146 | - | < 1 | 148 |
| CFC-114 | - | 2 | - | - | 4 | - | < 1 | 7 |
| CFC-115 | - | 12 | - | - | 0 | - | < 1 | 12 |
| HCFC-22 | - | 210 | - | - | 27 | - | 0 | 237 |
| Total 1991 | | 587 | | | 236 | | 17 | 842 |
| Peak Year | | 680 | | | 583 | | 44 | 1,307 |
| | ----------------------------(1991/Peak Year, %)---------------------------- | | | | | | | |
| CFC-11 (Peak-1987) | 72 | | 92 | 44 | | 10 | 32 | 56 |
| CFC-12 (Peak-1987) | 80 | | 31 | 16 | | 17 | 45 | 53 |
| CFC-113 (Peak-1989) | | 57 | | | 59 | | 143 | 59 |
| CFC-114 (Peak-1989) | | 84 | | | 35 | | 78 | 44 |
| CFC-115 (Peak-1989) | | 86 | | | - | | - | 86 |
| HCFC-22 (Peak-1989) | | 108 | | | 108 | | | 108 |
| Total | | 86 | | | 40 | | 39 | 64 |

Table 8 lists the absorption wavelength centers of several trace gases, including $CO_2$ and $H_2O$ vapor (Luther and Ellingson, 1985). This table provides more specific information on several of the CFCs and related compounds than given in Figure 1.

Radiative forcing functions have been derived for the radiatively-active trace gases as a function of present concentration and an incremental change in concentration:

$$\Delta F = f(C_o, C)$$

where F is the forcing in terms of W $m^{-2}$ for a change from an initial concentration, $C_o$, to an incremental

Table 6. CFC Production and Release Summary by Reporting Companies Cumulative to 1991. (Adapted from AFEAS, 1992.)

| | x$10^3$ Metric Tons | | | |
|---|---|---|---|---|
| | **Production** | **Released** | **Unreleased** | **% Released** |
| CFC-11 | 8,190 | 7,166 | 1,146 | 87% |
| CFC-12 | 10,631 | 10,116 | 781 | 95% |
| CFC-113 | 2,766 | 2,709 | 93 | 98% |
| CFC-114 | 492 | 475 | 24 | 97% |
| CFC-115 | 196 | 154 | 45 | 79% |
| HCFC-22 | 3,196 | 2,664 | 533 | 83% |
| Total | 25,471 | 23,284 | 2,622 | 91% |

NOTE: Released plus unreleased does not always equal production.

Table 7. Ozone depletion potential of CFCs, HCFCs, and HFCs calculated over their full lifetimes in the atmosphere. (Adapted from AFEAS, 1992.)

| CFC | | | | | | |
|---|---|---|---|---|---|---|
| 11 | 12 | 113 | 114 | 115 | | |
| 100 | 100 | 80 | 100 | 60 | | |
| **HCFC** | | | | | | |
| 22 | 123 | 124 | 141b | 142b | 225ca | 225cb |
| 5.5 | 2.0 | 2.2 | 11.0 | 6.5 | 2.5 | 3.3 |
| **HFC** | | | | | | |
| 32 | 125 | 134a | 152a | | | |
| 0 | 0 | 0 | 0 | | | |

Table 8. Absorption wavelength centers of several greenhouse-effect trace gases. (Recalculated from Luthur and Ellingson, 1985.)

| Trace Gas | Band Center, µm | | | | Mixing Ratio ppb |
|---|---|---|---|---|---|
| $N_2O$ | 7.78 | 8.56 | 16.98 | | 310 |
| $CH_4$ | 6.52 | 7.66 | | | 1,720 |
| $O_3$ | 4.70 | 9.07 | 9.61 | 14.27 | 20-100[a] |
| $CC\ell_3F$ (CFC-11) | 4.66 | 9.22 | 11.82 | | 280 |
| $CC\ell_2F_2$ (CFC-12) | 8.68 | 9.13 | 10.93 | | 484 |
| $CF_4$ | 7.93 | 8.06 | 15.82 | | |
| $CHC\ell F_2$ (HCFC-22) | 7.63 | 8.95 | | | 146 |
| $CC\ell_4$ | 12.89 | | | | |
| $CHC\ell_3$ | 8.20 | 12.92 | | | |
| $CH_2C\ell_2$ | 8.09 | 13.59 | 14.01 | | |
| $CH_3C\ell$ | 7.14 | 9.85 | 13.66 | | |
| $CH_3CC\ell_3$ | 9.23 | 14.14 | | | |
| $C_2H_2$ | 10.54 | | | | |
| $SO_2$ | 7.35 | 8.69 | 19.31 | | |
| $NH_4$ | 10.53 | | | | |
| $HNO_3$ | 5.90 | 7.50 | 11.76 | | |
| $CO_2$ | 4.30 | 15.0 | | | 360,000 |
| $H_2O$ | 6.4(5-8) | 12.5 | | | |

[a] Tropospheric ozone concentration

final concentration, C. Sometimes the forcing is expressed relative to $CO_2$ on a molecule per molecule basis. At today's atmospheric concentrations, an increase in $CH_4$, $N_2O$, and CFC-11 would produce a relative radiative forcing of 21, 206, and 12,400, respectively, on a molecule for molecule basis. The relative forcing for the CFCs and related gases range from 18,300 for CFC-114 to 6,590 for HFC-152a. Radiative forcing is computed in terms of the change in radiant energy flux density at the tropopause.

Of greater interest than the static molecule for molecule relative forcing is the time-integral of the effects of changes in concentration and radiative forcing of each of the greenhouse effect gases. The Greenhouse Warming Potential (GWP) has been defined as the ratio of the time-integral of the radiative forcing of a particular gas with respect to $CO_2$ as given below:

$$GWP = \int_0^n \Delta F_i C_i \, dt \; / \int_0^n \Delta F_{CO2} C_{CO2} \, dt$$

where n refers to the number of years.

Table 9 shows the GWP of various trace gases along with residence time and relative absorption capacity. The relative absorption capacity of CFCs is quite high relative to $CO_2$.

Table 10 provides an expansion of Table 9. In general, all HCFCs and HFCs offer some reduction in the GWP relative to CFCs. Much of the reduction in GWP is related to smaller estimated atmospheric lifetimes of the compounds.

The radiative forcing of greenhouse effect gases has been computed by Shine et al. (1990) over the past decades (Figure 2). These calculations show that, over the 1980-1990 decade, the radiative forcing of CFCs and HCFCs was second only to $CO_2$ itself (about 0.12 W $m^{-2}$ per decade for CFCs and HCFCs vs. about 0.30 W $m^{-2}$ per decade for $CO_2$). In terms of contributions of anthropogenetic to global warming US EPA (1990) estimated that CFCs contributed 17% to global warming, with 57, 15, 8, and 3% due to energy use and production, agricultural practices, land use modification, and other industrial sources, respectively. Thus, although UV radiation may have only a small effect on the greenhouse effect, the CFCs and related compounds released to the atmosphere is second only to rising levels of carbon dioxide. Wuebbles and Edmonds (1991) provide a summary of the greenhouse gases and impacts on global warming potential.

## CHANGES OF STRATOSPHERIC OZONE CONCENTRATION

The Total Ozone Mapping Spectrometer (TOMS) has been used to observe the amount and distribution of atmospheric ozone since November 1978 (Gleason et al., 1993). There has been a downward trend of stratospheric ozone estimated to be -0.27± 1.4% per year. Most of the depletion has occurred at mid and high latitudes, where reductions were estimated to be 3 to 5% and 6 to 8%, respectively, with essentially no losses near the equatorial latitudes. The September-October Antarctic springtime losses have reduced 1992 amounts to about 50% of the 1979 stratospheric ozone content over this area, creating the recurring "ozone hole." The eruption of Mount Pinatubo in the Philippines in June 1991 was followed by further ozone reductions in 1992 and continuing in 1993 (Gleason et al., 1993).

The effect of combined processes of increasing CFCs and reductions of stratospheric $O_3$ over the period 1979 to 1990 on radiative forcing of the surface-troposphere system have been assessed using the GFDL model and the Reading model (Ramaswamy et al., 1992). The increase of CFCs over that period was calculated to contribute an additional 0.10 W $m^{-2}$, whereas all other non-ozone gases ($CO_2$, $CH_4$, $N_2O$, stratospheric $H_2O$) were calculated to contribute 0.45 W $m^{-2}$ (Figure 3). Therefore, one point is that the use of CFCs has contributed greatly to the radiative forcing of climate during recent years.

Ramaswamy et al. (1992) further analyzed the effects of decreases in ozone concentration of the lower stratosphere during the 1979 to 1990 period. The decreases in ozone concentration has three distinct effects on the radiation balance. First, more solar radiation can reach the earth's surface and tropospheric gases. This factor should provide a global solar radiative forcing of about 0.10 W $m^{-2}$. Secondly, in the absence of decreases of stratospheric temperature, less long wave length radiation should be emitted into the surface-troposphere system, so this would reduce the effective net radiative forcing from 0.10 to 0.06 W $m^{-2}$. However, solar heating of the lower stratosphere should be reduced with less ozone, and this would further reduce the emission of radiation from the stratosphere into the troposphere. Ramaswamy et al (1992) found that this effect was large enough to yield a negative net radiative forcing due to the reduction in stratospheric ozone, with a

Table 9. Global warming potential (GWP) of various trace gases. (Adapted from Duxbury et al., 1993.)

| Gas | Residence time (yr) | Relative absorption capacity[a] (unit mass) | Relative GWP[b] 20 yr | 500 y |
|---|---|---|---|---|
| $CO_2$ | 120 | 1 | 1 | 1 |
| $CH_4$ | | | | |
| Direct[c] | 10 | 58 | 26 | 2 |
| Indirect | | | 37[d] | 7 |
| $N_2O$ | 150 | 206 | 270 | 190 |
| $CCl_3F$ (CFC-11) | 60 | 3970 | 4500 | 1500 |
| $CCl_2F_2$ (CFC-12) | 120 | 5750 | 7100 | 4500 |
| $CHCl_2F_2$ (HCFC-22) | 15 | 5440 | 4100 | 510 |
| $CHCl_2CF_3$ (HCFC-123)[e] | 1.6 | 2860 | 310 | 29 |

[a] Per unit mass change from present concentrations.
[b] Following addition of 1 kg of each gas.
[c] Direct = through direct radiative effect of $CH_4$; indirect = through the participation of $CH_4$ in chemical processes which lead to the formation of other radiatively active species.
[d] Estimated contributions of 24, 10, and 3 for generation of tropospheric $O_3$, stratospheric $H_2O$, and $CO_2$, respectively.
[e] One of the proposed CFC substitutes.

Table 10. Global Warming Potential (GWP) of carbon dioxide, methane, and various CFCs, HCFCs, and HFCs. (Adapted from AFEAS, 1992.)

| Compound | Estimated Atmospheric Lifetime | GWPs for Various Integration Time Horizons 20 years | 100 years | 500 years |
|---|---|---|---|---|
| Carbon Dioxide | * | 1 | 1 | 1 |
| CFC-11 | 55 | 4500 | 3400 | 1400 |
| CFC-12 | 116 | 7100 | 7100 | 4100 |
| CFC-115 | 550 | 5500 | 7000 | 8500 |
| HCFC-22 | 15.8 | 4200 | 1600 | 540 |
| HCFC-123 | 1.7 | 330 | 90 | 30 |
| HFC-125 | 40.5 | 5200 | 3400 | 1200 |
| HCFC-141b | 10.8 | 1800 | 580 | 200 |
| HCFC-225ca** | 2.7 | 610 | 170 | 60 |
| HCFC-225cb** | 7.9 | 2400 | 690 | 240 |
| HFC-134a | 15.6 | 3100 | 1200 | 400 |
| HFC-152a | 1.8 | 530 | 150 | 49 |
| Methane*** | 10.5 | 35 | 11 | 4 |

*The decay of carbon dioxide concentrations cannot be reproduced using a single exponential decay lifetime. Thus, there is no meaningful single value for the lifetime that can be compared directly with other values in this table.

**The HCFC-225ca/cb GWP values were calculated by Atmospheric and Environmental Research, Inc. and are based on rate constant measurements reported by Z. Zhang et al., Geophysical Research Letters, Vol. 18, January 1991, pp. 5-7, and the infrared energy absorption properties measured at Allied-Signal Central Laboratory.

***The GWP values include the direct radiative effect and the effect due to carbon dioxide formation, but do not include any effects resulting from tropospheric ozone or stratospheric water formed as methane decomposes in the atmosphere.

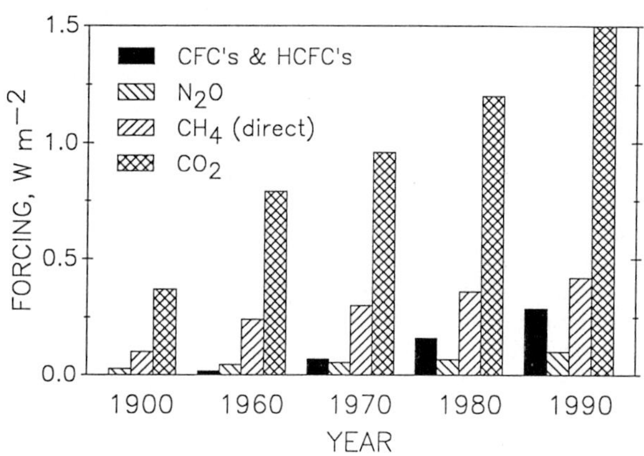


Figure 2. Changes in radiative forcing since 1765 to 1900, 1960, 1970, 1980, and 1990 by $CO_2$, $CH_4$ (direct), and CFCs & HCFCs. The indirect effect of $CH_4$ due to introduction of water vapor into the stratosphere is not shown, nor is the contribution of $N_2O$. (Derived from the data of Shine et al., 1990.)

value of -0.08 W m$^{-2}$ (Figure 3). Thus, these analysis shows that almost all of the positive climate forcing of added CFCs would be counteracted by the negative net radiative forcing due to a reduction in ozone in the lower stratosphere.

## SUMMARY

Ultraviolet radiation makes up only a small fraction of the total solar irradiance. The UV-B portion of extraterrestrial radiation is about 22.4W m$^{-2}$, or about 1.6% of total extraterrestrial irradiance. If the concentration of ozone decreases in the lower stratosphere, due to CFCs, then more radiant energy can pass through the stratosphere and arrive at the earth's surface, and thus cause a small amount of global warming. Furthermore, the CFCs also function as a greenhouse effect gas. However, recent studies predict that the loss of radiatively active ozone in the

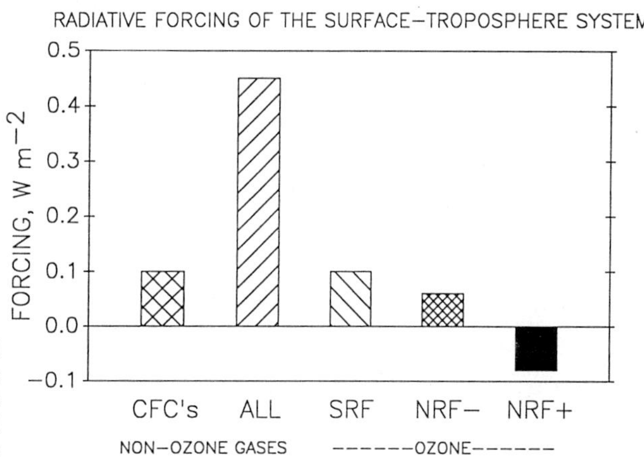


Figure 3. Contribution of CFCs, and the effects of ozone reductions due to CFCs, on the radiation forcing of the surface-troposphere system. (Derived from data of Ramaswamy et al., 1992.)

stratosphere would more than offset the radiative forcing of both direct solar heating and the greenhouse gas effect of the CFCs. Substitutes for CFCs (HCFCs and HFCs) have a shorter residence time and do not appear to have the potential to impact stratospheric ozone. The CFC substitutes may continue to contribute a small but significant amount to radiative forcing.

Although $CO_2$ will remain the primary greenhouse gas generated by industry and land use changes, the CFCs and CFC substitutes are likely to remain as important as methane and nitrous oxide in radiative forcing of the earth's climate. Direct UV-B radiation effects on climate are likely to decrease at some time in the future, unless volcanic eruptions inject massive amounts of chlorine into the stratosphere.

## REFERENCES


Alternative Fluorocarbons Environemntal Acceptability Study (AFEAS). 1992. Annual Report including Production, Sales and Atmospheric Release, & Summary Fact Sheets, September 1992. Washington, DC.

Duxbury, J.M., L.A. Harper, and A.R. Mosier. 1993. Contributions of Agroecosystems to Global Climate Change. p. 1-18. *In* L.A. Harper, A.R. Mosier, J.M. Duxbury, and D.E. Rolston (eds.), Agricultural Ecosystem Effects on Trace Gases and Global Climate Change. Proc. Symposium ASA-SSSA, 28 October 1991, Denver, CO. ASA Spec. Pub. No. 55. ASA-CSSA-SSSA, Madison, Wisconsin.

Elkins, J.W., T.M. Thompson, T.H. Swanson, J.H. Butler, B.D. Hall, S.O. Cummings, D.A. Fisher, and A.G. Raffo. 1993. Decrease in the growth rates of atmospheric chlorofluorocarbons 11 and 12. Nature 364:780-783.

Fröhlich, C. 1977. Contemporary measures of the solar constant. p. 93-109. *In* O. R. White (ed.) The Solar Output and its Variation. Colorado Associated University Press, Boulder, Colorado.

Gleason, J.F., P.K. Bhartia, J.R. Herman, R. McPeters, P. Newman, R.S. Stolarski, L. Flynn, G. Labow, D. Larko, C. Seftor, C. Wellemeyer, W.D. Komkyr, A.J. Miller, and W. Planet. 1993. Record Low Global Ozone in 1992. Science 260:523-526.

Handbook of Geophysics. 1960. p. 16-18, Chapter 16. The MacMillian Co., New York.

Heath, D.F., and M.P. Thekaekara. 1977. The solar spectrum between 1200 and 3000 Å. p. 193-212. *In* O. R. White (ed.) The Solar Output and its Variation. Colorado Associated University Press, Boulder, Colorado.

Luther, F.M., and R.G. Ellingson. 1985. Carbon dioxide and the radiation budget. p. 25-55. *In* M.C. MacCracken and F.M. Luther (eds.) Projecting the Climatic Effects of Increasing Carbon Dioxide. DOE/ER-0237, U.S. Department of Energy, Carbon Dioxide Research Division, Washington, DC.

Pierce, A.K., and R.G. Allen. 1977. The solar spectrum between .3 and 10 µm. p. 169-192. *In* O. R. White (ed.) The Solar Output and its Variation. Colorado Associated University Press, Boulder, CO.

Programme for Alternative Fluorocarbon Toxicity Testing (PAFT). 1992. Summary Fact Sheets, September 1992. Runcorn, Cheshire, United Kingdom.

Ramaswamy, V., M.D. Schwartzkopf, and K.P. Shine. 1992. Radiative forcing of climate from halocarbon-induced global stratospheric ozone loss. Nature 355:810-812.

Shettle, E.P., and A.E.S. Green 1974. Multiple scattering calculation of middle untraviolet reaching the ground. Appl. Optics 13:1567-1581.

Shine, K.P., R.G. Derwent, D.J. Wuebbles, and J-J. Morcrette. 1990. Radiative forcing of climate. p. 41-68. *In* J.T. Houghton, G.J. Jenkins, and J.J. Ephraums (eds.) Climate Change: The IPCC Scientific Assessment. World Meteorological Organization/United Nations Environment Prgramme, Intergovernmental Panel on Climate Change. Cambridge University Press, Cambridge, United Kingdom.

Thekaekara, M.P. 1974. Applied Optics 13:518-521.

Thekaekara, M.P., R. Krugewr, and C.H. Duncan. 1969. Applied Optics 8:1713-1716.

U.S. Environmental Protection Agency. 1990. Greenhouse gas emissions from agricultural systems. Vol. 1, Summary report, p. 19-24. *In* Proc. of IPCC-RSWG Workshop, Washington, DC. 12-14 Dec. 1989. USEPA Climate Change Div., Washington, DC.

Wallace, J.M., and P.V. Hobbs. 1977. Atmospheric science: An introductory survey. Academic Press, New York.

Wuebbles, D.J., and J. Edmonds. 1991. Primer on greenhouse gases. Lewis Publishers, Chelsea, Michigan.

# THERMOPHYSICAL PROPERTIES OF ENVIRONMENTALLY ACCEPTABLE REFRIGERANTS

C.A. Nieto de Castro,[1,3] U.V. Mardolcar,[2] and M.L. Matos Lopes[1]

[1]Faculdade de Ciências da Universidade de Lisboa, Departamento de Química and ICAT - Instituto de Ciência Aplicada e Tecnologia Campo Grande, Edif. C1- Piso 1, 1700 Lisboa, Portugal
[2]Departamento de Física, Instituto Superior Tecnico, 1096 Lisboa Codex, Portugal
[3]President of PROTECT

Key words: CFC, chlorofluorocarbons, HCFC, hydrochlorofluorocarbons, HFC, hydrofluorocarbons, refrigerants

**ABSTRACT**

It is well known that the generalized use of CFCs (chlorofluorocarbons) is one of the major causes contributing to the ozone depletion. Therefore it is highly recommended to banish these substances from use as soon as possible (1996), as agreed in the Copenhagen Meeting of November 1992.

The replacement of these compounds by HFCs (hydrofluorocarbons) should be gradual but easy and efficient. The implementation of HCFC (hydrochlorofluorocarbon) transition compounds should be restricted to the cases where their thermophysical properties will not necessitate the quick replacement of existing equipment such as new equipment for refrigeration, air conditioning, heat pumps, and expansion foams, as well as the construction of new plants for the manufacturing of new products. Important examples are HCFC 22 and its binary mixture with CFC 115, also known has R502.

It is of crucial importance to appeal for the governments, industry, and manufacturers of new equipment to invest in research of the thermophysical properties for the environmentally acceptable refrigerants. This investment should be channeled to university groups that have the expertise to obtain the best possible data in the shortest possible time.

To focus on this problems a Workshop was organized by PROTECT (**Pro**perties of Fluids **T**omorrow, **E**nvironmental **C**oncern **T**oday) in Ericeira, Portugal, 19-21 November 1992. The workshop gathered thirty-eight scientists from several countries, namely: Germany, the US, France, Greece, the Netherlands, Japan, Portugal, and the UK. They discussed the measurement and correlation of transport and equilibrium properties for the alternative refrigerants, as well as future international collaborations and their coordination.

In this contribution we will report the conclusions of this workshop as well as the research currently ongoing in Lisbon in this area.

## INTRODUCTION

Since the proposal in 1974 by Molina and Rowland [10] that chlorine atoms could be the leading element responsible for the destruction of the ozone layer, the replacement of CFCs has become a world issue. In the last five years there has been a considerable effort by many nations to move towards the use of refrigerant materials that are less hazardous to the Earth's environment than those currently in use. This initiative has produced an imprecedented levels of activity among members of the scientific community interested in the thermophysical properties of fluids. Naturally, this effort has been partially driven by the availability of

NATO ASI Series, Vol. I 18
Stratospheric Ozone Depletion/
UV-B Radiation in the Biosphere
Edited by R. H. Biggs and M. E. B. Joyner

funding to support the determination of the properties of fluids about which very little has been known so far. However, an altruistic motive behind the research has been apparent through the degree of international cooperation apparent among many of the laboratories concerned with these studies.

Knowledge of the thermophysical properties of the environmentally-acceptable refrigerants are necessary for the design and operation of equipment of the refrigeration and air conditioning industries, of heat pumps, and of expansion foams, as well for the construction of new plants for the manufacturing of new products.

Important examples these environmentally-acceptable refrigerants are HFC 134a, HFC 152a, HFC 32, HFC 141b, and HCFC 123 as well as the new generation of refrigerants based on the propane family.

Despite the intense efforts on both the national and international levels, an internationally-agreed set of values for any of the thermophysical properties of any of the new group of materials and, in particular, an agreed representation of them, is still lacking. From some properties, namely the transport properties, the effect of this intense activity has been to reinforce the view that the measurement of properties is only an exact science when there already exists a consensus about the result. The discrepancies between various independent measurements of the viscosity of some of the new refrigerants are as much as a factor of 10 more than the estimated measurements.

This situation of uncertainty led to the creation of PROTECT, **PRO**perties of Fluids **T**omorrow, **E**nvironmental **C**oncern **T**oday, an international group of scientists concerned with the current and future environmental impact of certain fluids, both on the atmosphere and on the hydrosphere, prepared to measure, predict, simulate, and correlate the thermophysical properties of those fluids and their possible replacements, presumed to be environmentally harmless. The group, founded in 1993, in Vienna during the 12th European Conference on Thermophysical Properties, has decide that its first meeting would be an workshop, dedicated to assessing the properties of hydroclorofluorocarbons (HCFCs) and hydrofluorocarbons (HFCs).

This workshop has therefore been organized to provide an opportunity for leading experts in the field to describe the current status of their work and their latest results. In addition, critical reviews have been presented of the experimental data already available, which form the basis of formulation of properties that may be put forward for international approval.

The workshop provided an informal atmosphere in which frank exchanges of views were encouraged in order to stimulate an elucidation of the problems that remain for the study of fluids already started in some detail. The same stimulus applied to the discussion of new measurements that are still required for the new materials and the pitfalls that were to be expected in the performance of those measurements, based on the recent experience.

It is the purpose of this paper to report on the conclusions of such a workshop. In addition, the paper will describe the work currently ongoing in Lisbon.

## THE SEARCH OF ALTERNATIVES TO CFCs

The numerous applications of CFCs in several fields of industry and products of domestic consumption create a difficult situation in the choice of alternatives. However, since 1988, several research projects where initiated to investigate the properties of the possible compounds that, we believe today, will not destroy the ozone layer and will not contribute significantly to global warming. The study of the usefulness of the new compounds is based on the determination of the following quantities:

ALT - atmospheric lifetime
ODP - ozone depletion potential
GWP - global warming potential
LFL - lower flammability limit
VLT - toxicity threshold limit value

The analysis of the possible alternatives to the CFCs suggests the existence of several possible compounds, but its application is limited by several factors [8]. We can illustrate these limitations with the case of refrigerants. The working fluid in a compressor must obey several requirements, from chemical and thermophysical properties, to safety and health and effects on the environment. These requirements can be summarized as follows [1,2,15]:
Chemical Properties

stable over long periods, no decomposition or deterioration;
noncorrosive of metals and other materials used in equipment;
nonreactive with lubricating oils in the presence of metals that could act as catalysts.

Thermophysical Properties
wide liquid range, with a low melting point;
adequate vapor pressure, normal boiling and critical points;
high heat of vaporization;
low heat capacity in the vapor phase;
low viscosity in both vapor and liquid phases;
high thermal conductivity in both vapor and liquid phases;
low surface tension.

Health, Safety, and Environmental Considerations
low toxicity and provoke no chronic diseases;
nonflammable and nonexplosive;
low ALT (<20 years);
zero ODP and low GWP.

Miscellaneous Considerations
lipophilic properties (low solubility in oils);
high dielectric constants and low dielectric losses;
easy leak detection;
low cost.

It is not easy to find a refrigerant that can fulfill all the above criteria. Some compromise can be made in the thermophysical properties. These compromises will depend on if we are going to use existing equipment with a new refrigerant or newly-designed equipment. However in the areas of Chemical Properties and Health, Safety, and Environmental Considerations, the degrees of freedom are small as we should not tolerate ecological disasters.

Figure 1 shows the triangular diagram developed by McLinden and Didion [8] where the vertices

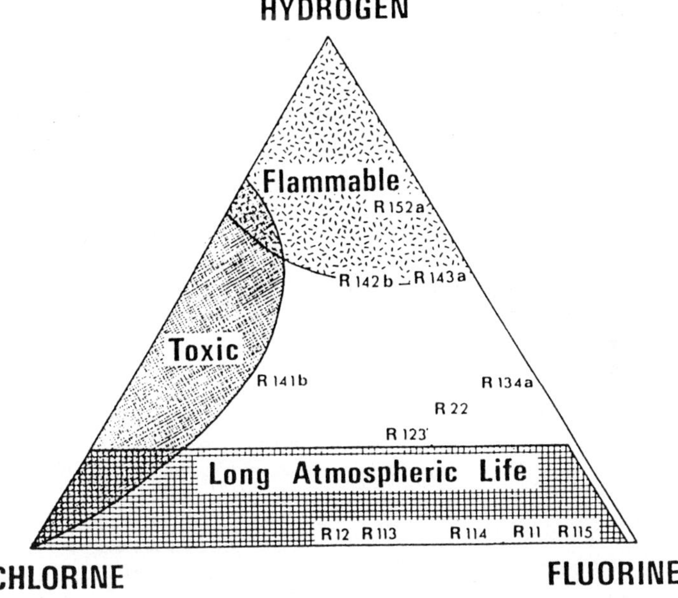


Figure 1. The triangular diagram for refrigerants choice. Adapted from McLinden and Didion, 1987.

Table 1. Summary of alternative refrigerants

| FLUID | FORMULA | MOLECULAR WEIGHT ($g\ mol^{-1}$) | NORMAL BOILING POINT(°C) | CRITICAL TEMPERATURE (°C) |
|---|---|---|---|---|
| HFC 23 | $CHF_3$ | 70.01 | -82.1 | 25.9 |
| HFC 32 | $CH_2F_2$ | 52.02 | -51.8 | 78.4 |
| HFC 125 | $CF_3CHF_2$ | 120.02 | -48.6 | 66.3 |
| HFC 143a | $CF_3CH_3$ | 84.04 | -47.4 | 73.1 |
| HCFC 22 | $CHClF_2$ | 86.47 | -40.8 | 96.2 |
| HFC 134a | $CF_3CH_2F$ | 102.03 | -26.1 | 101.1 |
| HFC 152a | $CHF_2CH_3$ | 66.05 | -24.2 | 113.3 |
| HFC 134 | $CHF_2CHF_2$ | 102.03 | -19.4 | 119.0 |
| HCFC 124 | $CHClFCF_3$ | 136.48 | -12.1 | 122.5 |
| HCFC 142b | $CClF_2CH_3$ | 100.50 | -9.3 | 137.1 |
| HCFC 123 | $CHCl_2CF_3$ | 152.93 | 27.9 | 183.8 |
| HCFC 141b | $CCL_2FCH_3$ | 116.95 | 32.0 | 204.7 |

correspond to each of the three elements, chlorine, hydrogen, and fluorine, that other than carbon constitute the molecule of the refrigerant*. We identify the toxicity limits (Cl, no F), flammability (high H content), long atmospheric life zone (low H), and the zone corresponding to the alternatives such as low CL and high H and F—the nonshadowed zone of the figure.

The best candidates are the compounds rich in F, with H and no Cl—the HFCs. We can select, by increasing order in normal boiling point, the compounds listed in Table 1. These fluids are also included in Table 2, grouped into classes according to their characteristics and environmental acceptability.[2] The properties necessary for the alternative refrigerants are the equilibrium properties such as density, vapor pressure, enthalpy, entropy, heat capacity, sound velocity, and dielectric constant, plus the transport properties such as viscosity and thermal conductivity, both for liquid and vapor phases, including superheated vapor.

[2] Recently the EPA (Environment Protection Agency) has identified eleven new compouds, seven derived from propane and two hidrofluroethers. In this last case the element oxygen has to be taken into account, specially because of the possible reactions with OH in the stratosphere.

The amount of data available for the thermophysical properties of these new refrigerants has increased substantially in recent years. From data available the summer of 1992 in the MIDAS data bank in the University of Stuttgart [6] there have been 393 publications about HFCs and 520 about HCFCs, making a total of 913 publications. The agreement between the data obtained in different laboratories and with different methodologies, however, do not agree as they do for other fluids, probably due to the following characteristics of these compounds:

high polarity (dipole moments up to 3.3 Debye);
extremely good solvent power;
difficulty of obtaining in purities greater than 99.5%;
strong absorbers of IR radiation.

These factors make the comparison of the results previously obtained very cumbersome, calling the attention of scientists to the need for international cooperation.

A survey conducted by the Advanced Heat Pump Program of the International Energy Agency, entitled Thermophysical Properties of Environmental Acceptable Refrigerants and published in 1991 by NIST [9], shows there was ongoing research in 15 countries

Table 2. Potentially environmentally-acceptable refrigerants shown with CFCs and $CO_2$ for comparison.

| Refrigerant | Formula | NBP (°C) | ALT (year) | ODP | GWP | LFL | TOX | Category |
|---|---|---|---|---|---|---|---|---|
| Ammonia | $NH_3$ | -33.3 | <1 | 0 | << | 15.0 | 25 | A |
| Propane | $C_3H_8$ | -42.1 | <1 | 0 | 3 | 2.1 | asphyx | A |
| n-Butane | $C_4H_{10}$ | -0.5 | <1 | 0 | 3 | 1.5 | 800 | A |
| i-Butane | $C_4H_{10}$ | -11.7 | <1 | 0 | 3 | 1.8 | 1000 | A |
| HFC 23 | $CHF_3$ | -82.1 | 310 | 0 | 12,000 | none | 1000 | A |
| HFC 32 | $CH_2F_2$ | -51.8 | 6 | 0 | 220 | 14.6 | ? | A |
| HFC 125 | $C_2HF_5$ | -48.6 | 28 | 0 | 860 | none | 1000 | A |
| HFC 134 | $C_2H_2F_4$ | -26.1 | 12 | 0 | ? | none | ? | A |
| HFC 134a | $C_2H_2F_4$ | -26.1 | 16 | 0 | 420 | none | 1000 | A |
| HFC 143a | $C_2H_3F_3$ | -47.4 | 41 | 0 | 1000 | 7.1 | ? | A |
| HFC 152a | $C_2H4F_2$ | -24.2 | 2 | 0 | 47 | 3.7 | 1000 | A |
| HFC 227 | $C_3HF_7$ | -17.3 | 30 | 0 | 110 | none | ? | A |
| HCFC 22 | $CHClF_2$ | -40.8 | 15 | 0.05 | 510 | none | 1000 | B |
| HCFC 123 | $C_2HCl_2F_3$ | 27.9 | 2 | 0.02 | 29 | none | 10 | B |
| HCFC 124 | $C_2ClF_4$ | -12.1 | 7 | 0.02 | 150 | none | 500 | B |
| HCFC 141b | $C_2H_3Cl_2F$ | 32.0 | 8 | 0.10 | 150 | 7.4 | 500 | B |
| HCFC 142b | $C_2H_3ClF_2$ | -9.3 | 19 | 0.06 | 540 | 6.9 | 1000 | B |
| CFC 11 | $CClF_3$ | 23.8 | 60 | 1 | 1500 | none | 1000 | C |
| CFC 12 | $CClF_2$ | -29.8 | 130 | 0.9 | 4500 | none | 1000 | C |
| CFC 13 | $CClF_3$ | -81.4 | 400 | 0.45 | ? | none | 1000 | C |
| CFC 113 | $C_2Cl_3F_3$ | 47.6 | 90 | 0.8 | 2100 | none | 1000 | C |
| CFC 114 | $C_2Cl_2F4$ | 3.8 | 200 | 0.7 | 5500 | none | 1000 | C |
| CFC 115 | $C_2ClF_5$ | -39.8 | 400 | 0.35 | 7400 | none | 1000 | C |
| Carbon dioxide | $CO_2$ | sublimes | 120 | 0 | 1 | none | 5000 | |

on 673 different projects for the refrigerants shown in Table I. Since then several other groups have started research in this field.

## THE WORKSHOP AT ERICEIRA, PORTUGAL

### GENERAL CONCLUSIONS

1) The workshop participants highly recommend to statesmen and politicians that CFCs, namely CFC 11 and CFC 12, be banished from use as soon as possible, that is, within a period of three to five years.

2) The replacement of these compounds by HFCs should be gradual but easy and efficient. The implementation of HCFC transition compounds should be restricted to the cases where their thermophysical properties will not necessitate the quick replacement of existing equipment such as new equipment for refrigeration, air conditioning, heat pumps, and expansion foams, as well as the construction of new plants for the manufacturing of new products. Important examples are HCFC 22 and its binary mixture with CFC 115, also known has R502.

3) It was decided to appeal to governments, in-

dustry, and manufacturers of new equipment to invest in research of new thermophysical properties for the environmentally-acceptable refrigerants. This investment should be channeled to university groups that have the expertise to obtain the best possible data in the shortest possible time.

4) It was decided that PROTECT, in conjunction with other organizations, would be responsible for alerting the public to the dangers of using CFCs. PROTECT should also take the responsiblity for creating a new awareness for the protection of the environment from other harmful compounds in addition to CFCs.

5) It was decided to collaborate with governments, industry, and manufacturers of equipment to attempt to solve the problem of the replacement of CFCs together.

6) The scientists were pleased with the results of the Copenhagen meeting which took place simultaneously to theirs, and were hopeful about its results.

7) Considering the increases in world population and the concomitant increases in energy consumption implied by these increases together with efforts to achieve a better way of life for the people of the world, changes in the environment are inevitable. PROTECT can contribute actively to solutions of these kinds of problems by obtaining reliable data to predict and simulate the global changing in the environment and by producing good data for the thermophysical properties necessary for an efficient use of the energy.

SPECIAL CONCLUSIONS

1) The properties of individual HFCs, namely HFC 32, HFC 125, HFC 134a, and HFC 152a, as well as their binary mixtures should be measured. The measurements should be systematic and as rigorous as possible. It is important to determine the following properties:

Equilibrium Properties: density, vapor pressure, entalphy, entropy, heat capacity, sound velocity, surface tension, and dielectric constant, plus their solubility in various oils.

Transport Properties: viscosity, thermal conductivity, and diffusion in air.

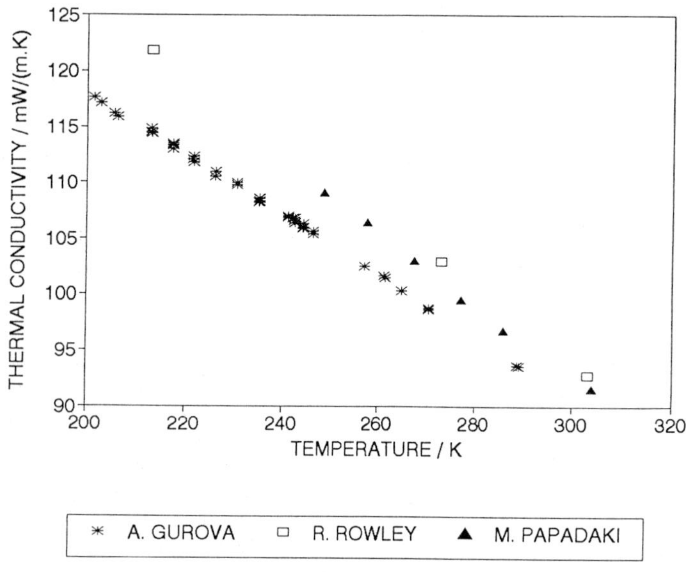


Figure 2. The thermal conductivity of HCFC 141b [3]. Comparison with the results obtained by Papadaki et al. [11] and Rowley et al [12].

Other: the range to cover of temperature and pressure should not be restricted to industrial needs. The purity of the samples used should be checked before and after the measurements.

2) The international cooperation of this group should be enlarged by establishing contacts with the International Energy Agency (EIA Annex 18) and the International Institute of Cold (IIF). The countries that do not belong to these institutions should be encouraged to do so and to collaborate in a systematic way, lending financial support in the international effort. It was decided to enlarge the PROTECT board by electing a president (Prof. Carlos Nieto de Castro from the University of Lisbon, Portugal) and a secretary (Prof. Akira Nagashima from the University of Keio, Japan) and by establishing a plan of action.

3) It was recommended to the members of the group that the pressures of industry to obtain new data should not be a reason for decreasing the quality of the measurements and correlations developed.

4) It was decided to publish the workshop proceedings, including the extracts of the scientific discussions that took place, as well as these conclusions.

## RESEARCH IN LISBON

The Portuguese group started its research in this field in 1990, collaborating with B. le Neindre and R. Tufeu of the LIMHP-CNRS, Villetaneuse [13,14], and R. Perkins of NIST-Boulder [7] in the fields of density and thermal conductivity measurement. The research in Lisbon concentrates on accurate measurement of the thermal conductivity and dielectric constant at temperatures from 170 to 370°K, and pressures to 20 MPa, and in prediction and correlation of densities of pure compounds and their mixtures.

The measurement of thermal conductivity is made using the polarized transient hot wire technique [4], while the dielectric constant uses the direct capacitance method [3]. The results, soon to be published [5], are presented in Figure 2, including results obtained for the thermal conductivity of HCFC 141b, at the pressure of 0.1 MPa, between 200 and 300°K. These results are compared to the only other results available, and it is shown that the discrepancies vary from 2% at the higher temperatures to 10% at 210°K. These results illustrate the difficulties in measuring the properties of these compounds, all polar and difficult to obtain in a pure state. During the workshop at Ericeira the possibility was discussed of performing measurements of thermal conductivity in ten samples from the same manufacturer, analyzed before and after the measurements. These samples, prepared by ICI UK, will be distributed to ten laboratories around the world and measurements will be performed through the end of 1993. The results of these measurements should be available shortly.

The discussion about the thermophysical properties of environmentally acceptable refrigerants has not yet finished. In the summer of 1993, during the 13th European Conference on Thermophysical Properties to be held in Lisbon August 30–September 3, another workshop on refrigerants will be held. The number of papers on refrigerants submitted to this second workshop covers about twenty percent of all contributions received a significant increase compared to the few papers presented just three years ago in Vienna. In our opinion, this reflects the importance that both science and industry, helped by governments, are giving to the problem of ozone-layer destruction by chlorine atoms. The PROTECT group hopes to continue this effort to eradicate all the CFCs now in use and replace them with HCFCs as soon as possible.

**Acknowledgments**

The authors would like to thank all the participants in the workshop at Ericeira as well as R. Krauss, University of Stuttgart. We would also like to thank the organizing committee of this NATO ARW for having invited this presentation, and NATO for the financial support. The work in Lisbon is supported by JNICT and the authors would like to express their gratitude.

## REFERENCES

1. Anon. 1985. 1995 ASHRAE Handbook, Fundamentals Volume, Chapter 16, ASHRAE, Atlanta, GA, USA.

2. Anon. 1991. Thermophysical properties of environmentally acceptable fluorocarbons—HFC 134a and HCFC 123, Japanese Association for Refrigeration and Japan Flon Gas Association.

3. Barao, M.T., U.V. Mardolcar, and C.A. Nieto de Castro. 1993. Dielectric constant of alternative refrigerants. First National Meeting on Physical Chemistry, Lisbon, June 1993.

4. Gurova, A.N., U.V. Mardolcar, and C.A. Nieto de Castro. 1993. Thermal conductivity of alternative refrigerants. First National Meeting on Physical Chemistry, Lisbon, June 1993.

5. Gurova, A.N., M.T. Barao, U.V. Mardolcar, and C.A. Nieto de Castro. 1993. The thermal conductivity and dielectric constant of HCFC 141b and HCFC 123. Submitted to High Temperatures–High Pressures. Paper to be presented at the 13th European Conference on Thermophysical Properties to be held in Lisbon, August 30–September 3, 1993.

6. Krauss, R. 1992. MIDAS Data Bank, University of Stuttgart, per. communic.

7. Laesecke, A., R.A. Perkins, and C.A. Nieto de Castro. 1992. Thermal conductivity of R134a. Fluid Phase Equilibria 80:263.

8. McLinden, M.O., and D.A. Didion. 1987. ASHRAE J. 32, December.

9. McLinden, M.O., V.M. Haynes, J.T.R. Watson, and K. Watanabe. 1991. A survey of current worldwide research on the thermophysical properties of alternative refrigerants, NISTIR 3969.

10. Molina, J.M., and F.S. Rowland. 1974. Nature 249:810.

11. Papadaki, M., M. Schmidt, A. Seitz, K. Stephan, B. Taxis, and W.A. Wakeham. 1993. Thermal conductivity of R134a and R141b within the temperature range 240–307°K at the saturation line. Int. J. Thermophys. 14:173.

12. R. Rowley et al. 1991.

13. Sousa, A.T., P.S. Fialho, C.A. Nieto de Castro, R. Tufeu, and B. le Neindre. 1992. The thermal conductivity of 1-chloro-1,1-difluoroethane HCFC 142b. Int. J. Thermophys. 13:383.

14. Sousa, A.T., P.S. Fialho, C.A. Nieto de Castro, R. Tufeu, and B. le Neindre. 1992. Density of HCFC 142b and of its mixture with HCFC 22. Fluid Phase Equilibria 80:213.

15. Threlkeld, J.L. 1990. *In* Thermal Environmental Engineering, 2nd Ed., Englewood Cliffs, NJ, Prentice Hall.

# PHYSIOLOGICAL AND BIOCHEMICAL CHANGES AS RELATED TO UV-B RADIATION

# PHYSIOLOGICAL CHANGES IN PLANTS RELATED TO UV-B RADIATION: AN OVERVIEW

Manfred Tevini

Botanical Institute II University of Karlsruhe,
Kaiserstraße 12, D-76128 Karlsruhe, Germany

## INTRODUCTION

Since local and global ozone depletion scenarios have given rise to the expectancy of increasing UV-B radiation (280-320 nm) on the earth's surface, biological impacts on plants have gained high scientific as well as political interest. Increases in UV-B radiation have already been observed in Antarctica during the ozone hole development as well as in the Northern hemisphere.[2-5]

The danger of increased UV-B to plants has been documented in the United Nations Environmental Program (UNEP) reports of 1989 and 1991, by SCOPE (1992), and by MacLochlainn (1993).[6-9] Reviews of UV-B effects in terrestrial plants have been published during the last five years by Caldwell et al. (1989), Bornman (1989), Tevini et al. (1989), Teramura et al. (1991), Tevini and Teramura (1989), Krupa and Kickert (1989), Stapleton (1992), Tevini (1993), and Bornman and Teramura (1993).[10-18]

UV-B radiation can be applied using artificial UV-B sources to biological material in growth chambers, greenhouses, or in the field supplementing solar UV-B radiation. In all cases, the biological effectiveness of the UV-B radiation has to be determined in order to compare the various simulated ozone depletion scenarios and to assess morphological, physiological, and biochemical consequences of solar UV-B radiation changes in plants due to ozone changes. UV-B can have damaging effects on DNA and/or physiological processes of the plant organisms as various mono- and polychromatic action spectra demonstrate.[19-24]

Figures 1 and 2 show action spectra in the UV range, all of which indicate increases in biological effectiveness with shorter wavelengths. Ozone depletion shifts the solar cut-off to shorter wavelengths and thus increases the irradiance at a given wavelength due to ozone absorption. Depending upon the shape of the action spectrum which is used as a weighting function for the radiation in use, the ozone simulation scenarios vary from drastic cases where a DNA action spectrum is used through moderately damaging cases when the generalized plant action spectrum is applied to lower effects when a polychromatic action spectrum for seedling growth is used.[19,20,25] Polychromatic action spectra include the beneficial effects of background white light on photorepair of DNA damage or on accumulation of target-shielding pigments and are thus closer to the natural radiation conditions.[26,21]

A new absolute action spectrum for pyrimidine dimer induction in intact alfalfa leaves also indicates that the slope in the short wavelength band is less steep when DNA is shielded. This consequently lowers the predicted impact of ozone depletion in comparison to DNA related responses.[27] (See Sutherland, this volume).

This impact could even be absent where DNA dimers are repaired by visible light, a process called photoreactivation. This emphasizes the importance of maintaining a realistic balance between UV-B, UV-A, and longer wavelengths which is largely not the case under artificial irradiation conditions. Furthermore, it is well known that physiological processes like photosynthesis are also negatively affected by

NATO ASI Series, Vol. I 18
Stratospheric Ozone Depletion/
UV-B Radiation in the Biosphere
Edited by R. H. Biggs and M. E. B. Joyner

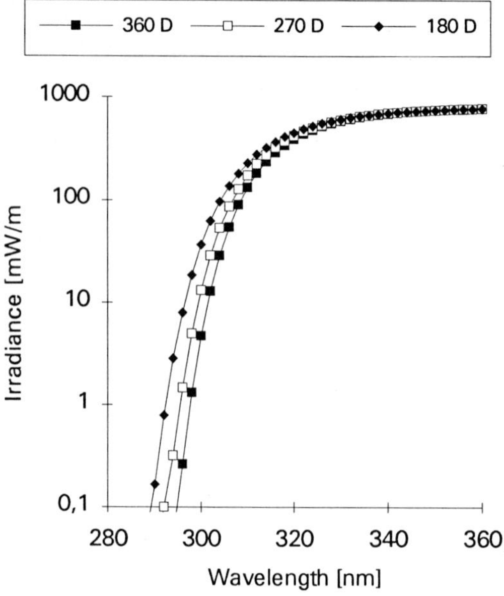


Fig.1: Solar UV-B irradiance at three ozone concentrations calculated for June 21 (52°N).

UV-B. However, the damage is much lower in the presence of high white light, indicating a modulating contribution of polychromatic radiation to the net effect of the UV-induced responses.[28,29] Thus, the use of only one weighting function for serious predictions is both very problematic and dangerous since action spectra are very complex and strongly related to the individual plant and to its different physiological responses which change with the environment. In addition, due to the absorption of many interacting compounds suvh as UV absorbing proteins and pigments, a photoreceptor molecule cannot be identified which clearly matches the response curve.

## EFFECTS AND PHYSIOLOGICAL CHANGES

The biological effects and physiological changes in terrestrial plants following enhanced UV-B radiation are numerous but nevertheless still insufficiently known. Most knowledge has come from crop plants mainly of temperate regions, whereas tropical plants and natural ecosystems have been more or less neglected. Studies of more than 300 plant species and cultivars (out of about 300,000 seed plant species) have been carried out and about 50% of these plants have been considered as UV-sensitive, where sensitivity is defined as any morphological, physiological, or biochemical change induced by UV-B. This definition also includes positive responses to enhanced UV-B such as the formation of UV-protective pigments. In the above-mentioned percentage only negative responses were subsumed.

Tree species, especially conifers in the seedling stage are also susceptible to enhanced UV-B radiation.[30] Out of 10 conifer species, four were susceptible to UV-B in terms of reductions in height or biomass.[31] Lower plants such as ferns and mosses

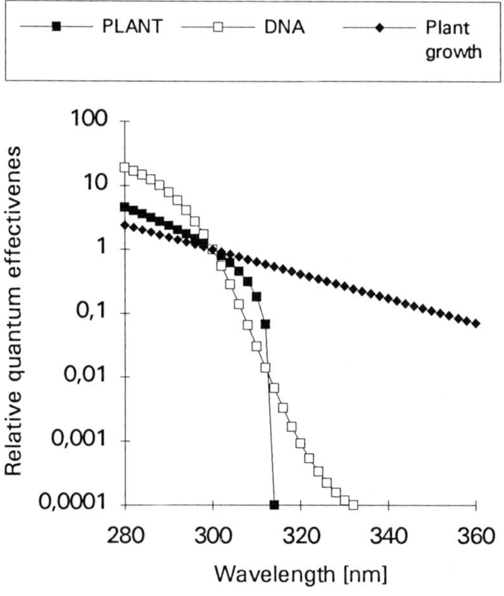


Fig. 2: Action spectra for DNA damage, [19] "general plant impact," [20] and plant growth. [25]

have not been studied with the interesting exception of *Bryum argenteum*.[32] Samples of this moss were collected between 1957 and 1989 and the flavonoid content measured as a possible indicator of fluctuations in solar UV-B radiation between those years.[32]

GROWTH

In many sensitive plant species such as wheat, rice, maize, rye, soybean, sunflower and cucumber[14], reduced leaf area and/or stem growth has been found. For leaf area and hypocotyl length of cucumber and sunflower seedlings grown in growth chambers under enhanced artificial UV-B radiation, a fluence response relationship could be demonstrated.[33] A reduction in stem length and leaf area was observed under conditions of increasing daily dose of artificial or natural UV-B radiation.

Recently, two comparable studies on rice cultivars of different geographical regions have been performed in greenhouses using 19 KJ $m^{-2} d^{-1}$ (plant weighted, normalization at 300 nm) as an enhanced daily dose and 15.7 KJ $m^{-2} d^{-1}$ respectively, compared to about 11 KJ $m^{-2} d^{-1}$ typically found in rice-growing areas near the Equator.[13,34,35]

From 16 rice cultivars native to the Philippines, India, Thailand, China, Vietnam, Nepal, and Sri Lanka, about one-third showed statistically significant decreases in total biomass and leaf area. Tiller number, highly correlated with yield, was reduced in 6 of these 16 cultivars. The Sri Lanka cultivar Kurkaruppan, however, showed increases in both total biomass and tiller number indicating that selective breeding might be a successful tool for obtaining UV-B tolerant cultivars.[13] In the second study on 4 cultivars from the Philippines based on total biomass changes different cultivar responses were reported, with IR74 the most sensitive and IR64 the least sensitive.[34] A further study on 4 Japanese rice cultivars revealed similar results, but unfortunately did not state the UV-radiation used in the phytotron in biological effective units since UV-C radiation also was emitted which is not present in solar radiation (Table 1).[36]

As stated before, proper irradiation conditions are the most crucial factors in evaluating and assessing effects of solar UV-B changes. One of the newer techniques to employ only solar UV-B was developed by Tevini et al. (1990).[33] Two identical growth chambers were located at a southern latitude with naturally high UV-B levels. These levels were attenuated by passing ozone through the plexiglas cuvette on top of one growth chamber. Plants beneath this cuvette received less than ambient UV-B which may be considered as a control UV-B radiation received at more northern latitudes. Plants in the second growth chamber were covered by a cuvette containing ambient air and thus received relatively enhanced UV-B compared to the control. With this ozone technique, a difference in solar UV-B radiation of 20% was simulated and growth, function, and composition of sunflower and corn seedlings have been intensively studied over several summer seasons.[33,37-39] In both sunflower and corn seedlings, growth parameters like plant size and leaf area were significantly reduced under higher UV-B radiation. These observations confirm results obtained in previous studies on other plant species with artificial UV-B applications in greenhouses and in the field or in growth chambers.[40,41]

Another possibility to provide plants with a different solar UV-B environment is given in greenhouses covered with UV-transmissive Plexiglass of different thickness. In a recent study two greenhouses located at a southern latitude (Portugal, 39°N) were covered with either 3 mm or 5 mm Plexiglass, simulating a 10% difference in solar UV-B. Even under these small UV-B differences, reductions in growth of bean plants can be observed under higher solar UV-B. Plant height is significantly reduced under higher UV-B throughout plant development in all cultivars. Leaf area and dry mass are affected more at an early developmental stage than later on. The reproductive phase was delayed as indicated by higher bud numbers and lower pod numbers as seen in 4 bean cultivars after 7 weeks (Table 2).[42] Growth reductions, however, can not always be correlated with lower biomass accumulation or photosynthesis. This was shown for wheat and wild oat which exhibited reduced plant height and leaf length when grown separately.[43] Growth reductions were also found in wild type and stable-phytochrome-deficient mutants of cucumber indicating that growth in the latter case is regulated by an unknown UV-B receptor, not by phytochrome.[44]

The molecular reasons for growth reductions can be attributed to changes in DNA and/or phytohormones. Phytohormones can affect growth by altering their concentration in the growth-sensitive tissue and by changing phytohormone-dependent processes. Elongation growth is related to indole acetic acid (IAA), which absorbs in the UV-B range and is readily photodestructed by UV-B *in vitro* and *in vivo* as shown in sunflower seedlings under low white light conditions. Furthermore the plastic epidermal cell wall extensibility, which is enhanced in auxin-induced elongation growth, was also reduced.[45] Peroxidase activity, which can reduce cell elongation at high activity by different mechanisms, was enhanced in UV-B irradiated sunflower and sugarbeet plants.[45,46] Another phytohormone, ethylene, which changes elongation to radial growth, is produced to a greater extent in UV-B irradiated sunflower seedlings.[45] In UV-B exposed cucumber and bean seedlings growth could be stimulated by gibberellins.[47,42]

## COMPETITION

One of the most important problems for agriculture is the growth competition between weeds and crops. Changes in morphological responses such as increased branching or number of leaves can lead to changes in competition for light.[48] In a competition study on wheat and wild oat, it was shown that under enhanced UV-B levels wheat had a competitive advantage over wild oat when they were grown together. By using a canopy microclimate and radiation interception model, these competitive differences can be quantitatively explained by shifts solely in the growth form of these species.[49]

In natural ecosystems, species pairs such as *Poa pratensis* and *Geum macrophylum* responded to a simulated 40% ozone depletion with a competitive advantage of *Geum* over *Poa*.[48] Further evidence for a shift of the competitive balance was derived from a previous experiment where seven competing pairs of agricultural crops and their associated weeds were grown in the field under increased levels of UV-B-radiation. It could be shown that the weeds *Avena*

Table 1: Effects of UV-radiation on growth of 6 rice cultivars.[13,34,36]

| Variety | Parameter | Control | Treated | | % change |
|---|---|---|---|---|---|
| **Norin 1** | Plant height, cm | 65,7 | 49,8* | a | -24,2* |
| | Dry weight, g/plant | 1,05 | 0,28* | a | -73,3* |
| | chlorophyll content, mg/g dw | 11,2 | 6,0* | a | -46,4* |
| **Sasanishiki** | Plant height, cm | 58,3 | 51,0* | a | -12,5* |
| | Dry weight, g/plant | 0,83 | 0,46* | a | -45,6* |
| | chlorophyll content, mg/g dw | 11,2 | 8,4* | a | -25,0* |
| **Kurkaruppan** | Plant height, % of control | - | - | | -6,2 |
| | Dry weight, g/plant | 1,42 | 1,74* | c | +22,8* |
| | chlorophyll content, mg/g dw | 9,9 | 8,2 | c | -17,2 |
| **N 22** | Plant height, % of control | - | - | | -14,8* |
| | Dry weight, g/plant | 2,57 | 4,56* | c | -39,3*Ü |
| | chlorophyll content, mg/g dw | 22,7 | 18,2 | c | -19,8 |
| **IR 64** | Plant height, cm | 80,1 | 73,9 | b | -7,3 |
| | Dry weight, g/plant | 3,17 | 3,08 | b | -2,8 |
| | chlorophyll content, mg/g fw | 2,95 | 3,40* | b | 15,3* |
| **IR 74** | Plant height, cm | 75,5 | 71,0* | b | -6,0 |
| | Dry weight, g/plant | 3,11 | 2,18* | b | -29,9* |
| | chlorophyll content, mg/g fw | 3,34 | 3,64 | b | 9,0 |

a) control: PAR (photosynthetic active radiation) 70 $\mu$mol m$^{-2}$ s$^{-1}$, treated: UV-B 0,74 W/m$^2$ not weighted, UV-C: 0,126 W/m$^2$

* = significantly different at 0,1% level using Students's test.[36]

b) control: PAR 1200$\mu$mol m$^{-2}$ s$^{-1}$, treated: UV-B 19,1 kJm$^{-2}$ d$^{-1}$ plant weighted

* = significantly different at $P<0{,}01$ using LSD test.[34]

c) control: PPF 20,7 mol m$^{-2}$ d$^{-1}$, treated: UV-B 15,7 kJm$^{-2}$ d$^{-1}$ plant weighted

* = significantly different at $P<0{,}01$ using LSD test.[13]

Table 2: Growth parameters of four bean varieties grown for 2 and 7 weeks in greenhouses covered with 5 mm (-UV) and 3 mm (+UV) plexiglas.[42]

| | | **Hilds Maja** | | **Primel** | | **Cropper Teepee** | | **Purple Teepee** | |
|---|---|---|---|---|---|---|---|---|---|
| | | 2 weeks | 7 weeks | 2 weeks | 7 weeks | 2 weeks | 7 weeks | 2 weeks | 7 weeks |
| **height** | 5 mm | 6.85 | 58.45 | 6.54 | 56.08 | 7.15 | 56.50 | 5.43 | 41.79 |
| (cm) | 3 mm | 5.57* | 50.50* | 5.67* | 52.29* | 5.69* | 50.50* | 4.51* | 36.04* |
| **% change** | | -18,7 | -13,7 | -13,3 | -6,8 | -20,4 | -11,6 | -16,9 | -13,8 |
| **dry weight** | 5 mm | 0.36 | 15.54 | 0.44 | 18.76 | 0.32 | 12.70 | 0.32 | 11.56 |
| (g) | 3 mm | 0.30* | 14.44 | 0.39* | 16.84 | 0.27 | 12.55 | 0.27* | 11.29 |
| **% change** | | -16,7 | -7,1 | -11,4 | -10,2 | -15,6 | -1,2 | -15,6 | -2,3 |
| **leaf area** | 5 mm | 62.6 | 2067.4 | 73.7 | 1676.4 | 68.48 | 1797.3 | 65.2 | 1866.1 |
| ($cm^2$/plant) | 3 mm | 50.7* | 1961.1 | 63.2 | 1581.2 | 50.1 | 1768.7 | 50.9* | 1794.1 |
| % change | | -19,0 | -5,1 | -14,2 | -5,7 | -26,8 | -1,6 | -22,9 | -3,9 |

* significantly different at $p < 0,05$ using LSD test

*fatua* or *Aegilops cylindrica* outcompeted *Triticum aestivum*.[50] These experiments imply significant ecological impacts on mixed-crop agriculture, species composition, and the biodiversity of natural ecosystems.[10]

FLOWERING

Several scientist investigating UV-B-effects on flowering demonstrated that exclusion of UV-B-radiation by Mylar plastic films or glass stimulated flowering as shown for *Melilotus*, *Trifolium dasyphyllum*, and *Tagetes*.[51-53] The inhibition of photoperiodic flower induction dependent on UV-B fluence rate and fluence could be demonstrated for the long day plant *Hyoscyamus niger* where a 20% decrease of flowering resulted from irradiation with 100 mW $m^{-2}$ UV-B, a 50% reduction from 300 mW $m^{-2}$ UV-B compared to plants merely irradiated with white light.[54] This suppression of flowering is correlated with changes in gibberellic acid content.

Investigations of UV-B effects on flowering also showed that the wall of anthers filters out over 98% of incident UV-B and thus protects pollen.[55] In addition, the pollen wall conteins UV-B absorbing compounds ensuring protection during pollination. Only after transfer to the stigma might the pollen be susceptible to UV-B, especially binucleate types of pollen with their longer time course for germination and penetration compared to trinucleate types (Flint, et al., 1984). At this stage, the assessment of the general impact of UV-B radiation on plant pollen is difficult as only few in vitro experiments have been carried out. It is thought that ovules offer sufficient protection for ovaries against solar UV-B radiation. Furthermore, experiments on bean plants suggest an influence of UV-B on the timing of flower production (Table 3).[42]

If the appearance of natural insect pollinators were not to coincide with the flowering of plants, marked consequences for the biodiversity of natural ecosystems could be the result.

Table 3: Distribution of buds, flowers, and pods (no./plant), as well as the yield potential (Σ of buds, flowers, and pods) and pod weight (g/plant) of 4 bean varieties grown under solar UV radiation in greenhouses covered with 5mm (-UV) and 3 mm (+UV) plexiglas.[42]

| | Hilds Maja | | Primel | | Cropper Teepee | | Purple Teepee | |
|---|---|---|---|---|---|---|---|---|
| | 5mm | 3mm | 5mm | 3mm | 5mm | 3mm | 5mm | 3mm |
| **buds** (no.) | 11.61 | 19.89 | 10.01 | 13.53 | 29.78 | 37.89 | 27.79 | 36.46 |
| **flowers** (no.) | 9.11 | 7.00 | 6.37 | 4.74 | 10.33 | 6.72 | 14.47 | 10.26 |
| **pods** (no.) | 18.56 | 9.72 | 7.16 | 3.63 | 5.66 | 1.06 | 5.68 | 1.68 |
| **Σ** (no.) | 39.44 | 36.61 | 23.53 | 21.89 | 45.78 | 45.67 | 47.95 | 48.26 |
| **pod weight** (g/plant) | 65.6 | 29.8 | 32.6 | 13.5 | 16.7 | 5.4 | 22.1 | 15.0 |
| **% change** | | -54.6 | | -58.6 | | -67.7 | | -32.1 |

PHOTOSYNTHESIS, RESPIRATION, AND TRANSPIRATION

*Photosynthesis.* This is one of the most examined physiological processes in UV-B studies and is accomplished mainly through growth experiments. When high UV-B irradiances were used in combination with low levels of white light (the usual conditions in growth chambers), effects on photosynthesis measured as $CO_2$-assimilation were generally deleterious in UV-B sensitive plants.[57] However, even in the presence of higher levels of white light found in greenhouses and the field, reductions in photosynthesis of up to 17% were reported in the UV-B sensitive soybean cultivar 'Essex' when supplied with UV-B equivalent to a 16% ozone depletion.[58,59]

Sunflower and maize seedlings grown under normal and enhanced solar UV-B radiation using the ozone filter technique also exhibited a significant decrease of about 17% in photosynthetic activity related to chlorophyll content and the different plant species when a 12% ozone depletion was simulated.[14] Based on leaf area, no differences were observed. This discrepancy is due to the strong effect of enhanced UV-B on growth resulting in a reduction of leaf area combined with an increase of chlorophyll content and probably chloroplast number per leaf area although total chlorophyll content per plant remained unaffected. Therefore, the leaf area is not always suitable as a relation parameter for estimating UV-B effects on photosynthesis.[39]

With respect to UV-B effects on photosynthesis, the PS II and the $CO_2$-fixing enzymes are the possible targets of UV-B radition. For sunflower and maize seedlings subjected to increasing UV-B fluence rates in a climate-controlled growth chamber under a 6000 W xenon lamp, a decrease in ribulosebisphosphat or phosphoenolpyruvat carboxylase activity as well as decreased photochemical capacity of the PS II and general vitality (RFD) could be proved after 4 weeks of growth.[39] Multiple sites of inhibition have been demonstrated, with the most sensitive around photosystem II.[11,60.62]

The reason for different action even in Photosystem II again may be the insufficient irradiation sources often emiting UV-C in addition to UV-B. From previous experiments it has been assumed that UV-C acts primarily on the water splitting system or on the oxidizing site of PSII.[61,63,64] However, UV-B acts on either the reaction center itself, producing dissipative sinks that quench the variable fluorescence or on the reducing site of PSII or both of these.[60,65] Evidence for the first action comes from the additional appearance of polypeptide fragments from the PSII reaction center under UV-B irradiation.[65,66,67] Recently, however, it was also shown that modification of $Q_A$ by UV-B can prevent electron transport from reduced pheophytin to $Q_A$, thus quenching the variable fluorescence by charge recombination reactions between $P680^+$ and $Pheo^-$.[65] Further targets seem to be the quinone $Q_B$ and /or the quinone pool inducing the degradation of the D1 protein (formerly called herbicide or $Q_B$-binding protein) as seen in photoinhibition.[66] However, UV-B damage is different from damage by visible radiation because the D1-degradation rate is much slower (5-7 h) than that during photoinhibition.[68] Trebst and Depka (1990) also found the degradation of the D1 protein subunit of photosystem II, but they attributed it to UV-C (254nm) radiation.[69] UV-C destroys plastoquinone and plastohydroquinone, whereas UV-B does not except when photosynthetic electron flow is already low as the result of prolonged irradiation periods with high UV-B irradiances, as demonstrated in radish leaves.[70-72] In the field, when longer UV-B irradiation periods occur, photoinhibition and UV-B damage may be additive and thus increase impact on net photosynthesis. This was also deduced from the inhibition of violaxanthin deepoxidation by UV-B, which is thought to have a protective role against photoinhibition. This enzyme synthesizes zeaxanthin (via antheraxanthin) which dissipates some of the absorbed excessive light energy.[73] Photosystem I, on the other hand, seems much more resistant to UV-B radiation.[60-62,74]

In contrast, ribulose 1,5-bisphosphate carboxylase (Rubisco), the key enzyme of the Calvin cycle, has recently been shown, at least in pea leaves, to be very UV-sensitive.[75] For example, one day of UV-B exposure reduced the enzyme activity by 40%, the large (LSU) and small subunits (SSU) of Rubisco by 10-15%.[76] Consistent with this observation are recent results on the gene expression of pea chloroplast proteins.[75,77] Both the nuclear-encoded SSU and chloroplast-encoded LSU-mRNA transcripts are severely reduced with a time delay for chloroplast transcripts. Chlorophyll *a*/*b* binding protein transcripts were also quantitatively reduced in pea and *Arabidopsis* mutants.[78] The reduced regulation of gene expression for chloroplast protein recovered to 60% after three days, which correlates with a similar pattern in the quantum yield of photosynthesis.[79]

*Plant respiration.* This was measured in sunflower and maize grown in growth chambers under increasing UV-B fluence rates in the range of 0.5 $mW/m^2$ to 1391.1 $mW/m^2$ for 23 days.[39] Independent of plant age, respiration was shown to increase with increasing fluence rates. Seedlings grown under low UV-B fluence rates showed similar respiration rates. Considering the whole plant a rise in respiration was seen only under high UV-B fluence rates.

With respect to the whole plant , the C4 species did not show differences in respiration. For leaf area only highest fluence rates had an effect on respiration which increased. However, with increasing duration of irradiation a rise in respiration occured also under low fluence rates.

*Transpiration.* The same experimental conditions were applied to study effects on transpiration. In sunflowers, transpiration decreased significantly from 353.3 $mW/m^2$ onwards, considering the whole plant.

An increase in transpiration of up to 50% occurred in maize seedlings with increasing UV-B exposure. In these plants, using the whole plant as a relation parameter, only during early development could a reduction in transpiration be observed. These initial differences soon disappeared and after 18 days transpiration was higher with the exception of plants grown under highest fluence rates. These showed transpiration rates similar to the control. Under solar radiation both respiration and transpiration remained unaffected in sunflowers and maize. This discrepancy is probably due to a higher portion of UV-A and/or PAR in the solar spectrum.

## YIELD

Since photosynthesis is essential for plant productivity, UV-induced damage to photosynthetic function will often manifest itself in less biomass or lower yield. Ideally, studies on plant yield ought to be conducted in carefully designed field studies since the supply of visible radiation is relatively low in growth chambers and greenhouses. It is well-documented that the effectiveness of UV-B radiation is magnified when given levels of radiation that are below the saturation level of photosynthesis. The reasons for this phenomenon are not completely known; however, it is thought that natural UV protective mechanisms do not become fully developed in low levels of visible light.

Earlier results on about 25 plant species illustrate the degree of technical difficulty and uncertainty surrounding the estimates of UV-B effects on yield. Many of these studies were conducted with filtered UV-B sources simulating relatively high ozone reduction rates.[80,81] In a German field study using filtered lamps and simulating 10% and 25% ozone reductions, no UV-B effects were demonstrated in 3 cabbage cultivars, or in lettuce or rape. Since flavonoid contents increased in most plants, there might have been a UV-B effect on food quality, although this was not specifically tested.[82] Biggs et al. (1978, 1984) grew 10 crop species in Gainesville, Florida, (29°N) and found yield reductions between 5% and 90% in half, among them wheat (-5%), potato (-21%), and squash (-90%), whereas rice, peanut, and corn were unaffected. Studies on six cultivars of soybean were conducted under conditions of a 16% ozone depletion simulation.[80,81,83] In all cultivars no change as a response to enhanced UV-B was observed. This brief summary indicates a high degree of interspecific variability complicated further by differences in artificial light sources, climate, soil quality, day length, and other growing conditions.

In addition to this interspecific variability, there is a high degree of intraspecific variability found among cultivars.[84,40] In the soybean cultivar 'Essex' a 25% reduction in photosynthesis was coupled with the same reduction in soybean yield when the plants were irradiated in a realistic arrangement with supplemental UV-B in the field.[85] A 20% reduction in yield was further established in a 6-year field study for 'Essex' when simulating a 25% ozone reduction (5.1 KJ $m^{-2}$ $d^{-1}$ supplemental UV-B), whereas in the cultivar 'Williams' the yield increased under the same conditions.[86] This again indicates differing cultivar sensitivities within the same species.

Currently not enough is known about the reasons for these large differences in UV-B responses among cultivars. Ideally, we need to understand the genetic bases (heritability) of UV tolerance and sensitivity. Once this is known, we could estimate the possibility of using conventional breeding practises to minimize the potential impacts of UV damage. At present, plant breeders have not yet considered UV sensitivity as a selective factor. This has resulted in the development of cultivar 'Essex' which appears to be quite sensitive to UV-B radiation. 'Essex' is a newer cultivar which is replacing the UV-tolerant cultivar 'Williams' in many areas of the US.

Bean cultivars grown in different greenhouses, one covered with 3 mm and the other with 5 mm Plexiglas (lower solar UV-B), showed significant reductions in yield under higher UV-B when harvested after 7 weeks. The reason for this was a delay in development of pods in favor of flowers and buds.[42]

## UV-PROTECTION AND ADAPTATION

Evolutionarily, plants have adapted to enhanced UV-B radiation mainly by developing two protection mechanisms: photoreactivation and accumulation of UV-absorbing pigments such as phenylpropanoids, flavonoids, and anthocyanines. Based on the flavonoid content of samples of *Bryum argenteum* collected from the Ross Sea area between 1957 and 1989, conclusions were extrapolated on fluctuating UV-B radiation conditions and ozone levels between those years.[32] Hence, protective substances can be an important indicator of environmental radiation conditions. Photoreactivation is directed towards the repair of thymine dimers of the cyclobutane type which may be responsible for most UV-induced damage of the DNA in plants. Recently it has been shown in alfalfa sprout leaves that the UV-B fluence response for the dimer formation is linear up to 690 J $m^{-}$.[24] The photolyase responsible for this repair uses visible light and is UV-B-inducible in *Arabidopsis thaliana* but phytochrome-regulated in bean hypocotyls.[87,88] Several studies have confirmed the importance of high white-light levels in the reduction of UV-B-induced damage normally found under low white-light levels.[28,29,36,89] Therefore, the repair capacity as far as photolyase is involved must be high in sunlight, since all active light qualities are present in high amounts. However, enhanced UV-B may have damaging effects on the enzyme activity itself (by destruction of UV-absorbing amino acids) and thus reduces the repair capacity, especially when the photolyase is located in the epidermal layer of the leaves.

UV-resitant species may be well protected by shielding pigments synthesized via the phenylpropanoid pathway in the epidermal layer or in leaf hairs.[90-92] Penetration of UV-B radiation through the epidermal layer into the mesophyll has been measured using a fibee-optic microprobe.[93] Two subalpine conifer species with increased growth under supplemental UV-B radiation as well as 11 species from the Rocky Mountains, had no measurable 300-nm radiation penetration into the mature needles.[31,94,95] In contrast, in postemergent and elongating leaves the photosynthetic mesophyll received small but measurable UV-B fluxes, indicating that there are smaller amounts of screening properties in early needle stages.[96] The protective function of flavonoids in the mechanism of photosynthesis has been clearly demonstrated in rye seedlings which accumulate isovitexins and other phenylpropanoids exclusively in the epidermal layer and thus damage to photosynthesis is prevented.[38]

The biosynthesis of these compounds is regulated by cis-trans isomerization of cinnamic acids.[97] The equilibrium between the cis and trans forms is shifted to the cis form by shorter wavelengths in the UV-B, thus lowering the feedback inhibition of the key enzyme L-phenylalanine ammonia lyase (PAL) and possibly also of the subsequent enzymes normally produceded by the trans form of cinnamic acids. The UV-regulation at the genetic level was intensively studied in cultured cells and leaves of *Petroselinum* by Hahlbrock's group,[91] demonstrating that the increase in mRNA transcripts for PAL, 4-coumarate:CoA ligase (4CL) and chalcone synthetase (CHS) is caused by UV-induced changes in transcription.[98-100] Schulze-Lefert et al. (1989) defined DNA sequences necessary for UV activation of the parsley CHS gene, and Block et al. (1990) further characterized a core of essential nucleotides containing ACGTGGC.[101,102] The cis-acting DNA sequence lies in unit 1 of the CHS promoter with two separate boxes necessary for UV inducibility. Three binding factors to the box II core sequence have been isolated and have been called common plant regulatory factors (CPRF), indicating that several other environmental factors apart from UV radiation can also induce CHS responses. This response was shown for pathogen challenge in soybean and snapdragon.[102,103] CPRF-1 accumulation induced by UV radiation is consistent over time with CHS mRNA accumulation, supporting the suggestion that this factor is involved in UV regulation.[105,106] The enzyme 4CL is also UV inducible and requires an additional cis-acting element which is distinct, however, from that controlling development, being located in the exonic region of the gene.[107]

In addition to UV radiation, blue and red/far-red light are also involved in the regulation of gene expression ,as shown for CHS in parsley cells, petunia, and mustard seedlings.[108-110] In parsley cells blue light abolishes the 2 h lag phase after UV induction, whereas phytochrome alters the extent of the CHS induction. In petunia, tissue-specific light regulation occurs since CHS expression is UV dependent in seedlings and controlled by red light in petals. In mustard, the UV-B and blue light influence on CHS expression were found in primary leaves but not in cotelydons demonstrating that the yet-unidentified UV-B photoreceptor either appears later in seedling development or that responsiveness for UV-B in the gene expression changes. In parsley cells it has been shown recently that flavin is an endogenous photoreceptive chromophore by feeding the cells with riboflavin, thus increasing the levels of CHS and flavonoids when irradiated for 6 h with UV-B plus white light and compared to unfed controls. Blue light had no effect in this respect.[111] In addition to flavonoids, polyamines which stabilize the membrane structure and inhibit lipid peroxidation may also play a role in protecting plants from UV-B stress.[112,113]

## COMBINATION EFFECTS

*$CO_2$ concentration and temperature.* Increases in $CO_2$ concentration and temperature due to the greenhouse effect are also anticipated in the future. Model calculations predict by the middle of the next century a temperature increase of 1° to 5°C and a doubling of the ambient $CO_2$ content. It is therefore of interest to investigate whether these joint stress factors show simply cumulative enhancing or compensative effects. Over the last three years progress has been made in answering this question.

Several studies describe the joint effects of enhanced UV-B and $CO_2$ (Table 4). Wheat, rice, and soybean were grown in greenhouses at 350 ppm and 650 ppm $CO_2$ at 8.8 and 15.7 KJ $m^{-2}$ $d^{-1}$UV-B. As expected for $C_3$ plants, an increase in $CO_2$ increases light saturated photosynthesis, apparent quantum efficiency, water use efficiency, biomass, and seed

Table 4: Seed Yield, light saturated photosynthetic rates and total biomass (g plant$^{-1}$) for wheat, rice and soybean grown in ambient and elevated $CO_2$ with ambient and supplemental levels of UV-B radiation. Percent change for each parameter is relative to ambient conditions.[114]

| | Seed Yield | Change % | Total Biomass | Change % | Photosynthesis \mol m$^{-2}$ s$^{-1}$ | Change % |
|---|---|---|---|---|---|---|
| **Wheat** | | | | | | |
| Control | 11.6 ca | | 27.1 b | | 23.4 b | |
| +UV-B | 10.7 c | (-8) | 27.4 b | (+1) | 23.3 b | (+0) |
| +$CO_2$ | 17.5 a | (+51) | 44.8 a | (+65) | 37.4 a | (+60) |
| +$CO_2$,+UV-B | 13.2 bc | (+14) | 37.1 a | (+37) | 36.1 a | (+54) |
| **Rice** | | | | | | |
| Control | 56.4 b | | 118.0 b | | 27.2 b | |
| +UV-B | 55.2 b | (-2) | 114.3 b | (-3) | 26.3 b | (-3) |
| +$CO_2$ | 66.6 a | (+18) | 130.8 a | (+11) | 32.7a | (+20) |

[a] Means with the same letter are not significantly different at the 95% level according to the Student-Newman-Keuls test.

yield in all three species. Enhanced UV-B had no significant damaging effect on seed yield, total biomass or photosynthesis of wheat or rice. However, increased UV-B radiation reduced the $CO_2$-induced increase in biomass and seed yield in both wheat and rice, whereas in soybean, biomass increased further and seed yield remained high under both environmental stresses.[114] As pointed out above, the reasons for UV-B-induced reductions may be both reduced PSII-activity and the biochemistry of $CO_2$ fixation. At least in rice, this damage to the photosynthetic apparatus can not be ameliorated by increased $CO_2$. Van de Staaij et al. (1992) reported that in *Elymus athericus,* a $C_3$ grass, the UV-induced dry weight reductions could not be significantly ameliorated by doubling of $CO_2$.[115]

The agricultural species pea and tomato also showed the same negative effect.[116] In both studies this lack of response may be due to unrealistic high UV-B-levels used in these experiments. In contrast to sunflower seedlings, maize seedlings grown in growth chambers using the ozone filter technique, doubling of $CO_2$ could not further increase biomass (over soil) normally found under increased solar UV-B.[39,117] An increase in temperature of 4°C over the normal daily temperature course can also compensate for biomass reductions normally found at lower temperatures and enhanced UV-B as also shown for sunflower and maize (Tables 5 and 6).[117] Since protective pigments are accumulated in UV-B-irradiated rye, bean, and maize plants, but not in sunflower, species-dependent protection mechanisms can be expected.

Ozone in combination with UV-B reduces pollen tube growth more than either stress factor alone. A UV-exposure of 3 Wm$^{-2}$ for 30 minutes followed by $O_3$ at 120 ppb for 3h reduced pollen growth of *Nicotiana tabacum* by 79% and *Petunia hybrida* by

Table 5. Growth, chlorophyll content, and gas exchange of 18-day-old sunflower seedlings grown at ambient and elevated temperature and $CO_2$ with ambient and reduced UV-B % = percentage change related to the control; means with the same letter are not significantly different at the 95% level according to an ANOVA.[117]

| | | | |
|---|---|---|---|
| **A:** | control | 28 °C | 340 ppm $CO_2$ |
| **B:** | + 25% UV-B | 28 °C | 340 ppm $CO_2$ |
| **C:** | + 25% UV-B | 32 °C | 340 ppm $CO_2$ |
| **D:** | + 25% UV-B | 32 °C | 680 ppm $CO_2$ |

| | A | B | % | C | % | D | % |
|---|---|---|---|---|---|---|---|
| **growth** | | | | | | | |
| fresh weight [g] | 1.77 a | 1.45 b | -18 | 1.98 c | +12 | 1.89 ac | +6 |
| dry weight [mg] | 136.70 a | 116.70 b | -14 | 160.60 c | +17 | 203.70 d | +4 |
| size [cm] | 8.64 a | 6.26 b | -27 | 8.42 a | -2 | 8.32 a | -3 |
| leaf area [$cm^2$] | 2.78 a | 24.67 b | -17 | 34.82 c | +16 | 33.06 c | +11 |
| **chlorophyll** | | | | | | | |
| chl [$mg/cm^2$] | 23.12 a | 27.47 b | +18 | 25.30 ab | +9 | 23.17 a | 0 |
| chl [mg/plant] | 775.80 a | 846.60 a | +9 | 955.30 a | +23 | 843.50 a | +8 |
| **photosynthesis** | | | | | | | |
| NP [mmol $CO_2$ $m^{-2}s^{-1}$] | 15.46 a | 17.32 b | +12 | 14.59 a | -5 | 15.39 a | -1 |
| NP [mmol $CO_2$ $plant^{-1}s^{-1}$] | 50.98 a | 46.06 b | -9 | 50.64 ab | -1 | 55.63 a | +9 |
| NP [mmol $CO_2$ $g^{-1}chl.^{-1}s^{-1}$] | 71.98 a | 57.93 b | -19 | 55.91 b | -22 | 72.03 a | 0 |
| **gas exchange** | | | | | | | |
| DR [mmol $CO_2$ $m^{-2}s^{-1}$] | 0.94 a | 1.08 b | +14 | 0.88 a | -6 | 0.91 a | -3 |
| TR [mmol $H_2O$ $m^{-2}s^{-1}$] | 2.33 a | 2.64 ab | +13 | 2.07 b | -11 | 1.91 b | -18 |
| TR [mmol $H_2O$ $plant^{-1}s^{-1}$] | 7.65 a | 7.15 a | -6 | 7.16 a | -6 | 6.87 a | -10 |
| **wue** | 6.77 a | 7.01 a | +3 | 7.54 a | +11 | 10.03 b | +48 |

NP= net photosynthesis; DR= dark respiration; TR= transpiration; wue= water use efficiency; chl= chlorophyll content

Table 6. Growth, chlorophyll content, and gas exchange of 18-day-old maize seedlings grown at ambient and elevated temperature and $CO_2$ with ambient and reduced UV-B; % = percentage change related to the control; means with the same letter are not significantly different at the 95% level according to an ANOVA.[117]

| | | | |
|---|---|---|---|
| **A:** | control | 28 °C | 340 ppm $CO_2$ |
| **B:** | + 25% UV-B | 28 °C | 340 ppm $CO_2$ |
| **C:** | + 25% UV-B | 32 °C | 340 ppm $CO_2$ |
| **D:** | + 25% UV-B | 32 °C | 680 ppm $CO_2$ |

| | A | B | % | C | % | D | % |
|---|---|---|---|---|---|---|---|
| **growth** | | | | | | | |
| fresh weight [g] | 2.54 a | 2.03 b | -20 | 2.78 a | +9 | 2.70 a | +6 |
| dry weight [mg] | .10 a | 221.00 b | -25 | 379.10 c | +31 | 382.30 c | +32 |
| size [cm] | 33.51 a | 28.30 b | -15 | 41.26 c | +23 | 37.78 d | +12 |
| leaf area [$cm^2$] | 83.42 a | 65.01 b | -22 | 88.70 a | +6 | 87.76 a | +5 |
| **chlorophyll** | | | | | | | |
| chl. [$mg/cm^2$] | 14.82 a | 17.13 b | +15 | 13.36 a | -9 | 15.32 ab | +3 |
| chl. [mg/plant] | 1046.30 a | 1055.50 a | 0 | 1066.60 a | +1 | 984.80 a | -6 |
| **photosynthesis** | | | | | | | |
| NP [mmol $CO_2$ $m^{-2}s^{-1}$] | 9.44 a | 8.98 a | -4 | 7.59 ab | -19 | 6.16 b | -32 |
| NP [mmol $CO_2$ $plant^{-1}s^{-1}$] | 72.29 a | 57.33 b | -20 | 70.55 a | -2 | 68.38 a | -5 |
| NP [mmol $CO_2$ $g^{-1}chl.^{-1}s^{-1}$] | 70.96 a | 50.45 b | -29 | 67.86 a | -4 | 71.12 a | 0 |
| **gas exchange** | | | | | | | |
| DR [mmol $CO_2$ $m^{-2}s^{-1}$] | 0.58 a | 0.66 b | +13 | 0.61 ab | +5 | 0.53 a | -8 |
| TR [mmol $H_2O$ $m^{-2}s^{-1}$] | 1.01 a | 0.95 a | -5 | 1.21 b | +19 | 0.45 c | -55 |
| TR [mmol $H_2O$ $plant^{-1}s^{-1}$] | 7.78 a | 6.22 a | -20 | 11.38 b | +46 | 4.90 c | -37 |
| **wue** 9.49 a | 9.44 a | -1 | 6.56 b | -30 | 14.74 | +56 | |

NP= net photosynthesis; DR= dark respiration; TR= transpiration; wue= water use efficiency; chl= chlorophyll content

Table 7: Growth parameters and chlorophyll content ($\mu g/cm^2$) of sunflower, bean, and rye grown for 7 weeks at different nitrogen and UV levels.

| | - UV 50mg | + UV 50mg | - UV 150mg | + UV 150mg | - UV 290 mg | + UV 290 mg |
|---|---|---|---|---|---|---|
| **sunflower** | | | | | | |
| height [cm] | 32.20 a | 29.30 b | 39.80 c | 35.00 d,e | 33.50 a,d | 36.20 e |
| dry weight [g/plant] | 1.56 a | 1.44 a | 3.15 b | 1.89 a,c | 2.25 c | 2.43 c |
| leaf area [$cm^2$/plant] | 97.43 a | 109.23 a | 355.31 b | 296.71 c | 379.04 b | 405.79 b |
| chlorophyll [$\mu g/cm^2$] | 22.02 a | 21.66 a | 37.41 b | 39.92 b | 57.61 c | 49.93 c |
| **maize** | | | | | | |
| height [cm] | 53.40 a | 48.30 b | 63.73 c | 60.60 d | 60.30 d | 60.00 d |
| dry weight [g/plant] | 1.12 a | 1.09 a | 1.79 b | 1.47 c | 1.59 b,c | 1.53 b,c |
| leaf area [$cm^2$/plant] | 201.83 a | 195.86 a | 340.70 b | 331.92 b | 351.16 b | 347.03 b |
| chlorophyll [$\mu g/cm^2$] | 7.53 a | 9.13 a,b | 15.00 b,c | 17.60 c | 19.38 c,d | 24.25 d |

control: PAR 400 $\mu mol\ m^{-2}\ s^{-1}$, UV-B 0,22 $kJm^{-2}\ d^{-1}$
treated: UV-B 5,0 $kJm^{-2}\ d^{-1}$ plant weighted norm: 300 nm

75%. The effect appeared to be additive, indicating that different targets may be affected by the two environmental factors.[118]

*Nitrogen supply.* Nitrogen has a great influence on UV-B effects on plant morphology and physiology. The sensitivity to UV-B is positive for all crops, although the degree of this sensitivity, however, depends on nitrogen supply. UV-B-dependent growth reductions were observed in rye plants at 50 mg N/kg soil in sunflower, bean, and maize plants at 150 mg N/kg soil. Sensitivity to UV-B dissapeared at the greatest nitrogen content (290 mgN/kg) in nearly all plant species. An increase in nitrogen supply is positively correlated to chlorophyll accumulation, whereas a decrease in protective pigments and soluble proteins is proportionate to nitrogen levels (Table 7).

*Heavy metals.* In most casesheavy metals are toxic to plants. Only one study describing the joint effects of 5 mM cadmium and UV-B radiation of only 6.17 KJ $m^{-2}$ $d^{-1}$ to *Picea abies* L. has appeared so far.[119] In general, the combined treatment accentuates the reductions of growth and photosynthesis normally found under separate treatments, showing that certain environ-mental factors reveal additive responses in plants, especially in the field where plants are usually subjected to more than one stress factor.

**Acknowledgments**

Work done in the laboratories of Prof. M. Tevini was supported by the German Ministry for Research and Technology. I am grateful to Thomas Hietzker (BSc hons) and Ernst Heene for their help in preparing the manuscript.

**REFERENCES**

1. Lubin, D., B.G. Mitchell, J.E. Frederick, A.D. Alberts, C.R. Booth, T. Lucas, and D. Neuschuler. 1992. A contribution toward understanding the biospherical significance of Antarctic ozone depletion. J. Geophys. Res. 97:7817-7828.

2. Lubin, D., and J.E. Frederick. 1990. Column ozone measurements from Palmer Station, Antarctica: Variations during the austral springs of 1988 and 1989. J. Geophys. Res. 95:13883-13889.

3. Lubin, D., and J.E. Frederick. 1991. The ultraviolet radiation environment of the Antarctic Peninsula: The roles of ozone and cloud cover. J. Appl. Met. 30:478-493.

4. Blumthaler, M., and W. Ambach. 1990. Indication of increasing solar ultraviolet-B radiation flux in Alpine regions. Science 248, 206-208.

5. Blumthaler, M. 1993. Solar UV Measurements. In: M. Tevini (ed.), UV-B radiation and ozone depletion: Effects on humans, animals, plants, microorganisms, and materials, Lewis Publishers Boca Raton, USA. p.71ff.

6. UNEP. 1989. Environmental effects panel report. In: J.C. van der Leun, M. Tevini, and R.C. Worrest (eds.), United Nations Environmental Program, Nairobi, Kenya.

7. UNEP. 1991. Environmental effects of ozone depletion: 1991 update. In: J.C. van der Leun, M. Tevini, and R.C., Worrest (eds.), United Nations Environmental Program, Nairobi, Kenya.

8. SCOPE. 1992. Effects of increased ultraviolet radiation on biological systems: A research implementation plan addressing the impacts of increased UV radiation due to stratospheric depletion. Proceedings, Budapest, February.

9. MacLochlainn, C., 1993. Report on stratospheric ozone and ultraviolet radiation. DG XII-Environment Program, Services of the EC Commission.

10. Caldwell, M.M., A.H. Teramura, and M. Tevini. 1989. The changing solar ultraviolet climate and the ecological consequences for higher plants. Trends Ecol. Evol. 4:363-366.

11. Bornman, J.F. 1089. Target sites of UV-B radiation in photosynthesis of higher plants. J. Photochem. and Photobiol. B Biology 4, 145-158.

12. Tevini, M., J. Braun, P. Grusemann, and J. Ros. 1989. UV-Wirkungen auf Nutzpflanzen. In: Lauffener Sem.beitr. 3/88, 38-51. Akad. Natursch. Landschaftspflege (ANL) Laufen/Salzach.

13. Teramura, A.H., L.H. Ziska, and A.E. Szetin. 1991. Changes in growth and photosynthetic capacity of rice with increased UV-B radiation. Physiol. Plant. 83:373-380.

14. Tevini, M., and A.H. Teramura. 1989. UV-B effects on terrestrial plants. Photochem. Photobiol. 50:479-487.

15. Krupa, S.V., and R.N. Kickert. 1989. The greenhouse-effect-impacts of ultraviolet-B (UV-B) radiation, carbon-dioxide ($CO_2$), and ozone ($0_3$) on vegetation. Env. Pollution 61:263-393.

16. Stapleton A.E. 1992. Ultraviolet radiation and plants: Burning questions. Plant Cell 4:1353-1358.

17. Tevini, M. 1993. Effects of enhanced UV-B radiation on terrestrial plants. In: M. Tevini (ed.), UV-B radiation and ozone depletion: Effects on humans, animals, plants, microorganisms, and materials, Lewis Publishers, Boca Raton, USA, pp.125ff.

18. Bornman, J.F., and A.H. Teramura. 1993. The effects of ultraviolet-B radiation on terrestrial plants. In: L.O. Björn (ed.), Environmental UV-Photobiology, in press.

19. Setlow, R.B. 1974. The wavelength in sunlight effective in producing skin cancer: A theoretical analysis. Proc. Nat. Acad. Sci. 71:3363-3365.

20. Caldwell, M.M. 1971. Solar UV irradiation and the growth and development of higher plants. In Photophysiology VI (edited by A.C. Giese) Academic Press, New York. pp.131-268.

21. Caldwell, M.M., L. B. Camp, C.W. Warner, and S.D. Flint. 1986. Action spectra and their key role in assessing biological consequences of solar UV-B radiation change. In: R.C. Worrest and M.M. Caldwell (eds.), Stratospheric Ozone Reduction, Solar Ultraviolet Radiation and Plant Life. NATO ASI Series G: Ecological Sciences, 8:87-112. Springer-Verlag Berlin.

22. Coohill, T.P. 1989. Ultraviolet action spectra and solar effectiveness spectra for higher plants. Photochem. Photobiol. 50:451-457.

23. Coohill, T.P. 1991. Photobiology School, Action Spectra Again? Photochem. Photobiol. 54:859-870.

24. Quaite, F.E., M.B. Sutherland, and J.C. Sutherland. 1992b. Quantitation of pyrimidine dimers in DNA from UVB-irradiated alfalfa (*Medicago sativa* L.) seedlings. Applied and Theoretical Electrophoresis 2:171-175.

25. Steinmüller, D. 1986. On the effect of ultraviolet radiation (UV-B) on leaf surface structure and on

the mode of action of cuticular lipid biosynthesis in some crop plants. Karls. Beitr. Entw. Ökophysiol. M. Tevini (ed.), 6:1-174.

26. Rundel, R.D. 1983. Action spectra and estimation of biologically effective UV radiation. Physiol. Plant. 58:360-366.

27. Quaite F.E., B.M. Sutherland, and J.C. Sutherland. 1992a. Action spectrum for DNA damage in alfalfa lowers predicted impact of ozone depletion. Nature 358:576-578.

28. Mirecki, R.M., and A.H. Teramura. 1984. Effects of ultraviolet-B irradiance on soybean. V: The dependence of plant sensitivity on the photosynthetic photon flux density during and after leaf expansion. Plant. Physiol. 74:475-480.

29. Cen Y.-P, and J.F. Bornman. 1990. The response of bean plants to UV-B radiation under different irradiances of background visible light. J. Experim. Bot. 41:1489-1495.

30. Sullivan, J.H., and A.H. Teramura. 1991. The effects of UV-B radiation on loblolly pine. 2: Growth of field-grown seedlings. Trees 6:115-120.

31. Sullivan, J.H., and A.H. Teramura. 1988. Effects of ultraviolet-B irradiation on seedling growth in the Pinaceae. Amer. J. Bot. 75:225-23O.

32. Glasgow, L. 1990. The history of the ozone layer. New Scientist, 1990:24 November.

33. Tevini, M., U. Mark, and M. Saile. 1990. Plant experiments in growth chambers illuminated with natural sunlight. In: H.D. Payer, T. Pfirrmann, and P. Mathy (eds.), Environmental Research with Plants in Closed Chambers, Air Pollution Res. Rep. 26:240-251.

34. Dai Q., P. V. P. Coronel, B.S. Vergara, P.W. Barnes, and A.T. Quintos. 1992. Ultraviolet-B radiation effects on growth and physiology of four rice cultivars. Crop Sci. 32:1269-1274.

35. Bachelet, P.W., D. Barnes, and M. Brown. 1991. Latitudinal and seasonal variations in calculated ultraviolet-B irradiance for rice-growing regions of Asia. Photochem. and Photobiol. 54:411-422.

36. Kumagai T., and T. Sato. 1992. Inhibitory effects of increase in near-UV radiation on the growth of Japanese rice cultivars (*Oryza sativa* L.) in a phytron and recovery by exposure to visible radiation. Japan J. Breed. 42:545-552.

37. Tevini, M., U. Mark, G. Fieser, and M. Saile. 1991b. Effects of enhanced solar UV-B radiation on growth and function of selected crop plant seedlings. In: E. Riklis (ed.), Photobiology, Plenum Publ., New York, pp.635-649.

38. Tevini, M., J. Braun, and G. Fieser. 1991a. The protective function of the epidermal layer of rye seedlings against ultraviolet-B radiation. Photochem. Photobiol. 53:329-333.

39. Mark, U. 1992. Zur Wirkung erhöhter artifizieller und solarer UV-B-Strahlung in Kombination mit erhöhter Temperatur und Kohlendioxidkonzentration auf das Wachstum und den Gaswechsel von ausgewählten Nutzpflanzen. (On the effect of increased artificial and solar UV-B radiation in combination with increased temperature and carbon dioxide concentration on growth and gas exchange of selected crop plants). In: M. Tevini (ed.), Karlsr. Beitr. Entw. Ökophys. 11:1-220.

40. Teramura, A.H., and N.S. Murali. 1986. Intraspecific differences in growth and yield of soybean Glycine max exposed to UV-B radiation under greenhouse and field conditions. Environ. Exp. Bot. 26:89-95.

41. Tevini, M., and W. Iwanzik. 1986. Effects of UV-B radiation on growth and development of cucumber seedlings. In: R.C. Worrest and M.M. Caldwell (eds.), Stratospheric Ozone Reduction, Solar Ultraviolet Radiation and Plant Life, NATO ASI Series G: Ecological Sciences, 8:271-286. Springer-Verlag Berlin.

42. Saile-Mark, M. 1993. Zur Beteiligung von Phytohermonen an Wachstum und Blütenbildung verschiedener Bohnenkulturvarietäten (*Phaseolus vulgaris* L.) in Abhängigkeit von artifizieller und solarer UV-B-Strahlung. Thesis, University of Karlsruhe.

43. Barnes, P.W., P.W. Jordan, W.G. Gold, S.D. Flint, and M.M. Caldwell. 1988. Competition morphology and canopy structure in wheat (*Triticum aestivum* L.) and wild oat (*Avena fatua* L.) exposed to enhanced ultraviolet-B radiation. Funct. Eco. 2:319-330.

44. Ballare, C.L., P.W. Barnes, and R.E. Kendrick. 1991a. Photomorphogenic effects of UV-B radiation on hypocotyl elongation in wild type and stable-phytochrome-deficient mutant seedlings of

cucumber. Physiol.Plant 83:652-658.

45. Ros, J. 1990. Zur Wirkung von UV-Strahlung auf das Streckungs wachstum von Sonnenblumenkeimlingen (*Helianthus annuus* L.). (On the effect of UV-radiation on elongation growth of sunflower seedlings (*Helianthus annuus* L.) (Thesis). Karls. Beitr. Entw. Ökophysiol. 8:1-157. M. Tevini (ed.), Bot. Inst. II, Karlsruhe.

46. Panagopoulos, I., J.F. Bornman, and L.O. Björn. 1990. Effects of ultraviolet radiation and visible light on growth, fluorescence induction, ultraweak luminescence and peroxidase activity in sugar beet plants. Photochem. and Photobiol. B: Biol., 8:73-87.

47. Ballare, C.L., J.J. Casal, and K.E. Kendrick. 1991b. Photochem. Photobiol. 54:819-826.

48. Barnes P.W., S.D. Flint, and M.M. Caldwell. 1990. Morphological responses of crop and weed species of different growth forms to ultraviolet-B radiation. Amer. J. Bot. 77:1354-1360.

49. Ryel, R., P.W. Barnes, W. Beyschlag, M.M. Caldwell, and S.D. Flint. 1990. Plant competition for light analyzed with a multispecies canopy model. I: Model development and influence of enhanced UV-B conditions on photosynthesis in mixed wheat and wild oat canopies. Oecol. 82:304-310.

50. Gold, W.G., and M.M. Caldwell. 1983. The effects of ultraviolet-B radiation on plant competition in terrestrial ecosystems. Physiol. Plant. 58:435-444.

51. Kasperbauer, L., and W. Loomis. 1965. Inhibition of flowering by natural daylight on an inbred strain of *Melilotus*. Crop Sci. 5:193-194.

52. Caldwell, M.M. 1968. Solar ultraviolet radiation as an ecological factor for alpine plants. Ecol. Monogr. 38, 243-288.

53. Klein, R., P. Edsall, and A. Gentile. 1965. Effects of near- ultraviolet and green radiations on plant growth. National Sci. Foundation and Contract AT(30-1-2587) from the Atomic Energy Commission.

54. Rau, W., H. Hoffmann, A. Huber-Willer, U. Mitzke-Schnabel, and E. Schrott. 1988. Die Wirkung von UV-B auf photoregulierte Entwicklungsvorgänge bei Pflanzen. Gesellschaft für Strahlen- und Umweltforschung mbH., München Abschluábericht.

55. Flint, S.D., and M.M. Caldwell. 1983. Influence of floral optical properties on the ultraviolet radiation environment of pollen. Amer. J. Bot. 7O:1416-1419.

56. Flint, S.D., and M.M. Caldwell. 1984. Partial inhibition of *in vitro* pollen germination by simulated ultraviolet-B radiation. Ecology 65:792-795.

57. Teramura, A.H. 1986. Interaction between UV-B radiation and other stresses in plants. In: R.C. Worrest and M.M. Caldwell (eds.), Stratospheric Ozone Reduction, Solar Ultraviolet Radiation and Plant Life, 375. NATO Advanced Science Institutes Series G, Ecological Sciences 8:327-344.

58. Murali, N.S., and A.H. Teramura. 1986. Effectiveness of UV-B radiation on the growth and physiology of field-grown soybean modified by water stress. Phytochem. Phytobiol. 44:215-219.

59. Murali, N.S., and A.H. Teramura. 1987. Intensity of soybean photosynthesis to ultraviolet-B radiation under phosphorus deficiency. J. Plant. Nutr. 10:501-515.

60. Iwanzik, W., M. Tevini, G. Dohnt, M. Voss, W. Weiss, P. Gröber, and G. Renger. 1983. Action of UV-B radiation on photosynthetic primary reactions in spinach chloroplasts. Physiol. Plant. 58:401-407.

61. Renger, G., H.J. Eckert, R. Fromme, P. Grüber, M. Volker, and S. Hohmveit. 1989. On the mechanism of photosystem II deterioration by UV-B irradiation. Photochem. Photobiol. 49:97-105.

62. Strid A., W.S. Chow, and J.M. Anderson. 1990. Effects of supplementary ultraviolet-B radiation on photosynthesis in Pisum sativum. Biochimica et Biophysica Acta, 1020:260-268.

63. Yamashita, T., and W.L. Butler. 1968. Inhibiton of Chloroplasts by UV-Irradiation and Heat-Treatment. Plant Physiol. 43:2037-2040.

64. Kulandaivelu, G., and A. Noorudeen. 1983. Comparative study of the action of ultraviolet-C and ultraviolet-B radiation on photosynthetic electron transport. Physiol. Plant. 58:389-394.

65. Melis A., J.A. Nemson, and M.A. Harrison, 1992. Damage to functional components and partial degradation of photosystem II reaction center proteins upon chloroplast exposure to ultraviolet-B radiation. Biochem. Biophys. Acta 1100:312-320.

66. Greenberg B.M., V. Gaba, O. Canaani, S. Malkin, A.K. Mattoo, and M. Edelman. 1989. Separate photosensitizers mediate degradation of the 32-kDa photosystem II reaction center protein in the visible and UV spectral regions. Proc. Natl. Acad. Sci. 86:6617-6620.

67. Nedunchezhian N., and G. Kulandaivelu. 1991. Evidence for the ultraviolet-B (280-320 nm) radiation induced structural reorganization and damage of photosystem II polypeptides in isolated chloroplasts. Physiol. Plant. 81:558-562.

68. Greenberg, B.M., A.K. Mattoo, V. Gaba, and M. Edelman. 1989. Degration of the 32kDa photosystem-II reaction center protein in UV, visible and far red-light occurs through a common 23.5 kDa intermediate. Zeitschrift für Naturforschung 44:450-452.

69. Trebst, A., and B. Depka. 1990. Degradation of the D-! protein subunit of photosystem II in isolated thylakoids by UV light. Z. Naturforsch. 45c:765-771.

70. Trebst, A., and Pistorius, E. 1965. Photosythetische Reaktionen in UV-bestrahlten Chloroplasten. Z. Naturforschg. 20b:885-889.

71. Lichtenthaler, H.K., and M. Tevini. 1969. Die Wirkung von UV Strahlen auf die Lipochinon-Pigment-Zusammensetzung isolierter Spinatchloroplasten. Z. für Naturforschung, Band 24b, Heft 6:764-769.

72. Tevini, M., and W. Iwanzik. 1983. Inhibition of photosynthetic activity by UV-B radiation in radish seedlings. Physiol. Plant. 58:395-400.

73. Pfündel, E.E., R.S. Pan, and R.A. Dilley. 1992. Inhibition of violaxanthin deepoxidation by ultraviolet-B radiation in isolated chloroplasts and intact leaves. Plant. Physiol., 98:1372-1378.

74. Prasil, O., N. Adie, and I. Ohad. 1992. Dynamics of photosystem II: Mechanisms of photoinhibition and recovery processes. In: J. Barber (ed.), The Photosystems: Structure, Function and Molecular Biology, Elsevier, Amsterdam, pp.295-348.

75. Jordan B.R., J. He, W.S. Chow, and J.M. Anderson. 1992. Changes in mRNA levels and polypeptide subunits of ribulose 1,5-bisphosphate carboxylase in response to supplementary ultraviolet-B radiation. Plant, Cell and Environ. 15:91-98.

77. Jordan B.R., W.S. Cow, A. Strid, and J.M. Anderson. 1991. Reduction in *cab* and *psb* A RNA transcripts in response to supplementary ultraviolet-B radiation. FEBS Letters, 284:5-8, Elsevier Science Publishers B.V.

78. Jordan, B.R. 1993. The molecular biology of plants exposed to ultraviolet-B radiation and the interaction with other stresses. In: M.B. Jackson and C.R. Black (eds.), Interacting Stresses on Plants in a Changing Climate, Nato Advanced Workshop, Measurements. Solar Energy. 49:535-540.

79. Chow, W.S., A. Strid, and J.M. Anderson. 1992. Short-term treatment of pea plants with supplementary ultraviolet-B radiation: Recovery time-courses of some photosynthetic functions and components. In: N. Murata (ed.), Proceedings of IX International Congress on Photosynthesis.

80. Biggs, R.H., and S.V. Kossuth. 1978. Effects of ultraviolet-B radiation enhancement under field conditions on potatoes, tomatoes, corn, rice, southern peas, peanuts, squash, mustard and radish. In: UV-B Biological and Climatic Effects Research (BACER). Final Report, US EPA, Washington, DC.

81. Biggs, R.H., P.G. Webb, T.R. Garrard, and S.H. West. 1984. The effects of enhanced ultraviolet-B radiation on rice, wheat, corn, soybean, citrus and duckweed. Year 3 Interim Report. 808075-03, US EPA. Washington, DC.

82. Dumpert, K., and T. Knacker. 1985. A comparison of the effects of enhanced UV-B radiation on some crop plants exposed to greenhouse and field conditions. Biochem. Physiol. Pflanz. 180:599-612.

83. Sinclair, T.R., O. N'Diaye, and R.H. Biggs. 1990. Growth and yield of field-grown soybean in response to enhanced exposure to ultraviolet-B radiation. J. Environ. Qual. 19:478-481.

84. Biggs, R.H., S.V. Kossuth, and A.H. Teramura. 1981. Response of 19 cultivars of soybeans to ultraviolet-B irradiance. Physiol. Plant. 53, 19-26.

85. Teramura, A.H., and J.H. Sullivan. 1988. Effects of ultraviolet-B radiation on soybean yield and seed quality: A six-year field study. Envir. Pollut. 53:466-468.

86. Teramura, A.H., J.H. Sullivan, and J. Lydon. Effects of UV-B radiation on soybean yield and seed quality: A 6-year field study. Physiol. Plant. 80:5-11.

87. Pang, Q., and J.B. Hays. 1991. UV-B-inducible and

temperature-sensitive photoreactivation of cyclobutane pyrimidine dimers in *Arabidopsis thaliana,* Plant. Physiol. 95:536-543.

88. Langer B., and E. Wellmann. 1990. Phytochrome induction of photoreactivating enzyme in *Phaseolus vulgaris* L. seedlings. Photochem. Photobiol. 52:861-863.

89. Lercari, B., F.S. Sodi, and M. Lipucci di Paola. 1990. Photomorphogenic responses to UV radiation: Involvement of phytochrome and UV photoreceptors in the control of hypocotyl elongation in *Lycopersicon esculentum.* Physiol. Plant. 79:668-672.

90. Jahnen W., and K. Hahlbrock. 1988. Differential regulation and tissue-specific distribution of enzymes of phenylpropanoid pathways in developing parsley seedlings. Planta 173:453-458.

91. Hahlbrock K., and D. Scheel. 1989. Physiology and molecular biology of phenylpropanoid metabolism. Annu. Rev. Plant. Physiol. Plant Mol. Biol. 40:347-369.

92. Karabourniotis G., D. Papadopoulos, M. Papamarkou, and Y. Manetas. 1992. Ultraviolet-B radiation absorbing capacity of leaf hairs. Physiol. Plant. 86:414-418.

93. Bornman, J.F., and T.C. Vogelmann. 1991. Effects of UV-B radiation on leaf optical properties measured with fiber optics. J. Exper. Bot. 42:237, 547-554.

94. De Lucia, E.H., T.A. Day, and T.C. Vogelmann. 1991. Ultraviolet-B radiation and Rocky Mountain environment: Measurement of incident light and penetration into foliage. In: D.D. Randall, D.G. Blevins, and C.D. Miles (eds.), Current Topics in Plant Biochemistry and Physiology. 10:32-48.

95. Day, T.A., T.C. Vogelmann, and E.H. DeLucia. 1992. Penetration of ultraviolet-B radiation in foliage of Rocky Mountain plants: Direct measurements with fiber-optic microrobes. Oecologia 83 (in press).

96. DeLucia, E.H., T.A. Day, and T.C. Vogelman. 1992. Ultraviolet-B and visible light penetration into needles of two species of subalpine conifers during foliar development. Plant Cell and Environ. 15:921-929.

97. Braun, J., and M. Tevini. 1993. Regulation of UV-protective pigment synthesis in the epidermal layer of rye seedlings (*Secale cereale* L. cv Kustro). Photochem. Photobiol. 57, 318-323.

98. Chappell J., and K. Hahlbrock. 1984. Transcription of plant defense genes in response to UV light or fungal elicitor. Nature 311:76-78.

99. Kuhn, D.N., J. Chappell, A. Boudet, and K. Hahlbrock. 1984. Induction of phenylalanine ammonia-lyase and 4-coumarate: CoA ligase mRNAs in cultured plant cells by UV light or fungal elicitor. Proc. Natl. Acad. Sci. USA 81:1102-1106.

100. Schmelzer, E., W. Jahnen, and K. Halbrock. 1988. *In situ* localization of light-induced chalcone synthase mRNA, chalcone synthase and flavonoid end products in epidermal cells of parsley leaves. Proc. Natl. Acad. Sci. 85:2989-2993.

101. Schulzelefert, P., W. Schulz, J.L. Dangl, M. Beckerandre, and K. Halbrock. Inducible *in vivo* DNA footprints define sequences necessary for UV-light activation of the parsley chalcone synthase gene. EMBO Journal 8:651-656.

102. Block A., J.L. Dangl, K. Hahlbrock, and P. Schulze-Lefert. 1990. Functional borders, genetic fine structure, and distance requirements of cis elements mediating light responsiveness of the parsley chalcone synthase promotor. Proc. Natl. Acad. Sci. USA 87:5387-5391.

103. Wingender R., H. Röhrig, C. Höricke, and J. Schell. 1990. Cis-regulatory elements involved in ultraviolet light regulation and plant defense. The plant Cell 2:1019-1026.

104. Fritze K., D. Staiger, I. Czaja, R. Walden, J. Schell, and D. Wing. 1991. Developmental and UV-light regulation of the snapdragon chalcone synthase promotor. Plant Cell 3:893-905.

105. Weisshaar, B., G.A. Armstrong, A. Block, O. da Costa e Silva, and K. Hahlbrock. 1991a. Light-inducible and constitutively expressed DNA-binding proteins recognizing a plant promoter element with functional relevance in light responsiveness. The EMBO Journal 10:1777-1786.

106. Weisshaar, B., A. Block, G.A. Armstrong, A. Herrmann, P. Schulze-Lefert, and K. Hahlbrock. 1991b. Regulatory elements required for light-mediated expression of the *Petroselinum crispum* chalcone synthase gene. In: G.I. Jenkins and W. Schuch (eds.), Society for Experimental Biology SEB Symposia Series 45:191-210.

107. Douglas, C., H. Hoffmann, W. Schulz, and K. Halbrock. 1987. Structure and elicitor of UV-

light-stimulated expression of two 4-coumarate: CoA ligase gene in parsley. EMBO. J. 6:1189-1195.

108. Bruns B., K. Hahlbrock, and E. Schäfer. 1986. Fluence dependence of the ultraviolet-light-induced accumulation of chalcone synthase mRNA and effects of blue and far-red light in cultured parsley cells. Planta 169:393-398.

109. Koes, R.E., J.N.M. Mol, and C.E. Spelt. 1989. The chalcone synthase multigene family of Petunia-Hybrida (V30)-differential, light-regulated expression during flower development and UV-light induction. Plant Molecular Bio. 12:213-225.

110. Batschauer, A., E. Bruno, and E. Schäfer. 1991a. Cloning and characterization of a chalcone synthase gene from mustard and its light-dependent expression. Plant Molecular Biology 16:175-185.

111. Ensminger, P.A., and E. Schäfer. 1992. Blue and ultraviolet-B light photoreceptors in Parsley cells. Photochem. Photobiol. 55:437-447.

112. Smith, T.A. 1985. Polyamines. Ann. Rev. Plant Physiol. 36:117ff.

113. Kramer, G.F., D.T. Krizek, and R.M. Mirecki. 1992. Influence of photosynthetically active radiation and spectral quality on UV-B induced polyamine accumulation in soybean. Photochem. 31:1119-1125.

114. Teramura, A.H., J.H. Sullivan, and L.H. Ziska. 1990. Interaction of elevated ultraviolet-B radiation and CO2 on productivity and photosynthetic characteristics in wheat, rice and soybean. Plant Physiol. 94:470-475.

115. van de Staaij, J.W., G.M. Lenssen, M. Stroetenga, and J. Rozema. 1992. The combined effects of elevated $CO_2$ levels and UV-B radiation on growth characteristics of *Elymus athericus* (=*E. pycnanathus*), Vegetatio 0:1-8.

116. Rozema J., G.M. Lenssen, and J.W.M. van de Staaij. 1990. The combined effect of increased atmospheric $CO_2$ and UV-B radiation on some agricultural and salt marsh species. In: J. Goudriaan, H. van Keulen, and H.H. van Laar (eds.), The Greenhouse Effect and Primary Productivity in European Agroecosystems, Pudoc, Wageningen, pp.68-71.

117. Tevini M., and U. Mark. 1993. Effects of elevated ultraviolet-B-radiation, temperature and $CO_2$ on growth and function of sunflower and corn seedlings. In: A. Shima, M. Ichahashi, Y. Fujiwara, and H. Takebe (eds.), Frontiers Of Photobiology: Proceedings of the 11th International Congress on Photobiology, Kyoto, Japan, 7-12 Sep., 1992, Elsevier Science Publishers B.V.

118. Feder, W. A., and R. Shrier. 1992. Combination of UV-B and ozone reduces pollen tube growth more than either stress alone. Environ. Bot. 30:451-454.

119. Dube, L.S., and J.F. Bornman. 1992. The response of young spruce seedlings to simultaneous exposure of ultraviolet-B radiation and cadmium. Plant Physiol. Biochem. 30:761-767.

# EXPOSURE RESPONSE CURVES ACTION SPECTRA AND AMPLIFICATION FACTORS

Thomas P. Coohill

Ultraviolet Consultants
Suite 106, 652 East 14th Street
Bowling Green, KY 42101 USA

Key words: Action spectra, effectiveness spectra, exposure response curves, ozone depletion, photobiology, radiation amplification factor, UV radiation

## ABSTRACT

Once it was established that in many places additional ultraviolet (UV) radiation was reaching the earth's surface due to the depletion of the earth's protective ozone layer, the primary question became: What does this mean to life forms? Surprisingly little is known about the biological effects of UV-B (290-320nm) radiation, the major region of the solar spectrum to be affected by ozone loss. This is partly because the UV-C spectrum, which DNA and proteins absorb heavily, and the UV-A and visible spectra, where most of photosynthesis takes place, attracted the greatest scientific attention. The dearth of information about the effect of UV-B on life forms is now being addressed. Among the first pieces of information needed to answer the question posed above is the construction of reliable UV-B action spectra (AS) for crucial biological endpoints. When available, these AS combined with an estimate of the ambient radiation spectra can give rise to effectiveness spectra (ES) that will allow us to estimate the effects of this increased UV-B on life.

## EXPOSURE RESPONSE CURVES (ERC)

In order to begin the analysis known as action spectroscopy one has to generate a series of accurate exposure response curves. It is important to realize that these curves often vary in shape and will be influenced by the amount and wavelength distribution of the incident radiation. If true monochromatic radiation is employed, the latter problem is alleviated. Some curves exhibit immediate exponential changes in biological effects at the lowest exposures utilized. Some responses are not exponential; others exhibit a variety of shapes that are highly dependent on the organisms being studied and the type of radiation being used. The literature attests to the complexity of ERCs [5]. However complicated, such curves usually show at least the additional amount of effect due to increased exposure, and thus aid in predicting the new level of response. An idealized example of one type of CRC is shown in Figure 1. This curve exhibits a typical "survival" curve shape, here measuring the degree of cell death by exposure to increasing amounts of UV-B. The initial flat portion of the curve is the "shoulder" region where no effect is noted until a certain threshold exposure is reached (here, 6 $Jm^{-2}$). The extent of this shoulder varies among different cells and is the result of the amount of damage sufficient to inactivate a cell and the ability of that cell to repair some of that damage. Cells normally being exposed to an amount of UV-B that is within the limits of the threshold may not show any increase in response in, for example, a situation of ozone depletion where 20% additional UV-B is reaching the exposed cells. This is shown in the figure by an initial exposure $A_0$ of 4 $Jm^{-2}$ compared to an exposure $A_1$ of 4.8 $Jm^{-2}$, an increase of 20%. Hence, cells under low levels of stress, as determined by their position on the ERC, would show no detrimental effects due to this amount of ozone loss. Again, as shown in Figure 1, the shoulder region is followed by a region of exponential

NATO ASI Series, Vol. I 18
Stratospheric Ozone Depletion/
UV-B Radiation in the Biosphere
Edited by R. H. Biggs and M. E. B. Joyner

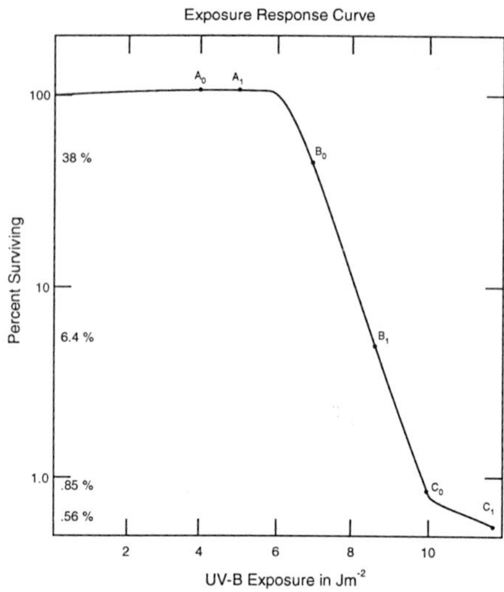


Figure 1. Idealized exposure response curve for cell killing by UV-B.
Exposure $A_0 = 4$ Jm$^{-2}$; $A_1 = A_0 + 20\% = 4.8$ Jm$^{-2}$
Exposure $B_0 = 7$ Jm$^{-2}$; $B_1 = B_0 + 20\% = 8.4$ Jm$^{-2}$
Exposure $C_0 = 10$ Jm$^{-2}$; $C_1 = C_0 + 20\% = 12$ Jm$^{-2}$

loss of function with increasing exposure. Here the cells obey the rules of target theory and respond to increasing amounts of radiation by losing function in an exponential manner. Additional exposures to cells on this portion of the curve give rise to significant increases in effect. As shown, an exposure of 7 Jm$^{-2}$ lowers cell survival to the 38% level. An additional 20% exposure (for a total of 8.4%) lowers survival down to a level of 6.4%, a six-fold increase. If organisms are at this position on the curve due to normal ambient radiation, then even small differences in ozone cover can have large effects. It should be mentioned that the point where the shoulder of the curve ends and the exponential portion begins is often called the "stress limit" of the organism for that type of radiation. At this time, it appears that some phyto-plankton in natural waters may be at their stress limit and could be severely impacted by further ozone loss [13], although ex-periments reported in the literature vary widely in measuring and predicting the UV-B responses of these crucial organisms. Finally, some survival curves exhibit a resistant "tail," such as the one shown in Figure 1, at exposures above 10 Jm$^{-2}$. The existence of this tail may be due to a resistant sub-population, changes in sensitivity due to position in the developmental cycle, or a variety of other causes. If the organisms in question are at this position on the CRC, then additional UV-B will cause moderate increases in effect. Here note that an exposure of 10 Jm$^{-2}$ (Co) lowered survival to a level of 0.85%; an additional 20% (12 Jm$^{-2}$) brought survival down to a level of 0.56%, an increase of less than a factor of two. Thus it is essential that a knowledge of the raw data used to generate the ERCs be included in the first paper that reports any set of new results. Only when the ERCs are accurate and well characterized can one go to the next step.

## ACTION SPECTRA (AS)

Once a series of monochromatic exposure response curves are available, a true AS can be constructed. There are a variety of ways to choose a parameter for comparing ERCs. These include a specific level of response (e.g.. the exposure to lower survival to the 10% level), the slope of the exponential portion of the ERCs, or other methods [16]. These choices are subject to a variety of constraints [18,7], one of which is that the ERCs should be superimposed upon one another if a suitable exposure modification factor is applied [14]. Usually an AS is then plotted as the relative effect per photon. Examples of this are shown in Figure 2. As shown, even for a complex set of reactions like photosynthesis, the wide variety of AS generated appear to have little agreement in the UV. An advantage of this type of plot is that it can readily be compared with other AS regardless of the exposures needed to elicit the different effects. A disadvantage is that this type of plot can obscure the absolute values needed to cause an effect. This becomes important in the construction of ES because it does not tell you if any effect will be observed. For example, if the amount of available ambient radiation is insufficient to effect a change in the response measured, even with the predicted extra UV-B, then a false appearance of extra effect is given. Only if both the AS and the solar radiation spectrum are in absolute units will a true effectiveness spectrum be generated.

## ANALYTICAL ACTION SPECTRA

If a number of carefully executed ERCs at several individual wavelengths are available, then a "true" or analytical action spectrum (AAS) can be con- structed. If this AAS corresponds closely to be the absorption spectrum of a molecule that can be shown to affected by exposure to radiation in the wavelength region being tested, and if that the molecule, within reason, can be thought of as being involved in the biological process being investigated, then a claim can be made that this molecule is primarily involved in the effect, and that photodamage within the molecule leads to the noted changes in response. This method was the first to point to chlorophyll as the target molecule for photosynthesis [9], and to nucleic acid as the genetic material [12]. In some cases the absorbing chromophore may not be the in the molecule which causes the effect, but may act as an intermediate in the transfer of energy to the eventual target molecule. The latter analysis is more difficult to prove, but has been observed [8].

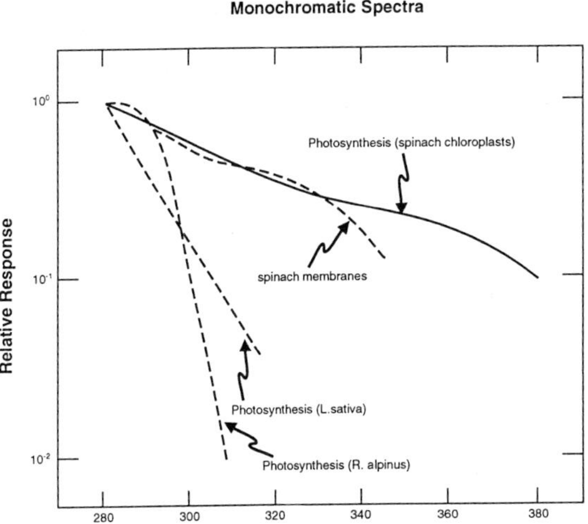


Figure 2. Monochromatic action spectra in higher plants. Spinach data from Jones and Kok (1966) and Bornman et al. (1984). Lower two curves from Bogenrieder (1982).

POLYCHROMATIC ACTION SPECTRA

It is sometimes impossible to utilize single wavelength radiation sources and to arrive at a true AAS. In addition, organisms live in an environment in which many UV wavelengths are present, from 290-380nm. Sometimes the response of individual isolated cells is not as desirable as a biological endpoint, but rather the response of the complex multicellular organism is the ultimate goal. Such is often the case for field studies. Therefore, many authors have produced polychromatic AS (PAS) that utilize a variety of sources, from polychromatic sources to single wavelengths added to a polychromatic ambient background. Such PAS are useful in two major ways: first, they combine in the measured responses the natural repair and ancillary reactions that occur in organisms in the natural setting; second, they are closest to the natural setting and more directly useful in anticipating such complex responses as plant yield. Their disadvantage is that they rarely point to the major chromophore for the individual response. Examples of PAS are shown in Figure 3. Once more, the agreement among authors is slight and reflects the paucity of data that can be relied upon for the choosing of a representative AS for this complex plant response.

## EFFECTIVENESS SPECTRA

Once an AS for a given effect is in hand, and if the ambient radiation background (here the total UV exposure and its wavelength distribution) is known, then the two can be combined into an effectiveness spectrum (ES). A typical ES is shown in Figure 4 [6].

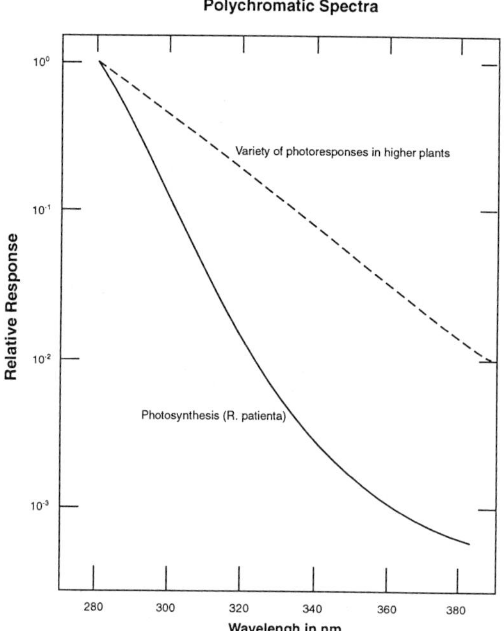


Figure 3. Polychromatic action spectra in higher plants. Dashed line data from Tevini and Steinmuller (1988); solid curve, Caldwell et al. (1986).

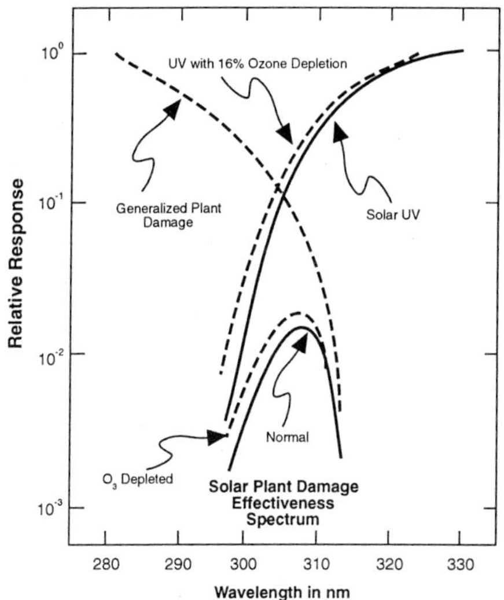


Figure 4. Solar effectiveness spectra from generalized plant damage action spectrum (Caldwell, 1977) and the solar spectrum with and without a 16% ozone loss.

Here one (of many) AS for generalized plant damage is combined with the known ambient UV radiation background and with one that would occur under an ozone layer that was depleted by 16%. This gives rise to two ES, one for the normal setting (solid line) and one for the ozone depleted setting (dashed line). The differences in the areas under these two curves gives an estimate of the additional damage to plant yield for a depleted ozone layer of 16%. For this figure, and only for this AS, the addi- tional loss of yield is about 15%. That is, for each 1% loss in ozone, a corresponding 1% loss in yield is calculated. Such estimates are useful only if one is aware of the limits of interpreting data from an AS. Figures 2 and 3 shown clearly that other AS would yield different ES. In other cases an ES is more exact. For example, if one combines the AS for damage to naked DNA [19] with a solar radiation spectrum, then a reasonably accurate ED is produced—for naked DNA only.

## BIOLOGICAL RADIATION AMPLIFICATION FACTORS (RAF)

It is sometimes desirable to assign a number called a "radiation amplifications factor" [17] to the differences in response for a biological system under different scenarios of UV exposure (e.g., due to various levels of ozone depletion). It thus becomes simple to state, e.g., that "for each 1% loss of ozone, a 2% increase in skin cancer is expected." The value of this terminology lies in its comprehensibility to laymen and scientists who are not familiar with action spectroscopy or even photobiology. A series of RAFs have been generated by different authors for different effects. For example, an RAF of between 2 and 3 for non-melanoma skin cancer [11]; between 1 and 2 for cataracts [20]; vaguely between 1 and 2 for certain immune responses [10]; about 1 for certain plant yields [6]; between 0 and 3 for oceanic phytoplankton production (various authors), etc. The availability of these RAFs is useful only if the user is aware of the limitations of the data that generated the RAF. Even if a number of studies and AS agree of the amount of effect, it should be remembered that RAFs, in general, will not be linear over a wide range of ozone depletion scenarios, as is to be expected from an analysis of ERCs. In addition, adaptation, pigmentation, avoidance response, change in behavior (and exposure to sunlight), mutation and other variables will limit the use of any RAF. If these caveats are taken seriously, and if the public is educated to expect RAFs to be updated as more data is collected, then their use is reasonable. So the challenge for the photobiologist is to provide these estimates so that policies and behaviors can be modified, but to couch them in terms that allow for their future change without public alarm. This may best be accomplished by the use of ranges of RAFs instead of single numbers. Although Lord Kelvin exclaimed that it was only science when a number could be assigned to some effect, a number, in and of itself, is often too powerful a thing, and gives the illusion of exact and immutable fact. We are not at the latter stage in photobiology, nor are we likely to get there.

## REFERENCES


1. Bogenrieder, A. 1982. Action spectra for the depression of photosynthesis by UV irradiation in *Lactu stativa* L. and *Rumex alpinus* L. *In* H. Baur, M.M. Caldwell, M. Tevini, and R.C. Worrest (eds.), Biologic Effects of UV-B Radiation, pp.132-139.

2. Bornman, J.F., L.O. Bjorn, and H.E. Akerlund. 1984. Action spectrum for inhibition by ultraviolet radiation of photosystem II activity in spinach thylakoids. Photobiochem. Photobiophys. 8:305-313.

3. Caldwell, M.M. 1971. Solar UV radiation as an ecological factor for alpine plants. *In* A.C. Giese (ed.), Photophysiology, Vol. 6, pp.131-177.

4. Caldwell, M.M., L.B. Camp, C.W. Warner, and S.D. Flint. 1986. Action spectra and their key role in assessing biological consequences of solar UV-B radiation. *In* R.C. Worrest and M. M. Caldwell (eds.), Stratospheric Ozone Reduction, Solar Ultraviolet Radiation and Plant Life, Springer, Heidelberg. pp. 87-111.

5. Coohill, T.P. 1984. Action spectra for mammalian cells *in vitro*. *In* K.C. Smith (ed.), Topics in Photomedicine, Plenum, New York, pp.1-37.

6. Coohill, T. P. 1989. Ultraviolet action spectra (280 to 380nm) and solar effectiveness spectra for higher plants, Photochem. Photobiol. 50:451-457.

7. Coohill, T.P. 1991. Action spectra again? Photochem. Photobiol. 54:859-870.

8. Coohill, T.P., M.J. Peak, and J.G. Peak. 1987. The effects of the ultraviolet wavelengths of radiation present in sunlight on human cells *in vitro*. Photochem. Photobiol. 46:1043-1050.

9. Daubeny, C. 1836. On the action of light upon plants and plants upon atmosphere. Phil. Trans. Roy. Soc. London, 1836, pp.149-179.

10. De Fabo, E.C., F.P. Noonan, and J.E. Frederick. 1990. Biologically effective doses of sunlight for immune suppression at various latitudes and their relationship to changes in stratospheric ozone. Photochem. Photobiol. 52:811-817.

11. De Gruijl, F.R., and J. van der Leun. 1992. Action spectra for photocarcinogenesis. *In* F. Urbach (ed.), Biological Responses to UV-A Radiation, Valdenmar Publishing, Overland Park, Kansas.

12. Gates, F.L. 1930. A study of the bacteriacidal action of ultraviolet light. III. The absorption of ultraviolet light by bacteria. J. Gen. Physiol. 14:31-42.

13. Hader, D.P. 1988. Ecological consequences of photomovement in microorganisms. J. Photo-chem. Photobiol. B: Biol. 1:385-414.

14. Jagger, J. 1967. Introduction to Research in Ultraviolet Photobiology. Prentice-Hall, Englewood Cliffs, NJ.

15. Jones, L.W., and B. Kok. 1966. Photoinhibition of chloroplast reactions. I: Kinetics and action spectra. Plant Physiol. 41:1037-1043.

16. Lipson, E.D. 1991. Action spectroscopy. *In* F. Lence, F. Ghetti, G. Colombetti, D.-P. Hader, and P.-S. Song (eds.), Biophysics of Photoreceptors and Photomovements in Microorganisms, Plenum, NY.

17. Rundel, R.D. 1983. Action spectra and estimation of biologically effective UV radiation. Physiol. Plant. 58:360-366.

18. Setlow, R.B. 1957. Action spectroscopy. *In* J.H. Lawrence and C.A. Tobias (eds.), Adv. Biol. Med. Physics, Vol. 5, Academic Press, New York.

19. Setlow, R.B. 1974. The wavelengths in sunlight effective in producing skin cancer. A theoretical analysis. Proc. Nat. Acad. Sci. USA 71:3363-3366.

20. Taylor, H.R. 1989. The biological effects of UV-B on the eye. Photochem. Photobiol. 50:489-492.

21. Tevini, M. and Steinmuller. 1988. Personal comminication.

# INTERACTION OF UV-B RADIATION AND THE PHOTOSYNTHETIC PROCESS*

Janet F. Bornman

Dept of Plant Physiology, Lund University S-220 07 Lund, Sweden

**ABSTRACT**

Alterations in morphology, ultrastructure, and pigments may be caused by exposure to UV radiation. This in turn may result in many indirect effects on cell processes and subsequent growth. One example of this is the altered distribution of photosynthetically-active radiation (PAR, 400-700 nm) with depth in a leaf after exposure of plants to UV-B radiation (Bornman and Vogelmann, 1991), with resultant change in the microenvironment for processes such as photosynthesis. Different protective mechanisms such as UV-B screening pigments modify the UV response, however. The effectiveness of these pigments is reflected in light gradients within a leaf as measured by fiberoptic microprobes.

The composition of photosystems (PS) I and II suggests that UV radiation would be effective in inducing changes in both systems, although PS II is generally more affected. The different components of the PS II reaction center seem to be variably influenced by UV-B radiation.

Simultaneous exposure to both UV-B radiation and high or low PAR was investigated with regard to the quantum yield (variable fluorescence/maximum fluorescence, $F_v/F_m$), photochemical ($q_P$) and non-photochemical quenching processes ($q_{NP}$), and D1 turnover. Low and high PAR (400 and 1600 mol $m^{-2}$ $s^{-1}$) were used with respective levels of UV-B radiation (13 and 3.25 kJ $m^{-2}$ $day^{-1}$), to simulate a period of cloudy weather followed by a period of high light. Photoinhibition and increased turnover of D1 were observed under high PAR and high UV-B radiation; no photoinhibition was observed with simulation of a cloudy day (400 mol $m^{-2}$ $s^{-1}$), and lower level of UV-B (3.25 kJ $m^{-2}$ $day^{-1}$), although increased turnover of the D1 protein still occurred under low PAR and UV-B radiation compared to low light conditions without UV-B radiation. Thus, even though small amounts of UV-B radiation did not enhance photoinhibition, the increased turnover of Dl may partly explain the often-reduced productivity in many studies (C. Sundby-Emanuelsson and J.F. Bornman, unpubl. data). However, photoinhibition by visilble light and UV inhibition may not always reflect damage via similar mechanisms.

Thus, although rates of $CO_2$ assimilation as a measure of net photosynthesis do not always follow the sometimes-larger changes due to UV radiation found within PS II, disturbances of these partial reactions may be reflected in more subtle changes of the carbon cycle.

Reference: Bornman, J.F., and Vogelmann, T.C. 1991 The effect of UV-B radiation on leaf optical properties measured with fibre optics. J. Exp. Bot. 42:547-554.

*Please see Dr. Bornman's full manuscript, p. 345, this publication.

NATO ASI Series, Vol. I 18
Stratospheric Ozone Depletion/
UV-B Radiation in the Biosphere
Edited by R. H. Biggs and M. E. B. Joyner

# BIOLOGICAL RESPONSES TO UV-B RADIATION DOSAGES

Eckard Wellmann

University of Freiburg, Germany

**ABSTRACT**

The effects of UV as a wide-spread phenomenon in sensitive plants have been identified. These effects result in growth inhibition and formation of UV-absorbing phenylpropanoin and flavonoid pigments. The responses occur with-in the framework of photomorphogenesis and represent protective mechanisms against UV damage. Though coactions of UV with longer wavelength light (via phytochrome and the blue light receptor) are typical, in all cases these UV effects show maximal spectral sensitivity in the UV-B range and are linearly dependent on the UV fluence (rate). Spectral fluence/effect relationships for UV-protective reactions and for DNA damage by pyrimidin dimer formation will be compared. Considering induction properties resulting from a further decay of the stratospheric ozone layer would not significantly contribute to an increase of the UV-protective reactions. This UV increase rather could inhibit these processes that depend on gene activation and, therefore, are sensitive to DNA damage, for example, by dimer formation.

NATO ASI Series, Vol. I 18
Stratospheric Ozone Depletion/
UV-B Radiation in the Biosphere
Edited by R. H. Biggs and M. E. B. Joyner

# TEMPORAL AND FLUENCE RESPONSES OF TREE FOLIAGE TO UV-B RADIATION

Joseph H. Sullivan

Department of Botany, University of Maryland
College Park, Maryland 20742

Key Words: leaf development, loblolly pine, photosynthesis, sweetgum, ultraviolet-B radiation

**ABSTRACT**

Although few studies have been conducted on the effects of UV-B radiation on trees, several field and glasshouse studies have demonstrated the importance of UV-B radiation to plant morphology and growth. However, in spite of these reductions in growth, consistent observations of reductions in photosynthesis have not been demonstrated. Some possible explanations for this lack of consistent effects on photosynthesis include transient UV-B effects, which may be in part a function of leaf age, chronic UV-B effects, which are not readily detected by gas exchange analysis,and direct effects of UV-B radiation on leaf growth and development which are independent of photosynthesis. This manuscript reviews the evidence for these hypotheses in loblolly pine (*Pinus taeda*), the tree species for which the most information exists. The possible extension of these results to other species is considered by an evaluation of the data available for sweetgum (*Liquidambar styraciflua*), a species with similar lack of shade adaptation but contrasting growth habit.

The direct effects of UV-B radiation on photosynthesis in loblolly pine appear to be transient, appearing only in specific age classes of needles. However, carbon isotope data also suggest that exposure to high levels of UV-B radiation may lead to small increases in internal $CO_2$ concentrations, which are indicative of reductions in photosynthesis over the life of the needle. In addition to photosynthetic depression, several studies have demonstrated reductions in needle biomass and length in response to UV-B radiation. Therefore the effects of UV-B radiation on growth in loblolly pine appear to be manifested by the combination of both reductions in photosynthetic leaf area and chronic reduction of photosynthetic capacity.

In contrast, preliminary studies on sweetgum suggest that UV-B radiation enhanced the rate of development of photosynthetic capacity in leaves and this was apparently the result of more rapid development of leaf chlorophylls and carotenoids. However, the accelerated leaf phenology may also result in early senescence as suggested by reduced photosynthetic capacity in mature leaves and reductions in green leaf area late in the growing season in sweetgum seedlings irradiated with supplemental UV-B radiation. Although biomass in sweetgum was only minimally affected by supplemental UV-B radiation, clearly both sweetgum and loblolly pine are capable of detecting and responding to alterations in light spectral quality in the ultraviolet region. However, additional research is needed in order to understand the role of ultraviolet radiation in leaf phenology, anatomy, and plant productivity.

## INTRODUCTION

Leaves must trap light to be effective as photosynthetic organs and in so doing must overcome possible damaging effects of infrared radiation and ultraviolet radiation [11]. Therefore, in addition to the often discussed photosynthesis-transpiration compromise, a related compromise also exists between the essential need to carry on metabolic processes (e.g., carbon assimilation) and the ability to avoid the

NATO ASI Series, Vol. I 18
Stratospheric Ozone Depletion/
UV-B Radiation in the Biosphere
Edited by R. H. Biggs and M. E. B. Joyner

damaging effects of radiation outside of the photosynthetically active region (400-700 nm). For this reason, several mechanisms of protection from and repair of damage from ultraviolet-B radiation (UV-B between 280 and 320 nm) have been attributed to higher plants as potential means of acclimation or adaptation to this environmental stress [3,7]. In spite of these protection mechanisms, many plants show adverse effects on growth or physiological processes when exposed to UV-B radiation (see [44,46] for recent reviews). However, much variability exists among plant species with respect to UV-B radiation sensitivity. This variability may in fact attest to the importance of UV-B radiation as a selective factor, but it also makes extrapolation of results between species extremely difficult. Therefore, plants representing a diverse range of life forms and habitats must be studied to enhance our understanding of the importance of UV-B radiation to terrestrial plant physiology, productivity and ecosystem processes.

Although forests may account for over two-thirds of global NPP, compared to about 11% for agricultural land [50], less then one dozen studies on the effects of UV-B radiation on growth or productivity of trees have been published. Basioumy and Biggs [2] demonstrated visual stunting of one-month-old peach seedlings under supplemental UV-B radiation in a greenhouse while Semeniuk [32] found no visual symptoms in other greenhouse studies on 6 ornamental species. Bogenreider and Klein [4], however, demonstrated in field studies that biomass accumulation in young seedlings of four tree species was increased when UV-B was excluded. Two other screening studies [25,37] found that roughly half of the coniferous species studied exhibited reduced growth under elevated UV-B irradiation. Further studies by Sullivan and Teramura [38] on loblolly pine in a greenhouse indicated that UV-B radiation reduced photosynthetic capacity, quantum efficiency, and growth even though the foliar concentration of UV-absorbing compounds increased.

Kaufmann [24] conducted the first UV supplementation study of trees in the field and irradiated seedlings of *Picea engelmannii* and *Pinus contorta* in the field for one season. Although he failed to detect any detrimental effects during that year, he found visible symptoms and increased mortality during the subsequent growing season following irradiation. This raised the question of whether UV-B damage might be cumulative in trees, especially evergreens. Therefore, Sullivan and Teramura [39] grew loblolly pine for three years in the field under supplemental UV-B irradiation and found, in general, growth reductions increased over the three year period and were on the order of 12 to 20%. Of particular interest was the fact that after three years, growth reductions in seedlings from two of the four seed sources were approximately equal between simulated 16% and 25% ozone depletion scenarios.

Growth of tree seedlings was affected by UV-B radiation in six of the eight studies cited above. However, few of the above studies evaluated the physiological processes which may be affected by UV-B radiation or experimentally addressed the potential mechanisms of UV-B radiation damage. This paper reviews some of the data available for trees and centers on studies on loblolly pine (*Pinus taeda*), the most widely studied of any tree species. It also includes some preliminary information on other species including sweetgum (*Liquidambar styraciflua*), which co-occurs with loblolly pine and is considered to be similarly shade intolerant, but exhibits a contrasting growth habit. Since leaves are the most important structures for plant productivity [49] this discussion will focus on UV-B effects on leaves and will consider two general modes of damage: reduction in photosynthetic carbon assimilation and direct effects on leaf growth.

## EPIDERMAL ATTENUATION OF UV-B BY TREE FOLIAGE

Any response of foliage to UV-B radiation is initiated by absorption of UV-B photons by chromophores. One means to avoid damage by UV-B radiation then is to prevent it from reaching sensitive targets. Leaf surface reflectance in the UV-B region is generally below 10% [10,20,23]. Although alterations in leaf epicuticular waxes [35,47] and leaf hairiness affect UV-B absorption [22,23], most UV-B radiation is presumed to be absorbed in the epidermis of foliar tissue. The accumulation of UV-absorbing compounds, such as flavonoids, in the epidermis accounts for much of this absorption [6,30,48] and is a protective response to UV-B radiation [3]. A recent survey of some 22 plant species [12] has shown a wide range of differences in the amount of UV-B radiation which penetrate through the epidermis, but essentially no UV-B radiation reached the mesophyll in the nine confers studied, and very little epidermal transmission (<10%) was observed in three woody dicots.

One conclusion to be drawn from these studies is that apparently the combination of screening mechanisms discussed above are quite effective in reducing the quantity of UV-B radiation which reaches potential photosynthetic targets in the mesophyll. However, four of the nine species screened by Day et al. [12] have been tested for UV-B sensitivity and, of these, two showed some sensitivity to enhanced levels of UV-B radiation. Ultraviolet-B radiation reduced the biomass of *Pinus ponderosa* and *P. contorta* in growth chamber [25] and glasshouse [37] studies although Day et al. [12] found essentially no transmission of UV-B radiation through the epidermis in native trees of these species. Also loblolly pine [39] and Scots pine (*P. sylvestris*) [18] appear to be deleteriously affected by UV-B radiation. Overall 8 of 16 conifer species tested, have shown some sensitivity to UV-B radiation. Therefore, physiologically significant levels of UV-B radiation are apparently present with enough frequency to cause damage in some tree species.

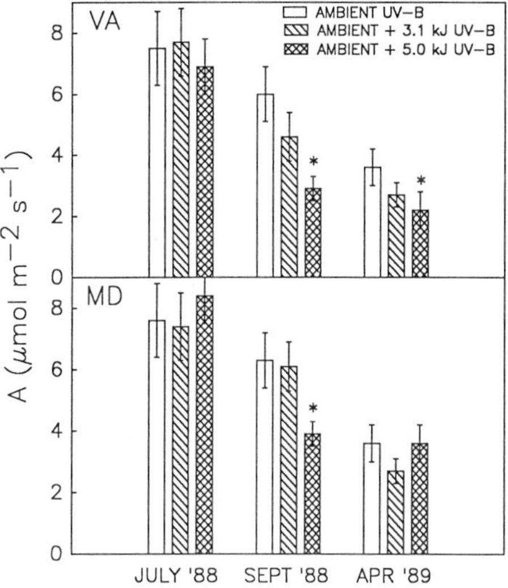


Figure 1. The effects of supplemental UV-B radiation on $CO_2$ assimilation of loblolly pine from two seed sources (northern Virginia [VA] and eastern Maryland [MD] grown for three years in the field under either ambient or ambient plus supplemental UV-B radiation. Each bar represents the mean of 6 plants and vertical bars are the standard error of the mean. An asterisk (*) represents a significant difference (P<0.05) from ambient UV-B controls.

## PHOTOSYNTHETIC DAMAGE IN LOBLOLLY PINE

In general, across all species studied, there are about as many reports of reductions in net photosynthesis by UV-B radiation as there are reports of no effects of UV-B radiation on net photosynthesis. Therefore, it is clear that response differences abound among species and even among genotypes [27,29]. UV-B radiation has been shown to reduce photosynthesis in loblolly pine under greenhouse [38,40] and field conditions [43] although measurements made on loblolly pine over an extended time period on the same needle cohort suggested that reductions in photosynthesis were not consistently detectable (Figure 1). Thus in an effort to explain the growth reductions observed for loblolly pine at least three hypotheses should be considered in future research. First, reductions in photosynthetic carbon assimilation may occur only during specific periods of needle development. Second, reductions in photosynthetic carbon assimilation may be chronic but not readily detectable by analysis of $CO_2$ uptake. Finally, growth reduction may be independent of direct effects of UV-B on net photosynthesis.

DuLucia et al. [13] demonstrated that in two high elevation conifers, UV-B penetration was dependent upon leaf age with significant UV-B penetration occurring only at very early developmental stages. In measurements made on several needle-age classes of field-grown loblolly pine [39], Sullivan [36] and Naidu et al. [28] observed age-dependant effects of UV-B radiation on the potential maximum photochemical efficiency of photosystem II (PS II) as measured by the ratio of variable to maximum chlorophyll fluorescence (Figure 2). Further studies that evaluated temporal effects of UV-B during needle development [40] also demonstrated that UV-B effects on photosynthesis may be transient since apparent damage to PS II was detected intermittently during needle development. The possible interactions between photosynthetic damage and changes in leaf

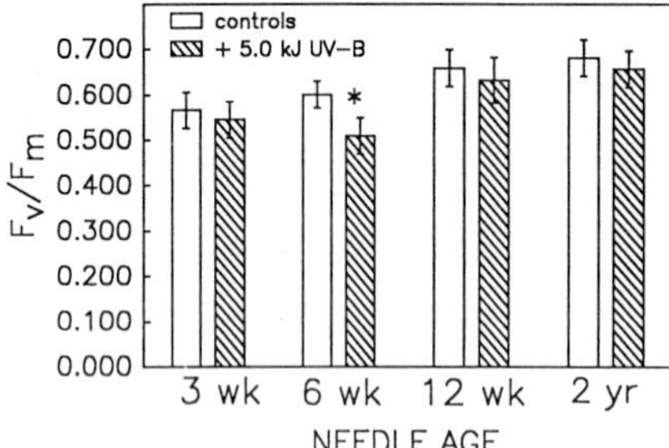


Figure 2. Potential maximum photosynthetic efficiency, as measured by the ratio of variable to maximum fluorescence ($F_v/F_m$). Measurement were made near the end of the photoperiod after 15 min. dark acclimation on plants grown in the field under either ambient or ambient plus supplemental UV-B radiation. Each bar is the mean of 12 plants and vertical bars are standard errors of the means. The asterisk (*) represents a significant difference ($P<0.05$) between treatments.

optical properties such as screening or in repair processes, which may be associated with needle development, have not been fully evaluated but further studies on this may contribute to our understanding of the nature of this transient UV-B damage.

The assessment of photosynthetic carbon assimilation by either oxygen evolution under saturating conditions of light and $CO_2$, or $CO_2$ uptake by infrared gas analyzers is highly variable. This is particularly true for the latter method where stomatal limitations may be quite important. For this reason small differences in photosynthesis between experimental treatments may go undetected since they could be hidden statistically within the large inherent variation of the measurements. However, the use of other techniques such as measurements of foliar nitrogen concentrations or $W^{13}C$ isotope ratios can provides a time integrated measure of photosynthetic function [16,17,19] and provide some useful insight into the photosynthetic performance of leaves. An analysis of $W^{13}C$ discrimination ratios in field-grown loblolly pines revealed that supplemental UV-B radiation resulted in statistically significant reductions (i.e. more negative $W^{13}C$ ratios) in two of three needle age classes measured [28]. This difference is indicative of slightly higher internal $CO_2$ concentrations over the life of the needle (Figure 3) which, in the absence of altered stomatal conductance, imply small and perhaps chronic reductions in photosynthesis.

## UV-B EFFECTS ON NEEDLE GROWTH

Leaf growth rate may be affects by many environmental stresses [11,31] and altered leaf growth may be an early response to stress observed in advance of gross physiological symptoms of stress [9]. Reduction in leaf growth may also represent a defense mechanism for plants exposed to increased levels of UV-B radiation [3]. Increased levels of UV-B radiation were found to reduce leaf expansion of several herbaceous species [13,26,33,34,41], and substantial evidence suggests that the growth of some conifer needles may also be reduced by UV-B radiation.

Reductions in needle biomass in loblolly pine have been reported in several studies [25,37-39] and Naidu et al. [28] also reported reduced needle lengths in response to UV-B radiation. The reduction in needle growth does not appear to be the result of reduced photosynthetic rate since $CO_2$ enrichment, which increased photosynthesis, did not eliminate reductions in needle growth by UV-B radiation [40]. Further evidence for reductions in needle elongation

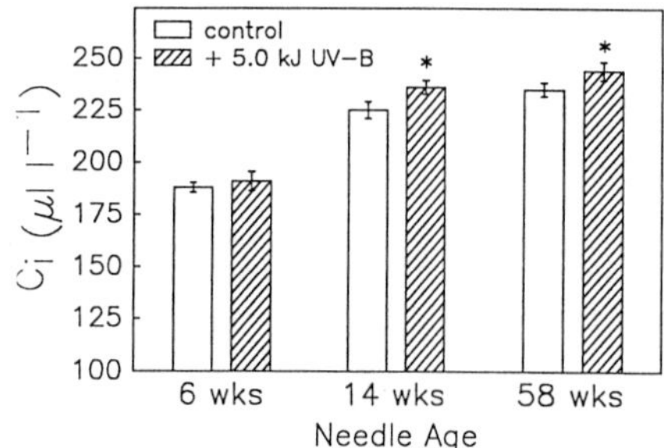


Figure 3. Calculation of internal $CO_2$ concentrations from carbon isotope ratios [28] in loblolly pine seedlings grown in the field for three successive years under either ambient or ambient plus supplemental UV-B radiation. Each bar represents the mean of 5 plants and vertical bars are the standard deviation of the mean. An asterisk (*) represents a significant difference ($P<0.05$) between treatments.

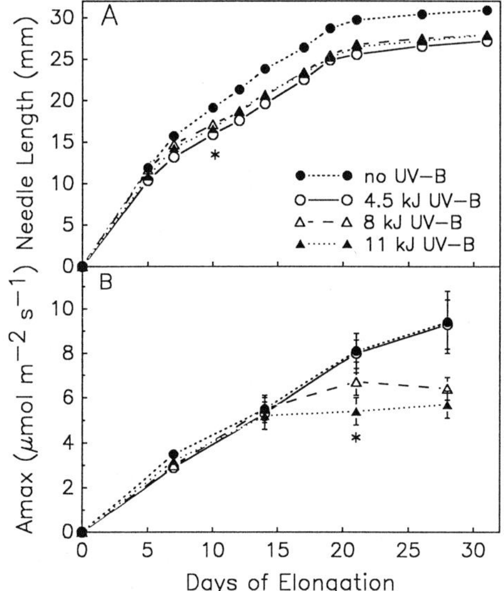


Figure 4. The effects of UV-B radiation on primary needle elongation (A) or photosynthetic capacity (Amax) as measured by oxygen evolution (B) in loblolly pine seedlings grown in greenhouse under either 0, 4.5, 8 or 11 kJ $m^{-2}$ of biologically effective UV-B radiation [5]. Each bar is the mean of 12 plants and vertical bars are standard errors of the means. The asterisk (*) represents the point at which significant differences ($P<0.05$) between supplemental and no UV-B radiation were detected.

rate were obtained in recent studies at the University of Maryland (J.H. Sullivan, L.R. Dillenburg, and A.H. Teramura [unpublished]) in which the elongation rate of primary needles of loblolly pine was reduced by UV-B radiation after 10 days of exposure (Figure 4A). This reduction was most likely independent of any photosynthetic effects since reductions in photosynthetic capacity were detected only after 21 days of exposure to UV-B radiation (Figure 4B).

In conclusion, it appears that growth reduction in loblolly pine as a result of UV-B radiation is likely to be due to a combination of factors including: 1) transient effects of UV-B radiation which may limit carbon gain during specific periods of needle development; 2) small chromic effects which reduce carbon gain over the life of the needle; and 3) direct effects of UV-B radiation on needle elongation which limit total plant carbon gain by reducing leaf surface area available for photosynthesis rather than by reducing photosynthetic rate per unit area.

## UV-B RADIATION EFFECTS ON SWEETGUM

Loblolly pine and sweetgum are important species which co-occur on successional sites and compete vigorously in commercial plantations throughout the southeastern United States. They are similarly shade intolerant and thus would be expected to be at least moderately tolerant to UV-B radiation. However, contrasting growth forms such as evergreen and deciduous provide a natural contrasting species pair for comparative studies. Due to differences in leaf longevity and costs of leaf construction, one might anticipate that deciduous and evergreen species would respond differently to increased levels of UV-B radiation. However, only a few studies have included hardwood deciduous trees [2,4,32], and of these studies only Bogenrieder and Klein [4] was conducted in the field.

Because of differences in background radiation,

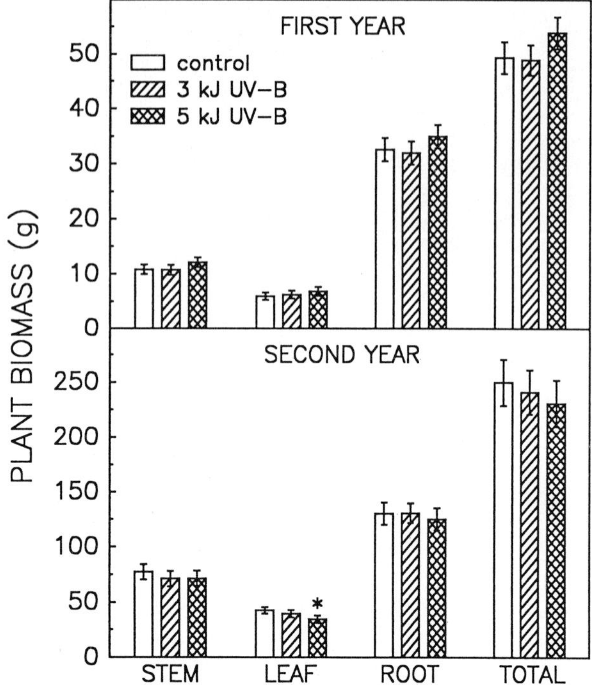


Figure 5. Biomass accumulation and allocation of sweetgum seedlings grown for two years in the field with either ambient or ambient plus supplemental UV-B radiation. Each bar is the mean of 12 plants and vertical bars are standard errors of the means. The asterisk (*) represents a significant difference (NP<0.05) between ambient UV-B and supplemental UV-B radiation

field and indoor studies very often generate contrasting results regarding sensitivity of plants to UV-B radiation. Much less sensitivity is usually found in field studies [42], and this may be due to repair mechanisms mediated by UV-A and visible radiation [1,8,18].

Sweetgum seedlings were grown in the field for two successive growing seasons under ambient and enhanced UV-B radiation. The supplemental UV-B radiation was provided by filtered fluorescent lamps as previously described [39,45] and the supplemental irradiance provided was calculated by the empirical model of Green et al. [21] to provide the amount of UV-B radiation anticipated with either a 16 or 25% depletion of stratospheric ozone over College Park, MD. This model predicted a maximum (clear sky on the summer solstice) daily supplementation of either 3.1 or 5.0 kJ $m^{-2}$ of biologically effective UV-B radiation when weighted with the generalized plant action spectrum of Caldwell [5].

In general, sweetgum appeared to be quite tolerant to UV-B radiation. The only growth reductions observed were reductions in leaf biomass following the second growing season (Figure 5). Interestingly, however, the production of photosynthetic pigments and the development of photosynthetic capacity was hastened by the highest supplemental UV-B irradiance [15]. For example, chlorophyll *a* content and photosynthetic capacity were increased by approximately 20% and 15%, respectively, by UV-B radiation during the first month of leaf development. However, photosynthetic capacity and chlorophyll content was found to be reduced by about 16% in mature (10 wk) leaves subjected to the same UV-B irradiation protocols. This suggests that leaf phenology may be altered by UV-B in sweetgum and once more points to the difficulty in interpretation of measurements made at single sampling periods. For example, contrasting conclusions would have been drawn had only one of the above measurements on photosynthetic capacity been conducted. These differences in the development of photosynthetic pigments and capacity imply that, while growth reductions were not observed, sweetgum was responsive to the spectral quality of incident light. This further suggests that the ultraviolet portion of sunlight or the ratio between visible and ultraviolet radiation may be important in the regulation of leaf development.

## CONCLUSIONS

Although the data available are minimal, it is apparent that both loblolly pine and sweetgum are responsive to the ultraviolet-B portion of the solar spectrum. In loblolly pine, this sensitivity includes reductions in needle elongation rate and slightly lower photosynthetic capacity. These effects, along with more pronounced reductions in photosynthesis during brief periods of needle development, were sufficient to cause significant losses in biomass over a three-year period.

Compared to loblolly pine, the effects of UV-B radiation on productivity in sweetgum appears to be reduced. However, some aspects of leaf development and canopy morphology such as potential reduction in leaf area duration (LAD) are influenced by UV-B radiation. Specifically, the development of photosynthetic pigments and capacity was hastened by UV-B radiation in sweetgum but this may have also resulted in early senescence and a loss of LAD.

Reductions in the productivity of some species as a result of an increase in solar ultraviolet radiation reaching the earth would obviously be important. Effects on productivity, per se, however, represent only one possible response to a changing solar radiation environment and essentially no data are available that adequately assess the likelihood of such effects at the level of forest ecosystems. Furthermore, subtle responses to UV-B radiation such as changes in leaf phenology, chemistry, and canopy structure could have ecological significance beyond simply alterations in biomass of a given species. For example, differential sensitivity to UV-B between species or changes in leaf phenology could alter plant competition, and changes in leaf chemistry may impact plant-insect interactions or leaf decomposition and nutrient cycling. It is clear that more information is needed on the role of current levels of UV-B radiation in the regulation of developmental and ecological processes before an adequate assessment of the likely consequences of a change in solar UV-B radiation can be made.

**Acknowledgments**

Grateful acknowledgment is made to A.H. Teramura, L.R. Dillenburg, S.L. Naidu, E.H. DeLucia, E. Pullman, A. McCoy, S. Luphold, and others who helped conceive and conduct many of the studies discussed in this review. This work was supported in part by USDA Competitive Grant No. 92-37100-7530.

## REFERENCES

1. Adamse, P., and S.J. Britz. 1992. Amelioration of UV-B damage under high irradiance: Role of photosynthesis. Photochem. Photobio. 56:645-650.

2. Basiouny, F.M., and R.H. Biggs. 1975. Photosynthetic and carbonic anhydrase activities in Zn-deficient peaches exposed to UV-B radiation. *In* Impacts of climatic change on the biosphere, CIAP Monograph 5, Part 1 (appendix B). Department of Transportation, Washington, DC.

3. Beggs, C.J., U. Schneider-Ziebert, and E. Wellmann. 1986. UV-B radiation and adaptive mechanisms in plants. *In* R.C. Worrest and M.M. Caldwell (eds.), Stratospheric Ozone Reduction, Solar Ultraviolet Radiation and Plant Life, Springer-Verlag: Berlin Heidelberg. pp.235-250.

4. Bogenrieder, A., and R. Klein. 1982. Does solar UV influence the competitive relationship of higher plants? *In* J. Calkins (ed.), The role of solar

ultraviolet radiation in marine ecosystems, Plenum Press: New York. pp.641-649.

5. Caldwell, M.M. 1971. Solar UV radiation and the growth and development of higher plants. *In* A. Giese (ed.), Photophysiology, vol. VI, 131-177. Academic Press, London.

6. Caldwell, M.M., R. Robberecht, and S.D. Flint. 1983. Internal filters: Prospects for UV-acclimation in higher plants. Physiol. Plant. 58:445-450.

7. Caldwell, M.M., A.H. Teramura, and M. Tevini. 1989. The changing solar ultraviolet climate and the ecological consequences for higher plants. Tree 4:363-367.

8. Cen, Y., and J.F. Bornman. 1990. The response of bean plants to UV-B radiation under different irradiances of background and visible light. J. Exp. Bot. 41:1489-1495.

9. Chapin, F.S. III. 1991. Integrated responses of plants to stress. Bioscience 41:29-36.

10. Clark, J.B., and G.R. Lister. 1975. Photosynthetic action spectra of trees. II: The relationship of cuticle structure to the visible and ultraviolet spectral properties of needles from four coniferous species. Plant Physiol. 55:407-413.

11. Dale, J.E. 1992. How do leaves grow? Bioscience 42:423-432.

12. Day, T.A., T.C. Vogelmann, and E.H. DeLucia. 1993. Are some plant life forms more effective than others in screening out ultraviolet-B radiation? Oecologia 92:513-519.

13. DeLucia, E.H., T.A. Day, and T.C. Vogelmann. 1992. Ultraviolet-B and visible light penetration into needles of two subalpine conifers during foliar development. Plant Cell and Environment 15:921-929.

14. Dickson, J.G., and M.M. Caldwell. 1978. Leaf development of *Rumex patientia* L. (Polygonaceae) exposed to UV irradiation (280-320nm). Amer. J. Bot. 65:857-63.

15. Dillenburg, L.R., J.H. Sullivan, and A.H. Teramura. 1994. Leaf expansion and development of photosynthetic capacity and pigments in *Liquidambar styraciflua*: Effects of UV-B radiation. Amer. J. Bot. (In review).

16. Farquhar, G.D., J.R. Ehleringer, and K.T. Hubick. 1989. Carbon isotope discrimination and photosynthesis. Ann. Rev. Plant. Physiol. Plant. Molec. Biol. 40:503-537.

17. Farquhar, G.D., M.H. O'Leary, and J.A. Berry. 1982. On the relationship between carbon isotope discrimination and the intercellular carbon dioxide concentration in leaves. Aust. J. Plant Physiol. 9:121-137.

18. Fernbach, E., and H. Mohr. 1992. Photoreactivation of the UV light effect on growth of Scots pine (*Pinus sylvestris* L.) seedlings. Trees 6:232-235.

19. Field, C., and H.A. Mooney. 1986. The photosynthesis-nitrogen relationship in wild plants. *In* T.J. Givinsh (ed.), On the Economy of Plant Form and Function, Cambridge University Press, New York. pp.25-55.

20. Gausman, H.W., R.R. Rodriguez, and D.E. Escobar. 1975. Ultraviolet radiation reflectance, transmittance and absorbance by plant leaf epidermises. Agron. J. 67:720-724.

21. Green, A.E.S., K.R. Cross, and L.A. Smith. 1980. Improved analytical characterization of ultra violet skylight. Photochem. Photobiol. 31:59-65.

22. Karabourniotis, G., A. Kyparissis, and Y. Manetas. 1993. Leaf hairs of *Olea europaea* protect underlying tissues against ultraviolet-B radiation damage. Env. Exp. Bot. 33:341-345.

23. Karabourniotis, G., K. Papadopoulas, M. Papamarkou, and Y. Manetas. 1992. Ultraviolet-B radiation absorbing capacity of leaf hairs. Physiol. Plant. 86:414-418.

24. Kaufmann, M.R. 1978. The effect of ultraviolet (UV-B) radiation on Engelmann spruce and lodgepole pine seedlings. UV-B Biological and Climatic Effects Research (BACER). Final Report EPA-IAG-D-0168, EPA: Washington, DC.

25. Kossuth, S.V., and R.H. Biggs. 1981. Ultraviolet-B radiation effects on early seedling growth of Pinaceae species. Canadian J. For. Res. 11:243-248.

26. Lindoo, S.J., and M.M. Caldwell. 1978. Ultraviolet-B radiation induced inhibition of leaf expansion and promotion of anthocyanin production. Plant Physiology 61:278-282.

27. Middleton, E.M., and A.H. Teramura. 1993. The role of flavonol glycosides and carotenoids in protecting soybean from UV-B damage. Plant Physiol. In press.

28. Naidu, S.L., J.H. Sullivan, A.H. Teramura, and E.H. DeLucia. 1993. The effects of ultraviolet-B radiation on photosynthesis of different needle age classes in field-grown loblolly pine. Tree Physiol. 12:151-162.

29. Reed, H.E., A.H. Teramura, and W.J. Kenworthy. 1993. Ancestral U.S. soybean cultivars characterized for tolerance to ultraviolet-B radiation. Crop Sci. 32:1214-1219.

30. Robberecht, R., and M.M. Caldwell. 1978. Leaf epidermal transmittance and of ultraviolet radiation and its implications for plant sensitivity to ultraviolet-radiation induced injury. Oecologia (Berl.) 32: 277-287.

31. Roden, J.E., V. Volkenburgh, and T.M. Hinckley. 1990. Cellular basis for limitation of poplar leaf growth by water deficit. Tree Phys. 6:211-219.

32. Semeniuk, P. 1978. Biological effects of ultraviolet radiation on plant growth and development in florist and nursery crops. In EPA-IAG-DG-0168, Research report on the impacts of ultraviolet-B radiation on biological systems: a study related to stratospheric ozone depletion. EPA, Washington, DC.

33. Sisson, W.B., and M.M. Caldwell. 1976. Photosynthesis, dark respiration, and growth of *Rumex patientia* L. exposed to ultraviolet irradiance (290 to 315 nanometers) simulating a reduced atmospheric ozone column. Plant Physiol. 58:563-568.

34. Sisson, W.B., and M.M. Caldwell. 1977. Atmospheric ozone depletion: Reduction of photosynthesis and growth of a sensitive higher plant exposed to enhanced UV-B radiation. J. Exp. Bot. 28:691-705.

35. Steinmuller, D., and M. Tevini. 1985. Action of ultraviolet radiation (UV-B) upon cuticular waxes of some crop plants. Planta 164:557-564.

36. Sullivan, J.H. 1992. UV-B effects on terrestrial plants. *In* Proceedings of the UV-B Monitoring Workshop: A Review of the Science and Status of Measuring and Monitoring Programs. Appendix C, pp. 64-82, Science and Policy Associates, Washington DC.

37. Sullivan, J.H., and A.H. Teramura. 1988. Effects of ultraviolet-B irradiation on seedling growth in the Pinaceae. Amer. J. Bot. 75:225-230.

38. Sullivan, J.H., and A.H. Teramura. 1989. The effects of ultraviolet-B radiation on loblolly pine 1. Growth, photosynthesis and pigment production in greenhouse-grown seedlings. Physiol. Plant 77:202-207.

39. Sullivan, J.H., and A.H. Teramura. 1992. The effects of ultraviolet-B radiation on loblolly pine 2. Growth of field-grown seedlings. Trees 6:115-120.

40. Sullivan, J.H., and A.H. Teramura. 1993. The effects of ultraviolet-B radiation on loblolly pine 3. Interaction with $CO_2$ enhancement. Plant Cell. In press.

41. Teramura, A.H., and M.M. Caldwell. 1981. Effects of ultraviolet-B irradiances on soybean IV. Leaf ontogeny as a factor in evaluating ultraviolet-B irradiance effects on net photosynthesis. Am. J. Bot. 68:934-941.

42. Teramura, A.H., and N.S. Murali. 1986. Intraspeciffic differences in growth and yield of soybean exposed to ultraviolet-B radiation under greenhouse and field conditions. Env. Exp. Bot. 26:89-95.

43. Teramura, A.H., and J.H. Sullivan. 1991. Field studies of UV-B radiation effects on plants: Case histories of soybean and loblolly pine. *In* Impact of Global Climate Changes on Photosynthesis and Productivity, Oxford and IBH Publishing Co. New Delhi. pp.147-161.

44. Teramura, A.H., and J.H. Sullivan. 1994. Effects of UV-B radiation on photosynthesis and productivity. Photosyn. Res. In press.

45. Teramura, A.H., J.H. Sullivan, and J. Lydon. 1990. The effectiveness of UV-B radiation in altering soybean yield: A six-year field study. Physiol. Plant. 80:5-11.

46. Tevini, M. (ed.) 1993. UV-B Radiation: Effects on humans, animals, plants, microorganisms and materials, Lewis, Boca Rotan, FL. 240pp.

47. Tevini, M., and D. Steinmuller. 1987. Influence of light, UV-B radiation and herbicide on wax biosynthesis of cucumber seedlings. J. Plant Physiol. 131:111-121.

48. Tevini, M., J. Braun, and F. Fieser. 1991. The protective function of the epidermal layer of rye seedlings against ultraviolet-B radiation. Photochem. Photobiol. 53:329-333.

49. Ticha, I. 1985. Ontogeny of leaf morphology and anatomy. *In* Z. Sestak (ed.), Photosynthesis during leaf development, W. Junk, Dordrecht. pp. 11-15.

50. Whittaker, R.H. 1975. Communities and Ecosystems. MacMillan Co., New York.

# ULTRAVIOLET-B IRRADIANCE MEASUREMENTS UNDER FIELD CONDITIONS[a]

Qiujie Dai, Arlene Q. Chavez, Shaobing Peng, and Benito S. Vergara

Agronomy, Plant Physiology, and Agroecology Division
International Rice Research Institute
P.O. Box 933, Manila 1099, Philippines

Key words: cellulose acetate filter, mylar polyester filter, rice, UV-B

**ABSTRACT**

Most of the studies reported on the effects of UV-B on plants were conducted using supplemental UV-B coming from UV-emitting lamps. The output of UV-B irrandance from the lamps, however, could be affected by different factors, a critical consideration in the study of UV-B effects on crops. An experiment was conducted at the IRRI field to determine UV-B irradiance at different voltages (160-220V), solar irradiance, and film photo-oxidation. Supplemental UV-B was provided by pre-aged (100h) UV-emitting fluorescent lamps that were enclosed by filters of either cellulose acetate (UV-B treatment) or mylar polyester (control treatment). UV-B irradiance was measured using an Optronic 752 Spectroradiometer.

UV-B output was negatively associated with electrical voltage. There was a 5-35% reduction in UV-B irradiance from 205 to 160 volts. To obtain better UV-B output from the lamps, the voltage must not be lower than 205 volts. At Los Baños, the ambient UV-B level during the dry season is about 6.0 kJ $m^{-2}$ day on a clear-sky day and very close to the simulated values of Björn and Murphy. This is the first measured value reported in tropical areas and is 172 percent higher than the simulated ambient UV-B level in Tokyo in March. A linear correlation ($r^2$=0.97) between ambient UV-B irradiance and total solar radiation was established. The photo-oxidation of cellulose acetate can reduce the UV-B output by about 25 percent after one week of exposure, suggesting that CA films must be replaced within 7 days to ensure appropriate UV-B output. Uniform distribution of UV-B irradiance under the lamps could be successfully evaluated by portable UVX radiometer.

## INTRODUCTION

A major consequence of ozone depletion is an increase in ultraviolet-B (UV-B; 280-320 nm) radiation which has profound effects on both plants and human beings [1]. Most studies reporting on the effects of UV-B on plants were conducted using supplemental UV-B from UV-emitting lamps [2,5]. Under field conditions with artificial lamps as enhanced UV-B source, the total UV-B irradiance the plant receives is equal to the UV-B emitted by the lamps plus the ambient UV-B. Supplemental UV-B irradiance from the lamps is affected by electrical voltage and photo-oxidation of the cellulose acetate (CA) filters (transmission down to 290 nm) while ambient UV-B is affected by solar radia-tion. A study needs to be conducted to quantify the effects of these factors to accurately provide supplemental UV-B radiation under field conditions.

## OBJECTIVES

To evaluate the effect of electrical voltage, solar irradiance, and filter photo-oxidation on UV-B irradiance received by plants and to measure the ambient

[a]Although the information in this document has been funded wholly or in part by the U.S. Environment Protection Agency, under agreement number CR 817426 to the International Rice Research Institute, it has not been subjected to the Agency's peer and administrative reviews. It does not necessarily reflect the views of the Agency and no official endorsement should be inferred.

NATO ASI Series, Vol. I 18
Stratospheric Ozone Depletion/
UV-B Radiation in the Biosphere
Edited by R. H. Biggs and M. E. B. Joyner

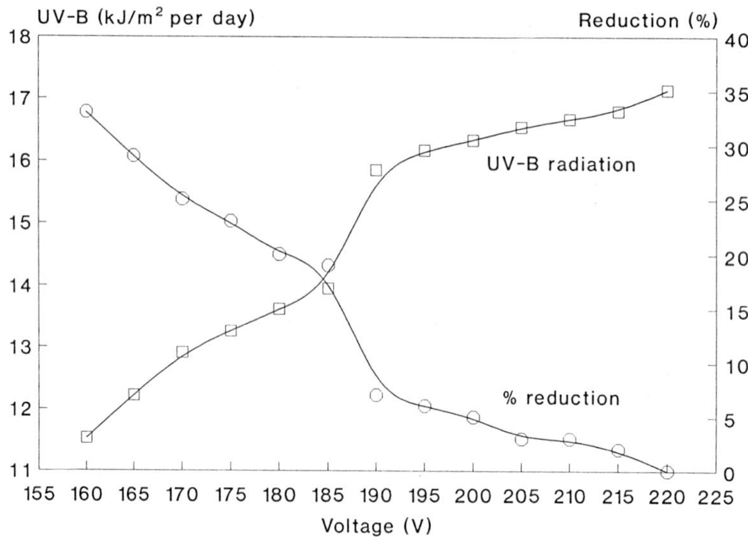


Figure 1. Biologically effective UV-B irradiance emitted by UV lamps at different voltage supplies (160–220V).

Figure 2. Biologically effective UV-B irradiance emitted by UV lamps covered with cellulose acetate filters photooxidized at different durations.

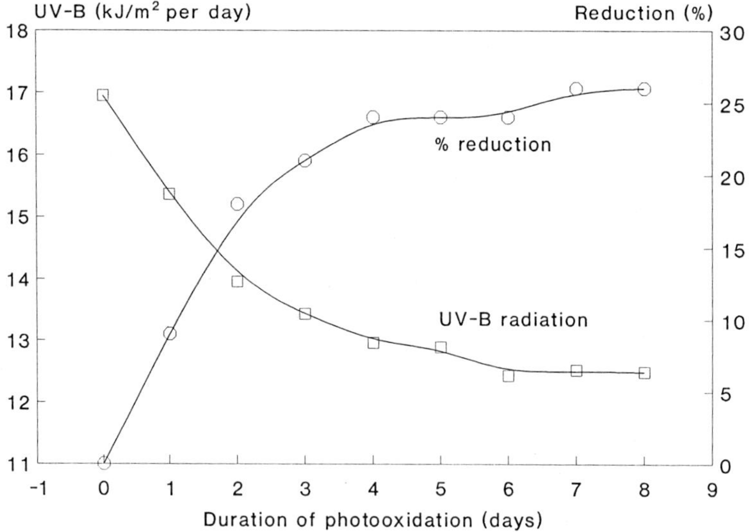

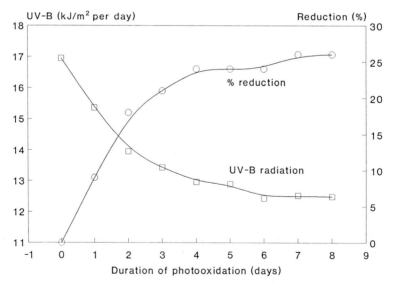

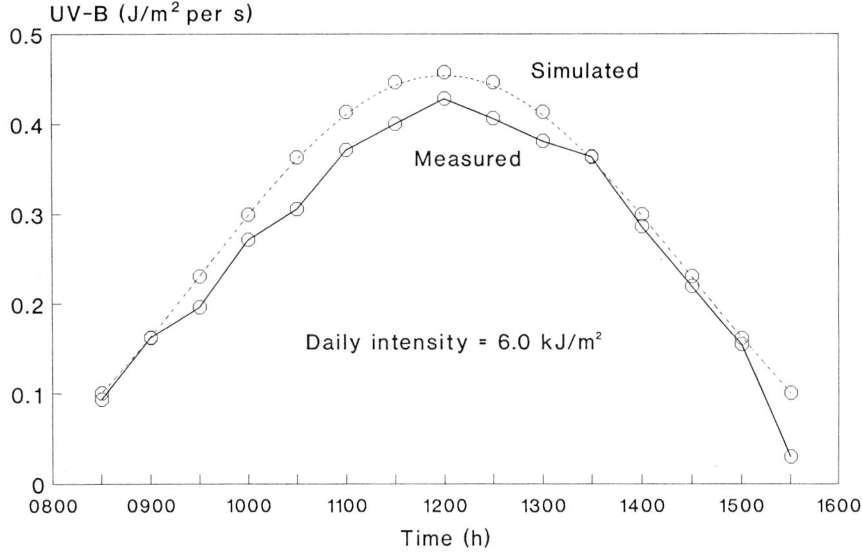


Figure 3. Measured and simulated (using Bjorn and Murphy's model) ambient UV-B irradiance at Los Banos in March 1993 on a clear sky day.

UV-B radiation at tropical area, especially at Los Baños, Laguna, Philippines.

## MATERIALS AND METHODS

The experiment was conducted at the International Rice Research Institute (IRRI) field in dry season, 1993. Supplemental UV-B was provided by pre-aged (100 h) UV-emitting fluorescent lamps (UV-B 313, Q-Panel, Cleveland, OH) that were enclosed by either CA (UV-B treatment) or mylar polyester (control treatment) filters. UV-B was measured with a double-monochromator spectroradiometer (Optronic Model 752, Orlando, Fl.). The reading was then converted into biologically effective UV-B irradiance by normalizing to unity at 300 nm according to the general plant-action spectrum of Caldwell [3]. Measurements were done at different voltages (160-220V) with the use of a variac. The solar irradiance was monitored simultaneously with UV-B irradiance using a Sunfleck ceptometer. Ambient UV-B was determined from 0830 to 1530 h with a 30-min interval in March 1993. The photo-oxidation of CA was evaluated daily with the CA filters exposed to UV-B from 0900 to 1500 h daily.

## RESULTS AND DISCUSSION

Ultraviolet-B output was negatively associated with electrical voltage (Figure 1). There was a 5-35% reduction in UV-B irradiance from 205 to 160 volts. The finding pointed out that to obtain better UV-B output from the lamps, the voltage must be maintained above 205 volts after loading the current.

The photo-oxidation of CA reduced the UV-B output by about 25% after one week of exposure (Figure 2). This result suggests that CA filters must be replaced within 7 days to ensure appropriate UV-B output.

At IRRI, Los Baños, Philippines, the ambient biologically effective UV-B level in the dry season is about 6.0 kJ/m$^2$ per day on a clear sky day (Figure 3), which was very close to the simulated values of Björn and Murphy [1]. This is the first measured value reported in the tropical areas and is 172% higher than

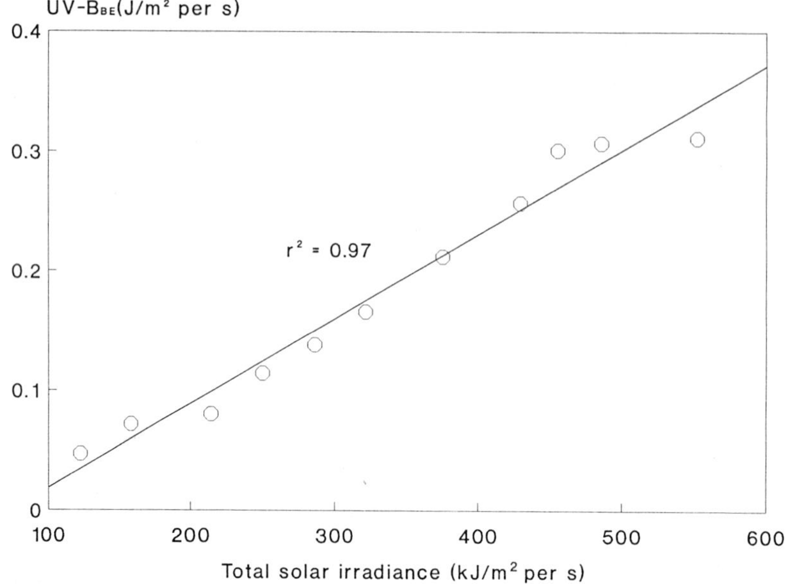


Figure 4. Relationship between ambient UV-B and total solar irradiance.

the simulated ambient UV-B level in Tokyo in March. A linear correlation ($r^2$=0.97) between ambient UV-B irradiance and the total solar radiation was established (Figure 4), indicating that UV-B radiation received by plants is significantly varied with sky conditions.

## CONCLUSION

By maintaining an adequate electrical voltage and regularly replacing the filters, supplemental UV-B can be accurately provided to experimental materials. The ambient UV-B must also be considered to determine the total UV-B radiation that the crop will receive during the growing season.

## REFERENCES

1. Björn L.O., and T.M. Murphy. 1985. Computer calculation of solar ultraviolet radiation at ground level. Physiologie Vègètal 23:555-561.

2. Booker, F.L., E.L. Fiscus, R.S. Philbeck, A.S. Heagle, J.E. Miller, and W.W. Heck. 1992. A supplemental ultraviolet-B radiation system for open-top field chambers. J. Environ. Qual. 21:56-61.

3. Caldwell, M.M. 1971. Solar ultraviolet irradiation and the growth and development of higher plants. In A.C. Giese (ed.), Photophysiology VI, Academic Press, New York. pp.131-177.

4. SCOPE Report. 1992. Effects of increased ultraviolet radiation on biological systems. Proceedings of a workshop on the effect of increased ultraviolet radiation on biological systems. February 17-22, 1992. Budapest, Hungary.

5. Webb, A.R. 1991. Solar ultraviolet-B radiation measurement at the earth's surface: Techniques and trends. Proceedings of the Indo-US workshop, Impact of Global Climate Changes on Photosynthesis and Plant Productivity. January 8-12, 1991, New Delhi, India. pp.23-38.

# EFFECTS OF STRATOSPHERIC OZONE DEPLETION AND INCREASED LEVELS OF ULTRAVIOLET RADIATION ON SUBANTARCTIC FORESTS AND WESTERN PATAGONIAN STEPPE: A RESEARCH PROJECT

Hector L. D'Antoni,[1] Roy Armstrong,[2] Joseph Coughlan,[2] Jay Skiles,[2] and Gustavo R. Daleo[3]

[1] NASA/Ames Research Center, Mail Stop 242-4, Moffett Field, CA 94035-1000.
[2] Johnson Controls World Services, Ames Operation, NASA/Ames Research Center.
[3] Facultad de Ciencias Exactas y Naturales, Universidad Nacional de Mar del Plata, Funes 3250, 7600 Mar del Plata, Argentina.

Key words: forest, ozone depletion, PAR, photosynthesis, steppe, subantarctic, UV-B radiation

**ABSTRACT**

The depletion of stratospheric ozone is a well established fact, supported by remote sensing data. A corresponding increase of solar UV-B radiation reaching the surface of the Earth is being documented. The effects of this increment of UV radiation on living organisms are uncertain. Most of our knowledge of this problem is the result of laboratory experiments and studies on marine phytoplankton of Antarctica.

The purpose of this project is to identify and measure indicators of the effect of increased solar UV-B radiation on terrestrial plants. Our goal is to establish these indicators so that we may use them to predict the effects of further increased UV-B radiation on terrestrial ecosystems.

We propose a three year study of two parallel ecosystem types (forest and steppe) that run N-S over a 15° latitudinal gradient, along a gradient of ozone depletion over the mid-latitudes in Argentina.

Two stations to measure UV radiation and photosynthetically active radiation (PAR) will be installed along the gradient. We will use the same type of radiometer that is operated simultaneously with the spectroradiometer of the NSF Polar Program installed at Ushuaia, in Tierra del Fuego. The data of the latter will also be used in this project. Samples and *in situ* measurements will be taken at 10 sites on the gradient in the austral spring in each of the three years. The tasks we expect to perform at these 10 sites are 1) photosynthesis measurements, 2) chlorophyll concentration measurements, 3) leaf area and leaf area index (LAI) measurements, 4) epidermal pigments and soluble proteins extraction.

Intensive work will be performed at 3 sites (northern end, center, and southern end of the transect). The tasks described above will be performed, but we will also take a) productivity measurements, b) spectroradiometry, and c) collection of seedlings (and/or seeds) of *Nothofagus* spp., *Festuca gracillima*, *Poa* spp. *Rumex acetosa* and *R. acetosella*. A greenhouse experiment will be carried out in the southernmost site (Ushuaia). The seedlings (and seeds) collected at the three instrumented sites will be grown under ambient conditions and under UV filters. All the above measurements will be performed on these plants at regular intervals by the research team while in the study area and by local research staff for the remainder of the year.

We expect to produce and calibrate biochemical, botanical (LAI, and others), and spectroradiometric data. Our ground level measurements of UV radiation will be correlated to remote sensing data of ozone. The data will be used to generate response surface and other models of those variables controlled by UV radiation. In turn, these data will be used in ecosystem models in order to simulate the current conditions in both ecosystems and predict future developments under different scenarios. A point model will be developed which will then be applied at regional scale at selected locations along the gradient.

NATO ASI Series, Vol. I 18
Stratospheric Ozone Depletion/
UV-B Radiation in the Biosphere
Edited by R. H. Biggs and M. E. B. Joyner

This research will complement NASA's commitment to the study of stratospheric ozone. Southern Argentina is the most appropriate place for study as the largest UV effects on terrestrial ecosystems are expected there. Our findings may be relevant for human and all other forms of life. Furthermore, this research addresses several issues highlighted in the Working Document of April 4, 1993 on the U.S. Federal Government-Wide Ultraviolet-B (UV-B) Research Activity.

## INTRODUCTION

*"Little research has been done in this area, despite the potential for declines in agricultural yields as ozone concentrations continue to fall and UV-B radiation consequently continues to increase into the next century."*

—NIE Network News, April 1993

The discovery of the Antarctic ozone hole in 1985 was a surprise to many researchers because it had not been predicted by any photochemical model and there was no immediately obvious explanation for it [50, 68]. However, the mechanism for this ozone loss was first postulated by Molina and Rowland in 1974 [41] who said that chlorofluorocarbons (CFCs) dissociate and liberate chlorine in the stratosphere where the chlorine catalyzes the breakdown of ozone. The consequent depletion of the ozone layer in the atmosphere causes an increase in incident solar ultraviolet (UV) radiation (280-400 nm) that impinges on the world's surface. (See Appendix for a summary of the biological effects of UV radiation). This mechanism has resulted in a large decrease in global ozone since about 1978 and a pronounced ozone hole in the southern hemisphere [27].

Natural and man-made causes coupled to produce a record low global ozone in 1992. Last year's ozone measurements were 2% to 3% lower than the lowest values recorded in earlier years. The lowest values of November and December 1992 were three to four standard deviations below the 12-year daily mean [21]. Although CFCs emissions have been sharply reduced in acceptance of the Montreal Protocol, a recent report [46] suggests that some CFCs may persist from three centuries to thousands of years. Further, harmful UV radiation may increase after volcanic eruptions [8,68] because volcanic plumes contain high concentrations of sulfuric acid droplets which can also reduce ozone column amounts either by enhancing chlorine-catalyzed chemistry or inducing convective lifting. The cause of the 1992 low ozone values is uncertain, but it is intuitively linked to the continuing presence of aerosol from the Mount Pinatubo eruption [21,35].

The 12% increase in harmful UV radiation resulting from a 9% decline in ozone at mid-latitudes [35] is currently linked to the augmented occurrence of biological phenomena such as skin erythema and cancer, cataracts, unusual pigmentation, and deformities and death of vascular plants. Since 1985, several phenomena directly attributable to this increased UV radiation have been found for phytoplankton [1,14,34,56], for terrestrial plants, and humans [37,17,29] in the Southern Hemisphere. A thinning ozone shield over the mid-latitudes and the Antarctic "ozone hole" are believed to be responsible for higher rates of human skin cancer in Australia, an apparently high incidence of cataracts in Patagonian sheep and Antarctic penguins, and lately, an increased red component in the canopies of the subantarctic forest of Patagonia (See Appendix, Part E).

Some side effects of increased UV with potentially large economic consequences include the possible collapse of complex aquatic food chains and lower crop yields which may force changes in the spatial location of agriculture. Studies conducted in Germany [18] have shown effects that impair the marketability of crops. These effects include overall size reduction, leaf size reduction, color changes, and morphology changes. (See Appendix C.)

Living systems have adapted to UV radiation mostly by filtering out the excess radiation with screening pigments [4,48]. (See Appendix, Part E). However, when these screening mechanisms fail, UV radiation causes different degrees of damage in plants, animals, and humans. In a recent statement in the

April, 1993 NIE Newletter the author said, "The only field data available on agricultural crops show a 20% decrease in yield in one soybean strain under increased UV-B radiation. In addition, laboratory studies appear to indicate that UV-B radiation alters the metabolism and growth of at least 50% of terrestrial plants." Investigated problems may have dramatic consequences on biotic productivity, survivability, and health of plants, animals, and man. Long-term ecological studies are needed to quantify the effects of increased UV radiation in the continents, assess the risks, and produce reliable data for prediction.

In response to increased UV radiation, the synergistic relationship among ecosystem components will produce effects that are nonlinear which will make it difficult to single out UV effects in field experiments [59]. Most of the regional data available on stratospheric ozone have been obtained by remote sensing studies of the thinning ozone layer so it is necessary to identify reliable terrestrial indicators of increased UV radiation. The thinned stratospheric ozone layer of the Southern hemisphere reaches as far north as approximately 34°S latitude [38,39,58]. Responses must be identified that can be easily measured and be unequivocally linked to increased UV radiation. If the production of screening pigments is an adaptive mechanism in plants, then a higher concentration of screening pigments in leaves may be interpreted as a natural response to increased UV radiation. If such higher concentrations of pigments filters out the excessive UV radiation, no damage will occur. If the screening effect is not sufficient then genetics, photosynthesis, growth, leaf shape and size, and pollen viability as well as pollen grain shape, size, and production may be affected by excessive UV radiation. It is necessary to monitor selected terrestrial ecosystems in sufficient detail to permit detection and interpretation of changes attributable to global climate change [13]. A model-guided design allows the prediction of systemic scenarios. At the same time, analytical models of ecosystem dynamics are necessary to determine the limits to which resilience of ecosystems can be stretched before they change, perhaps irreversibly [13].

## OBJECTIVE AND HYPOTHESES

**Objective**: To quantify the responses of terrestrial vegetation to UV radiation by studying pigment concentrations, photosynthesis, and spectral parameters of selected plant species in the field and in greenhouses.

**Hypothesis 1**: Terrestrial plants respond to increased solar UV radiation by changing the concentration of pigments in leaves.

**Hypothesis 2**: UV-induced changes in the amount of plant pigment can be inferred from variations in spectral reflectances.

**Hypothesis 3**: If UV radiation increases beyond the filtering capacity of foliar pigments, changes will occur in foliar composition and fluxes of carbon.

## PROPOSED RESEARCH

To evaluate the effects of increased UV-B radiation on land vegetation due to a thinning ozone shield, we propose a three-year study of forest and steppe of southern Argentina.

### THE REGION

The deciduous formations of the subantarctic forests extend from 39°50'S to 55°S on the eastern slopes of the Andes (Figure 5). This narrow stripe enclosed by the 700 and 1,500 mm precipitation isopleths (Figure 1) is more dense from 41° to 46°S, then it occurs in patches down to 52°S and becomes dense again in Tierra del Fuego (54° to 55°S). This forest, which encompasses 15 degrees of latitude, is located on the continental landmass closest to Antarctica, and therefore is well suited as a field laboratory for the study of UV radiation effects on terrestrial plants. In contact with the forest, the western end of Patagonia is a narrow belt of open vegetation growing between the precipitation isopleths of 700 mm (forest limit) and 200 mm (Patagonia regime). Both formations extend under a gradient of 50% annual cloudiness in the north to 70% in the south (Figure 2). These features are relevant because the forest and the neighboring grass steppe run parallel to the gradient of ozone depletion in the southern Hemisphere reaching well into region affected by Antarctic ozone depletion, and the UV radiation on the Earth's surface may be either lower or higher with incomplete overcast [39] and this ecosystem extends under zones of 50, 60 and 70% cloud cover.

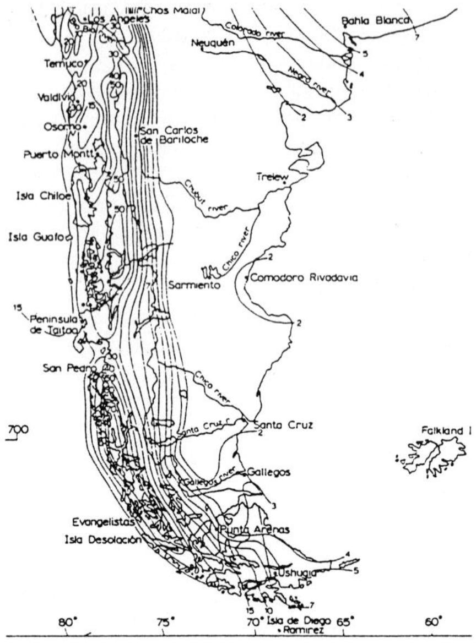


Fig. 1. Annual precipitation in dm. (from Prohaska [45], modified).

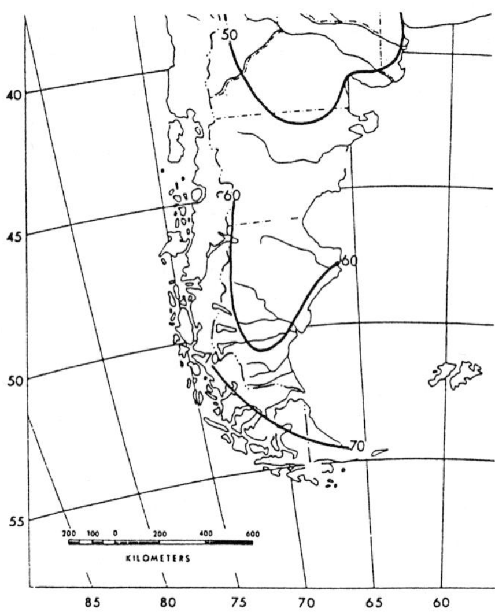


Fig. 2. Annual cloudiness in percent of sky cover (from Prohaska [45], modified).

UV RECORDING STATIONS

Two stations (equipped with five-channel radiometers designed to monitor UV irradiance by sensing four critical spectral regions in the UV-B and UV-A ranges), will produce a continuous record of UV radiation received at the surface. One station is already in Ushuaia (the site of one of NSF's UV-monitoring stations). At the other two stations, Bariloche and Lago Viedma, we will establish ground-based UV radiometers manufactured by Biospherical Instruments. These instruments measure and record UV-A, UV-B, and PAR (Photosynthetically Active Radiation). University personnel will check the instruments daily to ensure that all equipment is operating properly. Failures will be reported to the University of Mar del Plata for prompt servicing.

GREENHOUSE EXPERIMENT

As is known from the classic studies of Clausen et al. [12] and Hiesey et al. [28], plants of the same species collected across a transect can show marked ecotypic and phenotypic variation. To account for these differences, seedlings from selected points along the latitudinal gradient will be brought to Ushuaia. *Nothofagus antarctica* (Forster f.) Oersted, *N. pumilio* (Poeppig and Endl.) Krasser, the evergreen species, *N. betuloides* (Mirbel) Oersted, and steppe species including *Festuca gracillima* Hooker f., *Poa* spp., *Rumex acetosa* L., and *R. acetosella* L. will be grown in a greenhouse under total UV screening (control) and natural conditions (treatment). A robust experimental design (latin square and the like) will be used to allow application of statistical techniques for the analysis of data on growth and pigmentation. This experiment will play a major role in the testing process of Hypothesis 1. Non-destructive hemispherical reflectance measurements will be obtained with a spectroradiometer during the control period. Large numbers of seedlings will be grown in each experiment in order to provide sufficient materials for the necessarily-destructive techniques of pigment analysis. A sub-sample of the leaves will be removed for chemical analysis of pigments. This experiment will be performed in real time in connection with the NSF Polar Program* recording stations in order to obtain a simultaneous record of biological variables and UV radiation. The measurements will include UV and PAR modules as parts of complete and automatic weather stations.

HYPOTHESIS TESTING

We will test our hypotheses using forest and grassland species that span the latitudinal belt. Two deciduous arboreal species, *Nothofagus antarctica* and *N. pumilio*, found from 36°S to Tierra del Fuego, and an evergreen species, *N. betuloides* (46°40'S to Tierra del Fuego) will be studied to evaluate the responses of forest vegetation to increased solar UV radiation. *Festuca gracillima*, *Poa* spp., *Rumex acetosa* L., and *R. acetosella* L. will also be studied in this project. In the south, the Magellan sector of Patagonia (51-54°S) is a steppe that grows on the soils developed on the moraines of the last Ice Age.

TEST OF HYPOTHESIS 1

*Biochemistry.* To verify if the ozone depletion gradient generates parallel responses in plants, pigment analysis of leaves of *Nothofagus* spp. growing on the east Andean slope will be performed at 10 sites from 36° to 55°S. The neighboring steppe will serve to test the effects of increased UV radiation on open vegetation. Plants to be tested are autoctonous grasses (*Festuca* spp. and *Poa* spp.), and introduced weeds, *Rumex* spp. [41]. Extractions will be performed in the field with 70-80% methanol (with the addition of 0.1% hydrochloric acid to preserve anthocyanins). A study of the crude extract will be performed by two-dimensional paper or thin-layer chromatography [22]. Qualitative and quantitative analyses will be performed with a combination of paper chromatography, UV spectroscopy, and, specially, by high performance liquid chromatography (HPLC) in reversed-phase columns (C18), eluted in a gradient of methanol or acetonitrile [25]. Flavonoids will be detected by fluorescence or by absorbance at 355 nm. Additionally, the content of soluble proteins in leaves will be

* The National Science Foundation's (NSF) Polar Program has established a UV monitoring network in the western hemisphere. There are presently five stations in operation: Three in Antarctica, one in Ushuaia, Argentina, and one in Barrow, Alaska. The ground stations utilize spectroradiometers built by Biospherical Instruments, Inc. (San Diego, California). Data scans are conducted on an hourly basis when the sun is above the horizon are distributed, after calibration, on CD-ROM. These stations were established to provide independent confirmation of the role of the ozone layer in moderating UV irradiance [6].

analyzed. Leaf tissue will be homogenized and extraction with trichlorhydric buffer, 50 mmolar at pH 8.0 will be done. Quantification will be based on the measurement of peptide bonds. Results will be related to dry weight and other parameters.

TEST OF HYPOTHESIS 2

*Leaf-level measurements*. The green leaf reflectance curve in the visible region of the electromagnetic spectrum is characterized by absorption of incident radiation by plant pigments, particularly chlorophyll. Leaf spectra from various plant species have been used to estimate leaf nitrogen content [65,66], chlorophyll concentrations [66], and leaf water content [64]. Light absorption by water in intracellular spaces of leaves dominates reflectance in the short wave infrared region. A shift in plant pigment production, from chlorophyll to accessory pigments, can result in a corresponding spectral response in the visible region of the spectrum.

We will use individual leaf spectrometry for assessing the early response of *Nothofagus* spp. to UV-induced stress. A Spectron SE-590 spectroradiometer with an integrating sphere attachment will be used in the field and in the greenhouse (under two treatments: ambient and UV-restricted) to acquire hemispherical reflectance data from individual leaves. The leaves subsequently will be analyzed for chlorophyll, anthocyanins, flavonoids, carotenoids, and other pigments and secondary metabolites. Statistical analysis will determine the relationship of spectral data to variations in foliar pigment concentrations. [We are aware that there are procedural problems with the calibration of different SE-590 instruments (S. Goward, personal comm.).] Our use of the SE-590 will ensure that spectral measurements will be consistent within the proposed study.)

*Leaf Area Index*. Variations in plant biophysical parameters such as leaf area index (LAI) have been used for spectrally separating healthy from stressed plant canopies [36]. The LAI has been correlated with near-infrared to red wavelength ratios [15,43,57]. The LAI of *Nothofagus* forests will be estimated in the field using a Decagon ceptometer [44]. These measurements will be related to spectral ratios obtained from remote sensing data used in other research work. Remotely sensed estimates of LAI will be used in the ecosystem simulation models (see below).

TEST OF HYPOTHESIS 3

*Modeling*. Ecosystem modeling is a method to integrate and balance the opposing or compounding effects of increased solar UV radiation on plant molecular, cellular, and production processes. Our modeling will be carried out in two interdependent phases: (1) development of a point model and (2) application of the model on a regional scale, extending along the UV radiation gradient.

*Forest vegetation model*. Greenhouse measurements taken for the six species will provided data to test Hypothesis 3 and data to establish parameters of physiologically detailed models of forest and grassland ecosystems. The models can extrapolate greenhouse results to predict whole plant and ecosystem level responses to UV-B chronic exposure. Experimental measurements include leaf physical characteristics, composition, pigmentation, flux rates, and environmental conditions needed for testing Hypothesis 3.

Leaf physical characteristics are measured to quantify if either leaf area or specific leaf area changes with UV-B exposure as has been reported by Tevini [60,61]. The data also are needed to scale up flux measurements to plant levels because ecosystem models use a leaf biomass-to-area relationship in scaling up modeled leaf processes [50].

Leaf composition includes nitrogen and lignin concentrations which should correspond to enzymatic activity and leaf construction costs [52]. Leaf nitrogen is a critical parameter for both photosynthesis and respiration modeling [51,52]. Increased nitrogen usually causes an increase in both modeled respiration and photosynthesis rates. UV-B damaged leaves can have high leaf nitrogen concentrations but low photosynthesis rates [61]. Biochemical parameters in detailed photosynthesis models [19] can be altered to reflect changes in photosynthesis rates per leaf nitrogen concentrations. For estimating growth respiration and for long-term plant and ecosystem modeling, changes in leaf lignin are important because lignin has higher construction costs than carbohydrates (2.2 vs. 1.2 g/g) [53] and as litter, lignin slows decay which in turn slows soil nitrogen cycling rates and lowers soil nitrogen availability.

Carbon fluxes will be measured both at optimum photon flux densities and at night for each of the six species using the LICOR photosynthesis meter. The maximum photosynthesis and night respiration estimates will be made before the destructive leaf pigmentation testing in Hypothesis 1 and 2. Photosynthesis rates will be related to leaf composition and leaf structure for whole plant totals. Appropriate photosynthesis model adjustments reflect UV-B damage that can be made to mechanistic rubisco-nitrogen relationships or made by lowering maximum photosynthesis rates. Plant maintenance respiration, defined as a model parameter (g/c per g/living tissue), will be measured for leaves and estimated for stem and root tissues keeping in mind that respiration is temperature sensitive (Q10) and respiration coefficients are a function of total nitrogen [52] and UV-B exposure.

Ecosystem model extrapolations will integrate the leaf and tissue changes to show long-term changes in LAI, NPP, and litter decomposition. For the forest ecosystems, modifications will be made to the Forest-BGC ecosystem model [50,51] including sections of Biome-BGC [31] to replace the less detailed leaf physiology as needed by this experiment.

*Steppe Vegetation Model.* This model will be used in the Magellan steppe to simulate hydrology, snow accumulation and melt, plant growth, and mortality. Ecological differentiation between the simulated species will be accomplished by using different parameters for the different plant species simulated by the model. Thus grasses, forbs, and shrubs in the Patagonian steppe can be simulated by using a standard set of parameters modified to distinguish between the functional groups or plant species. Specific biomass levels required for input to the model will be measured during the field data-gathering efforts.

By studying grasses, critical differences between growth cycles of species groups that occur during the UV increase episodes can be seen. The ozone hole opens over the Antarctic in August and closes in November. This is the time some grasses, including *Festuca* spp., *Poa annua,* and *P. trivialis*, (the early flowering group or EFG) initiate flowering. At about the time the hole closes, other species (*Festuca purpurascens, Lolium multiflorum, Poa scuberula,* the late flowering group, LFG) initiate the flowering cycle. This might mean that the overt/direct effects of UV will be exerted primarily on the EFG of the grassland communities. These effects can be simulated by modifying the parameters that control the production of photosynthate [4,10] for the EFG while keeping the LFG parameters unchanged. Measurement of rates of photosynthesis for steppe vegetation is therefore important, since this will indicate carbon flux of above-ground plant components and thus give an estimate of steppe production. To test these ideas, direct and continuous recording of the UV radiation reaching the surface at the site of the experiments is needed.

Further, it has been postulated that the growth rates of plant parts will also change in response to higher UV as will some competitive interactions [10]. The model to be used will need a plant component with explicit relationships for allocation and competition. It will be used to simulate these changes in the steppe vegetation of the study area. The changes in the state variables will then give an indication as to how the grassland ecosystem will change.

We can postulate at this point that plant community composition might change because of the different amounts of time the EFG and LFG components of the grassland are exposed to UV. Because different plants have different optimal photosynthetic rates, we can expect that these rates (for the EFG) will change with UV exposure [10], changing the water uptake and transpiration rates of these plants. This will affect the water balance for the site and favor one group over the other to a greater extent than at present, thereby altering the species composition of the community. Although conducted under simulated severe ozone depletion (40%), the experiments of competing pairs of weeds and crops that were cited by Tevini [62] well illustrate our ideas regarding the Magellan steppe.

## CREATION OF DATABASES

*TOMS Data.* The Total Ozone Mapping Spectrometer (TOMS) has acquired daily global ozone data from October 1978 to May 1993. An historical database is available from NASA's Climate Data System (NCDS) and from CD-ROM. Figure 3 shows total column ozone data computed using a 100-day moving average for Ushuaia, Argentina [6]. This figure depicts a decreasing trend in ozone between 1979 and 1991, suggesting a possible parallel trend in the short-wavelength UV radiation incident on the Earth at this location. The amount of UV radiation reaching the Earth's surface is not only dependent on the amount of stratospheric ozone but is also a

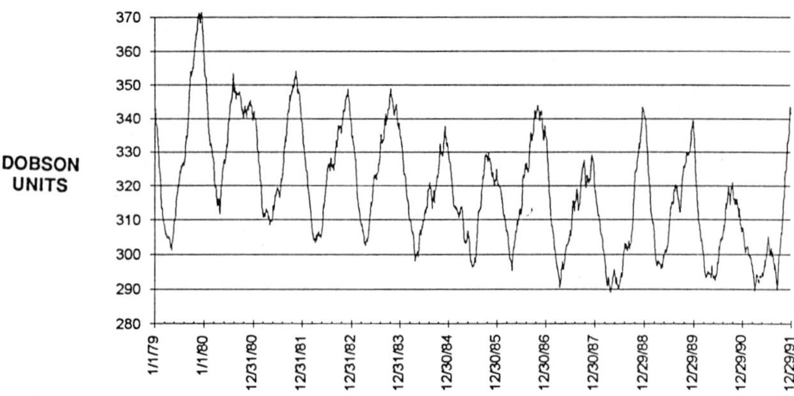


Fig.3. Total column ozone over Ushuaia (1978-1991) computed with a 100-day moving average (from Booth et al. 1992 [6], modified).

function of solar angle and cloud cover. This is particularly relevant for the longer wavelengths of UV-A. For the shorter wavelengths of UV, however, dramatic differences in UV irradiance have been correlated with ozone depletion [6].

*Other Databases*. Large amounts of data will be produced by this research work. We will file those data in ASCII and in common database programs to make the dataset available to future users. Data to be filed are: UV and PAR, climate, photosynthesis, respiration, chlorophyll concentration, leaf area, leaf area index, leaf weight, total leaf nitrogen, pigments, soluble proteins, productivity (including pollen), radiometry, and spectroradiometry.

## EXPERIMENTAL PLAN AND TIME TABLE

We will establish a transect in Argentina that runs north to south for approximately 1,500 km. The transect will have three stations (Table 1) in an area where we have been promised the use of facilities. The southernmost station on the transect will be at Ushuaia in Tierra del Fuego and will be our base station and the site of our greenhouse experiments.

Field work must start in beginning of October and extend into November. The Antarctic "ozone hole" begins to form in August and is largest in early October then dissipates in November. It is springtime there, soil moisture has been replenished by winter precipitation, and the trees are ready to produce flowers and new leaves.

If we assign two days per sampling site and two days to setup the group of collaborators and equipment in Mar del Plata plus a minimum loss due to unfavorable weather conditions, three weeks will be devoted to sampling along the transect. The remaining time will be used to work in the greenhouse experiment in Ushuaia. In addition, a greenhouse facility will be established in Tierra del Fuego which O.A. Bianciotto will supervise after the field season and take the responsibility for recruiting students of National University of Patagonia. These students will be responsible for the control and maintenance tasks of the greenhouse experiments.

## FIELD WORK

We will proceed from the north to the south along the margin of the forest, starting at Bariloche and traveling the distance to the southern end of the transect, sampling at regular intervals, one team in the forest and one team in the steppe or grassland areas. This will require two vehicles, one of which will be provided by the University of Mar del Plata. As shown in Figure 4, ten prospective sites have been selected for this study. Each site will be marked by a portable global positioning system (GPS) to enable us to retrace it in each of the subsequent years of the study. At each location, both the forest team and the steppe team will perform:

- photosynthesis measurements (LICOR photosynthesis meter)
- chlorophyll concentration measurements using a SPAD 502 instrument
- leaf area measurements
- pigments and soluble proteins extraction

Intensive work will be performed in three of these sites (~40°; ~50° and ~55°S) where precise and continued UV radiation and weather measurements will be taken. This work will be in addition to the standard measurements (listed above), and include

- meteorological measurements
- productivity measurements
- spectroradiometer measurements
- seeds and/or seedlings of *Nothofagus antarctica*, *N. pumilio* and *N. betuloides* (in Bariloche and Ushuaia) as well as *Festuca gracillima* (native), and *Rumex acetosa*, *Rumex acetosella*, *Poa annua* and *P. pratensis* (alien) will be collected and trans ported for growth in the greenhouse at Tierra del Fuego.
- Further analyses of all studied materials include weight, total leaf nitrogen, lignin.

## SCHEDULE

*July 1993-June 1994*:

1) Cooperative agreements between NASA and Argentine institutions (Government, universities, hospitals, schools, private citizens)
2) Cross-calibration of UV radiometers with NSF unit at Ushuaia.
3) Installation of UV recording stations. Initial database construction
4) Pigment and soluble protein analyses.
5) Preparation and initiation of greenhouse experiments in Tierra del Fuego
6) Spectral measurements—leaf level (mature trees)
7) Cross-calibration of UV radiometers with NSF unit at Ushuaia
8) Analysis of TOMS data

*July 1994-June 1995*:

1) Repetitions of 1993 experiments (adjusted in the light of the acquired experience)
2) Spectral measurements—leaf level (seedlings)

Table I. Proposed UV Study Sites.

| Site No. | Location | Latitude | Longitude | Elevation (m) |
|---|---|---|---|---|
| 1 | S.C. de Bariloche | 41° 06' S | 71° 10' W | 836 |
| 2 | Lago Viedma | 49° 30' S | 72° 30' W | 350 |
| 3 | Ushuaia | 54° 48' S | 68° 19' W | 6 |

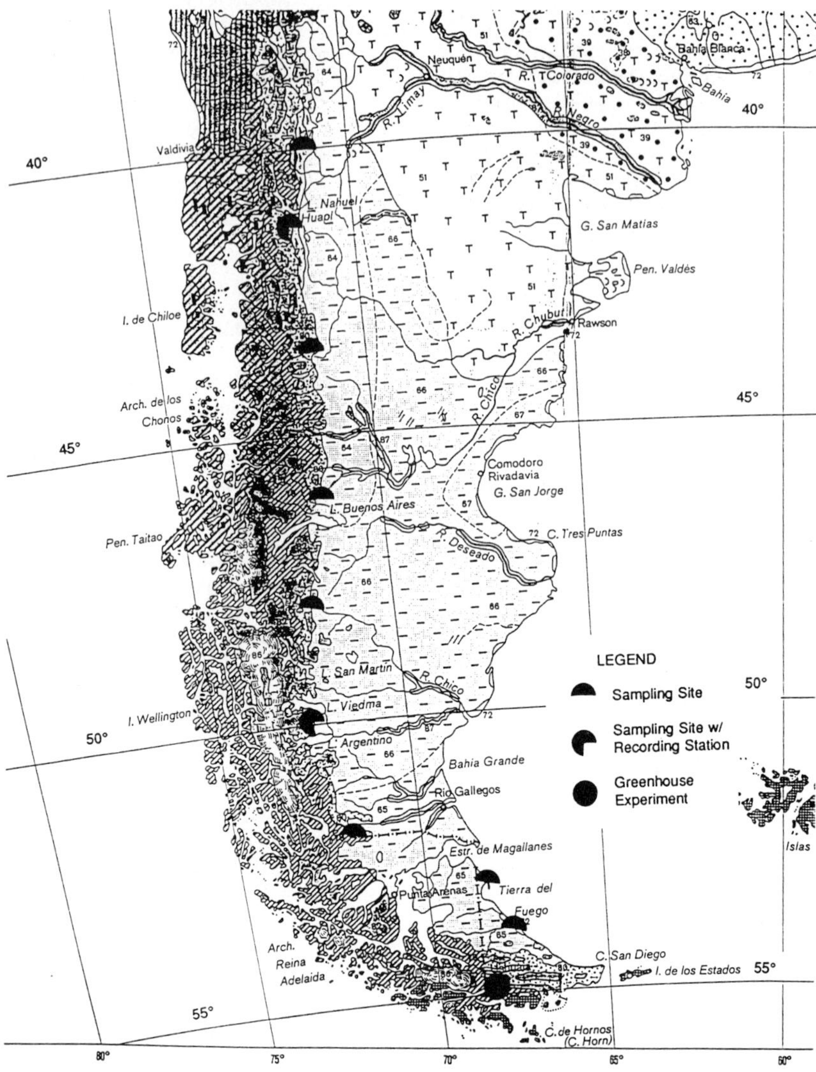


Fig.4. The ten sites selected for this study. The extent of studies to be undertaken at each site is indicated by the legend. (From Hueck & Seibert 1981 [30], modified).

*July 1995-June 1996*:

1) Repetitions of 1993, 1994 experiments
2) Use of steppe vegetation models on Magellan steppe
3) Use of forest vegetation models on Subantarctic forest

EDUCATION OUTREACH

Students of American, Argentine, and German universities will be admitted for technical work in the project that may eventually be used for M.S. theses or Ph.D. dissertations.

## ANCILLARY AND FUTURE RESEARCH

1) *Pollen studies*. It has been suggested that pollen viability is impaired by increased UV-B radiation [20]. South American *Nothofagus* expose their pollen-producing primordia as the Antarctic ozone hole opens and enter into anthesis during the maximum ozone hole and persist until the hole closes. The entire Subantartic Forest Ecosystem is located under the ozone depleted zone (since 1983 in the range of 350 Dobson units). Increased solar UV-B radiation resulting from such ozone depletion may be affecting not only the viability of the pollen but also its ontogenic process and hence its production, morphology, and size distribution. This general assumption has an impact on several areas of research and a number of hypotheses can be formulated and tested.

2) *Leaf morphology and anatomy studies*. Some cell structural components absorb in the UV-B range. Plants may respond to increased solar UV-B radiation by increasing the proportion of those components. Classical anatomic studies of leaves of both evergreen and deciduous species of *Nothofagus* and on leaves of *Rumex* spp. from 10 sampling sites may help understand other UV effects on plants.

3) *Plant embryology studies*. Plants may be affected by increased UV radiation in early stages of development because IAA is a UV target. We will invite Alfredo Cocucci (University of Cordoba) to initiate a research work on embryons of *Nothofagus* spp. and *Rumex* spp.

4) *DNA alteration studies*. A major target of UV radiation, DNA from samples taken at all 10 sites and the Ushuaia Greenhouse Experiment will be studied at the ARC by a post-doctoral fellow (Charles Cockell), under the guidance of Lynn Rothschild. Experiments in growth chambers will be conducted at the ARC under controlled (realistic) conditions.

5) *Public health*. During the field work season we expect scientific interaction with physicians in Patagonia and Tierra del Fuego. We will encourage them to keep a record of patients treated for erythema, skin cancer, and vision ailments (cataracts and the like). We will encourage joint research efforts taking advantage of our direct UV-B measurements and our capability to perform statistical analysis. We have already taken some steps in this direction.

**Acknowledgments**

We would like to acknowledge with thanks J.Brass, E.Condon, P.Matson, L.Morrissey, D.Peterson, P.Russell, J.Sharp and O.Toon (NASA/Ames Research Center) who reviewed an earlier version of this article and made valuable suggestions. The comments by R. Hilton Biggs (University of Florida) are also greatly appreciated.

# APPENDIX

## BIOLOGICAL EFFECTS OF UV RADIATION

The damaging effects of UV radiation occur at two major levels of organization:

*Molecular Level*. UV radiation affects nucleic acids, membranes, and phytohormones. Due to the aromatic-electron system, purine and pyrimidine bases of nucleic acids have a strong absorption near 260 nm. The aromatic amino acids included in protein molecules bring protein maximum absorption down to 280 nm. Abscisic acid (AA) and indole acetic acid (IAA) also absorb in the UV range (Figure A1). The action spectrum for killing bacteria peaks near 265 nm which is close to the absorption spectrum of DNA, suggesting that DNA is the target. However, wavelengths shorter than 280 nm do not reach the Earth's surface.

*Cellular Systems.* The mutagenic power of UV radiation is exercised on DNA molecules of viruses, prokaryotic, and simple eukaryotic organisms. Experiments on advanced eukaryotic cells have shown a still higher sensitivity to UV radiation [23]. The possible targets are DNA and proteins. The uncovered epidermis of many vertebrates is the main target for UV radiation. Short-term radiation can induce tanning and synthesis of vitamin D. Longer exposures result in erythema (sunburn) and can induce degenerative skin diseases and tumors [23].

EFFECTS OF INCREASED UV RADIATION ON TERRESTRIAL PLANTS

The effects of increased UV radiation on terrestrial life forms are probably best recorded by plants. In fact, increased UV radiation affects:

A. *Photosynthesis*. Increases of UV-B radiation (280-320 nm) equivalent to atmospheric ozone reductions of 50% or less adversely affect various component reactions of photosynthesis and carbon assimilation rates. UV-B has been linked to disruptions of the chloroplast envelope in *Beta vulgaris* [2,7,55], *Glycine max*, and *Rumex patientia* [11]. In all these studies, damage appeared to accumulate with duration of dose (Sisson 1986). The action of moderate levels of UV-B radiation on photosynthesis seems to concentrate on photosystem II (PS II), while photosystem I (PS I) remains unaffected. The impairment of PS II activity by UV-B radiation is due to blockage of PS II reaction centers rather than an inhibition of the water-splitting enzyme system [32,42]. Under a regional or even global perspective, the initial question to be addressed is whether chlorophyll concentrations are altered in plants exposed to UV-B radiation levels equivalent to a reduction in atmospheric ozone of less than 20% under field conditions [54]. Plants with photosynthetic C3 pathways are significantly affected by increased UV-B radiation while those with C4 pathway remain relatively unaffected [3]. Several authors [18,62,69] have found increased concentrations of soluble proteins in leaves of plants exposed to several levels of UV-B radiation [57].

B. *Carbon assimilation*. The detrimental effects of UV-B radiation on carbon assimilation have been extensively documented. Although physiologically sound, much of the available information on carbon assimilation has been obtained under conditions that were too artificial to be considered valid (i.e., under low visible light flux that is known to enhance the deleterious effects of UV-B radiation). It is therefore necessary to carry out studies under ambient conditions equivalent to a reduction of 20% or less of atmospheric ozone. A central question in carbon assimilation is whether the repression of photosynthesis is dose-dependent and cumulative and, hence, if reciprocity is maintained. The latter was demonstrated in several experiments [47,54]. Reduction of photosynthesis was evident during the early stages of leaf ontogeny.

C. *Growth*. Indole acetic acid (IAA) in plant meristems is a target of UV-B radiation [4,63]. Tevini and Iwanzik [63] observed that growth is slower as increased UV-B is experimentally applied to *Cucumis sativus* plants. They conclude that moderate enhancements of UV-B radiation (such as that resulting from small stratospheric ozone depletions) will cause significant reductions in growth of sensitive cucumber seedling. *Hordeum* sp. seedlings irradiated with UV-B have disturbed vertical growth, suggesting destruction of IAA molecules in the leaf tips. Leaf structure [7], nucleic acids, and pollen [20] are also affected. The experiment described by Tevini [61] reinforces these views. The conditions of such experiments (25% increase of solar UV-B radiation equivalent to a 12% ozone depletion) are similar to the ones prevalent at the latitude of Ushuaia [38,39]. This fact gives additional meaning to our LAI measurements along the gradient of ozone depletion.

D. *Plant Pathology*. Biggs and Webb [5] investigated yield and disease incidence and severity for wheat (*Triticum aestivum*) in connection with enhanced UV-B radiation in field conditions. They concluded that susceptibility to leaf rust (*Puccinia recondita* f. sp. *tritici*) is increased by enhanced UV-B radiation.

E. *Protective Mechanisms*. As stated by Beggs et al. [4], protective mechanisms can be grouped in three main classes: (1) UV damage reparation or effects negation, (2) reduction of the amount of UV radiation actually reaching sensitive targets, and (3) responses that minimize the negative effects or damages. These mechanisms may always be present or appear as the plants develop normally, without any special outside stimulus [4]. Structural attenuation plays little role in most plants [10]. Cuticles and cell walls do not absorb UV radiation. Sometimes cuticles (often thickened cuticles) are covered by external secretions of powders with flavonoids as well as waxes, oils, and some alcohols, all of which absorb UV radiation efficiently (4,16,33,72]). The mechanisms of reduction of the amount of UV radiation reaching sensitive targets are relevant for this project.

SCREENING PIGMENTS

It has been suggested [4,9,71] that, as one of their mayor functions, anthocyanins and flavonoids absorb UV radiation that might otherwise cause damage to the plant. These views are based on the fact that flavonoid synthesis is often stimulated by radiation, via phytochrome and/or the blue light receptor or induced by UV-B radiation [71]. As anthocyanins are not efficient UV screens, they need to be either esterified with cinnamic acids or to be present in very high concentrations [61]. This fact leads us to believe that some changes may be observed in the spectral signature of vegetation under UV stress. There are three further reasons to consider this as a protective mechanism: (1) the highest quantum efficiency occurs at the most damaging wavelengths reaching the Earth's surface (about 300 nm), (2) there is a linear fluence-effect relationship, and (3) the response after UV induction is quick [70,71]. Anthocyanins and flavonoids are located in epidermal cell vacuoles, although flavonoids also occur in chloroplasts. UV-B radiation stimulates the synthesis of epidermal pigments (flavone glycosides and anthocyanins) that absorb in the UV range and protect inner foliar tissues, but UV-induced reduction of photosynthesis shows that some UV reaches the inner tissues of the leaf [23]. Experiments in high latitudes have shown that increased UV-B irradiance inhibits photosynthesis and increases accumulation of UV-absorbing pigments in the leaves [26]. UV-induced decrease in mitosis frequency in *Rumex* spp. shows that UV damages the DNA molecule. It has been shown that the synthesis of flavonoids is activated by stress, including UV irradiation [24]. This indicates the existence of UV-activated promoters. Although the enzymes for flavonoid synthesis are involved in other pathways, conducting to phytoalexins and lignin, the increase in intermediates of pigment biosynthesis can explain the detected accumulation of UV-absorbing pigments in leaves [26]. Some flavonoids and anthocyanins also absorb in the visible. Therefore, it might be possible to identify changes in spectral signatures of vegetation exposed to enhanced UV-B radiation.

**REFERENCES**


1. Agrawal, S.B. 1992. Effects of supplemental UV-B radiation on photosynthetic pigment, protein and glutathione contents in green algae. Environ. Exper. Bot. 32:137-143.

2. Barnes, P.W., S.D. Flint, and M.M. Caldwell. 1987. Photosynthesis damage and protective pigments in plants from a latitudinal Arctic/Alpine gradient exposed to supplemental UV-B radiation in the field. Arctic and Alpine Res. 19:21-27.

3. Basiouny, F.M., T.K. Van, and R.H. Biggs. 1978. Some morphological and biochemical characteristics of C3 and C4 plants irradiated with UV-B. Physiol. Plant. 42:29-32.

4. Beggs, C.J., U. Schneider-Ziebert, and E. Wellmann. 1986. UV-B radiation and adaptive mechanisms in plants. *In* R.W. Worrest and M.M. Caldwell (eds.), Stratospheric ozone reduction, solar ultraviolet radiation and plant life. pp.235-250. Springer-Verlag Berlin, Heidelberg, New York, Tokyo. 374 p.

5. Biggs, R.H., and P.G. Webb. 1986. Effects of enhanced ultraviolet-B radiation on yield, and disease incidence and severity for wheat under field conditions. *In* R.W. Worrest and M.M. Caldwell (eds.), Stratospheric Ozone Reduction, Solar Ultraviolet Radiation and Plant Life. pp.303-311. Springer-Verlag Berlin, Heidelberg, New York, Tokyo. 374 p.

6. Booth, C.R., T. Lucas, T. Mestechkina, J. Tusson IV, D. Neuschuler, and J. Morrow. 1992. NSF Polar Programs UV Spectroradiometer Network 1991-1992 Operation Report. Biospherical Instruments Inc. San Diego. 152p.

7. Bornman, J.F., R.F. Evert, R.J. Mierzwa, and C.H. Bornman. 1986. Fine structural effects of UV

radiation on leaf tissue of Beta vulgaris. *In* R.W. Worrest and M.M. Caldwell (eds.), Stratospheric ozone reduction, solar ultraviolet radiation and plant life. Springer-Verlag, Berlin, Heidelberg, New York, Tokyo. 374p.

8. Brasseur, G. 1992. Volcanic aerosols implicated. Nature 359:275.

9. Caldwell, M.M. 1981. Plant response to solar ultraviolet radiation. *In* Lange, Nobel, Osmond, and Ziegler (eds.), Encyclopedia of Plant Physiology, New series, Physiological Plant Ecology I, vol 12A. Springer Verlag, Berlin, Heidelberg, New York, 169p.

10. Caldwell, M.M., A.H. Teramura, and M. Tevini. 1989. The changing solar ultraviolet climate and the ecological consequences for higher plants. Tree 4:363-366.

11. Campbell, W.F. 1975. Ultraviolet-radiation-induced ultrastructural changes in mesophyll cells of soybean (*Glycine max* (L.) Merr.). *In* Nachtwey, Caldwell & Biggs (eds.), Impacts of Climatic Change on the Biosphere, Assessment Program, US Dept. of Transportation Report No DOT-TST-75-55, NTIS, Springfield, Virginia, pp.4-79.

12. Clausen, J., D.D. Keck, and W.M. Hiesey. 1940. Experimental studies on the nature of species: I. Effect of varied environments on western North American plants. Carnegie Inst. Washington Pub. 520.

13. Crawley, M.J. 1990. The responses of terrestrial ecosystems to global climate change. *In* G. MacDonald and L. Sertorio (eds.), Global Climate and Ecosystems Change. Pleum, New York. p.141.

14. Cullen, J.J., P.J. Neale, and M.P. Lesser. 1992. Biological weighing function for the inhibition of phytoplankton photosynthesis by ultraviolet radiation. Science 258:646-650.

15. Curran, P.J., and H.D. Williamson. 1987. Estimating the green leaf index of grassland with airborne MSS data to estimate GLAI. Internatl. J. Remote Sensing 8:57-74.

16. Egger, K., Wollenweber, E., and M. Tissot. 1970. Freie Flavonol-Aglykone im Knospensekret von *Aesculus* Arten. Z. Pflanzenphysiol. 62:464-466.

17. El-Sayed, S.Z. 1988. Fragile life under the ozone hole. Natural History, October:72-80

18. Esser, G., D. Franz, M. Haeser, B. Hugemann, and H. Schaaf. 1979. Einfluß einr Schadstoffimmission vermehrten Einstrahlung von UV-B-Licht auf den Ertrag von Kulturpflanzen (FKW 22). Bericht für Strahlen- und Umweltforschung m.b.H. Nueherberg. Battelle-Institut E.V. Frankfurt am Main. 59p.

19. Farquhar, G.D., and S. von Caemmerer. 1982. Modelling of photosynthetic response to environment. *In* O.L. Lange, P.S. Nobel, C.B. Osmond, and H. Ziegler (eds.), Encyclopedia of Plant Physiology, NS vol 12B: Physiological Plant Ecology II. Springer-Verlag, Berlin, Heidelberg, New York, Tokyo. pp.549-587.

20. Flint, S.D., and M.M. Caldwell. 1986. Comparative sensitivity of binucleate and trinucleate pollen to ultraviolet radiation: A theoretical perspective. *In* Worrest and Caldwell (eds.), Stratospheric ozone reduction, solar ultraviolet radiation and plant life. Springer-Verlag, Berlin, Heidelberg, New York, Tokyo. 374p.

21. Gleason, J.F., P.K. Bhartia, J.R. Herman, R. McPeters, P. Newman, R.S. Stolarski, L. Flynn, G. Labow, D. Larko, C. Seftor, C. Wellemeyer, W.D. Komhyr, A.J. Miller, and W. Planet. 1993. Record low global ozone in 1992. Science 260:523-526.

22. Goodwin, T.W. 1976. Chemistry and biochemistry of plant pigments. 2nd. Ed. Vol. 2. Academic Press. London, New York, San Francisco. 373p.

23. Häder, D-P., and M. Tevini. 1987. General Photobiology, Pergamon Press, Oxford. 323p.

24. Hahlbrock, K., J. Chappell, and D.N. Kuhn. 1984. Rapid induction of mRNAs involved in defense reactions in plants. *In* Annual Proceedings of the Phytochemical Society of Europe. Vol 23 (P.J. Lea, and G.R. Stewart, eds.). Clarendon , Oxford.

25. Harborne, J.E. 1992. Chromatography. 5th Edition. Part B: Applications. E. Heftmann (ed.). Chap. 19, B363-B393. Elsevier, Amsterdam.

26. Hardy, T.J., M. Beherenfeld, H. Gucinski, and A Wones. 1992. AGU 1992 Ocean Sciences Meeting: 66. # 041c-7.

27. Heath, D.F. 1988. Non-seasonal changes in total column ozone from satellite observations, 1970-86. Nature 332:219-227.

28. Hiesey, W.M., J. Clausen, and D.D. Keck. 1942. Relations between climate and intraspecific variation in plants. Am. Naturalist 76:5-22.

29. Hively, W. 1989. How bleak is the outlook for ozone?

Am. Scientist 77:219-223.

30. Hueck, K., and P. Seibert. 1981. Vegetationskarte von Südamerika. 2. Auflage. Gustav Fischer Verlag, Stuttgart-New York. 90p. + 1 map.

31. Hunt, E., and S. Running. 1992. Simulated dry matter yields for aspen and spruce stands in the North American boreal forest. Canadian J. Remote Sensing 18(3):126-133.

32. Iwanzik, W., M. Tevini, G. Dohnt, M. Voss, W. Weiss, P. Gräber, and G. Renger. 1983. Action of UV-B radiation on photosynthetic primary reactions in spinach chloroplasts. Physiol. Plant. 58:401-407.

33. Johnson, N.D. 1983. Flavonoid aglycones from *Eriodyction californicum* resin and their implication for herbivory and UV screening. Biochem. System. and Ecol. 11:211-215.

34. Karentz, D. and L.H. Lutze. 1990. Evaluation of biologically harmful ultraviolet radiation in Antarctica with a biological dosimeter designed for aquatic environments. Limnol. and Ocean. 35:549-561.

35. Kerr, R.A. 1993. Ozone takes a nose dive after the eruption of Mount Pinatubo. Science 260:490-491.

36. Knipling, E.B. 1970. Physical and physiological basis for the reflectance of visible near-infrared radiation from vegetation. Remote Sensing of Environ. 1:155.

37. Lemonick, M.D. 1992. The ozone vanishes. Time, February, pp.60-63.

38. Madronich, S. 1992. Implications of recent total atmospheric ozone measurements for biologically active ultraviolet radiation reaching the earth's surface. Geophys. Res. Letters 19:37-40.

39. Madronich, S. 1993. UV Radiation in the natural and perturbed atmosphere. *In* Tevini, M. (ed.), 1993. UV-B radiation and Ozone Depletion: Effects on Humans, Animals, Plants, Microorganisms, and Materials. Lewis, Boca Raton. 248p. pp.17-70.

40. Molina, M.J., and F.S. Rowland. 1974. Stratospheric sink for chlorofluoromethanes: Chlorine atomic catalyzed destruction of ozone. Nature 249:810-812.

41. Moore, D.M. 1983. Flora of Tierra del Fuego. Anthony Nelson, England and Missouri Botanical Garden, USA. 396p.

42. Noorudeen, A.M., and G. Kulandaivelu. 1982. On the possible site of inhibition of photosynthetic electron transport by ultraviolet-B (UV-B) radiation. Physiol. Plant. 55:161-166.

43. Peterson, D.L., M.A. Spanner, S.W. Running, K.B., and A. Teuber. 1987. Relationship of thematic mapper simulator data to leaf area index of temperate coniferous forests. Remote Sensing of the Environ. 22:323-341.

44. Pierce, L.L., and S.W. Running. 1988. Rapid estimate of coniferous forest leaf area index using a portable integrating radiometer. Ecology 69:1762-1767.

45. Prohaska, F. 1976. CLimate of Argentina, Paraguay and Uruguay. *In* W. Schwerdtfeger (ed.), Climates of Central and South America. World Survey of Climatology, Vol. 12, pp.13-112. Elsevier, Amsterdam.

46. Ravishankara, A.R., S. Solomon, A.A. Turnipseed, and R.F. Warren. 1993. Atmospheric lifetimes of long-lived halogenated species. Science 259:194-199.

47. Robberecht, R., and M.M. Caldwell. 1978. Leaf epidermal transmittance of ultraviolet radiation and its implications for plant sensitivity to ultraviolet-radiation-induced injury. Oecologia 32:277-287.

48. Robberecht, R., and M.M. Caldwell. 1986. Leaf UV optical properties of *Rumex patientia* L. and *Rumex obtusifolius* L. in regard to a protective mechanism against solar UV-B radiation injury. *In* R.C. Worrest and M.M. Caldwell (eds.), Stratospheric ozone reduction, solar ultraviolet radiation and plant life. Springer-Verlag, Berlin, Heidelberg, New York, Tokyo. 374p.

49. Rogers, C. 1988. Global ozone trends reassessed. Nature 332:201.

50. Running, S., and J. Coughlan. 1988. A general model of forest ecosystem processes for regional applications. I. Hydrologic balance, canopy gas exchange and primary production processes. Ecol. Model. 442:125-154.

51. Running, S.W., and S.T. Gower. 1991. FOREST-BGC, A general model of forest ecosystem processes for regional applications II. Dynamic carbon allocation and nitrogen budgets, Tree Phys. 9:147-160.

52. Ryan, M.G. 1991. Effects of climate change on plant respiration. Ecol. Applic. (1), 2:157-167.

53. Sisson, W.B., and M.M. Caldwell. 1976. Photosynthesis, dark respiration, and growth of *Rumex patientia* L. exposed to ultraviolet irradiance (288-315 nm) simulating a reduced ozone column. Plant Phys. 58:563-568.

54. Sisson, W.B., and M.M. Caldwell. 1977. Atmospheric ozone depletion: Reduction of photosynthesis and growth of a sensitive higher plant exposed to enhanced UV-B radiation. J. Exper. Bot. 28:691-705.

55. Sisson, W.B. 1986. Effects of UV-B Radiation on Photosynthesis. *In* R.C. Worrest and M.M. Caldwell (eds.), Stratospheric ozone reduction, solar ultraviolet radiation and plant life. pp.161-170. Springer-Verlag, Berlin, Heidelberg, New York, Tokyo. 374p.

56. Smith, R.C., B.B. Prézelin, K.S. Baker, R.R. Bidigare, N.P. Boucher, T. Cooley, D. Karentz, S. MacIntyre, H.A. Matlick, D. Menzies, M. Ondrusek, Z. Wan, and K.J. Waters. 1992. Ozone depletion: Ultraviolet radiation and phytoplankton biology in Antarctic waters. Science 225:952-959.

57. Spanner, L., L. Pierce, D.L. Peterson and S.W. Running. 1990. Remote sensing of temperate coniferous forest leaf area index: The influence of canopy closure, understory vegetation and background reflectance. Internat. J. Remote Sensing 11:95-111.

58. Stolarski, R., R. Bojkov, L. Bishop, C. Zerefos, J. Staehelin, and J. Zawodny. 1992. Measured trends in stratospheric ozone. Science 256:342-349 (17 April 1992).

59. Teramura, A.H. 1986. Interaction between UV-B radiation and other stresses in plants. *In* R.C. Worrest and M.M. Caldwell (eds.), Stratospheric ozone reduction, solar ultraviolet radiation and plant life. NATO ASI Series, Vol. G8. Springer-Verlag, Berlin, Heidelberg, New York, Tokyo. pp.327-343.

60. Tevini, M., (ed.). 1993a. UV-B Radiation and Ozone Depletion: Effects on Humans, Animals, Plants, Microorganisms, and Materials. Lewis Publishers, Boca Raton, Ann Arbor, London, Tokyo. 248p.

61. Tevini, M. 1993b. Effects of enhanced UV-B radiation on terrestrial plants. *In* Tevini, M. (ed.), UV-B Radiation and Ozone Depletion: Effects on Humans, Animals, Plants, Microorganisms, and Materials. Lewis Publishers, Boca Raton, Ann Arbor, London, Tokyo. 248p. Ch.5:125-153.

62. Tevini, M., W. Iwanzik, and U. Thoma. 1981. Some effects of enhanced UV-B irradiation on the growth and composition of plants. Planta 153:388-394.

63. Tevini, M., and W. Iwanzik. 1986. Effects of UV-B radiation on growth and development of cucumber seedlings. *In* R.C. Worrest and M.M. Caldwell (eds.), Stratospheric Ozone Reduction, Solar Ultraviolet Radiation and Plant Life. Springer-Verlag, Berlin, Heidelberg, New York, Tokyo. 374 p. pp.271-285.

64. Thomas, J.R., L.M. Namken, G.F. Oerther, and G. Brown. 1971. Estimating leaf water content by reflectance measurements. Agron. J. 63:845.

65. Thomas, J.R., and G.F. Oerter. 1972. Estimating nitrogen content of sweet pepper leaves by leaf reflectance measurements. Agron. J. 64:11.

66. Tsay, M.L., D.H. Gjerstad, and G.R. Clover. 1982. Tree leaf reflectance: A promision technique to rapidly determine nitrogen and chlorophyll content. Canadian J. Forest. Res. 12:788.

67. Vitousek, P.M. 1992. Global environmental change: An Introduction. Annu. Rev. Ecol. Syst. 23:1-14.

68. Vogelmann, A.M., and T.P. Ackerman. 1993. Harmful UV radiation may increase after volcanic eruptions. Amer. Geophy. U. EOS. January 12, p.25.

69. Vu, C.V., L.H. Allen, and L.A. Garrard. 1982. Effects of supplemental UV-B radiation on growth and leaf photosynthetic reactions of soybean (Glycine max). Physiol. Plant. 52:353-362.

70. Wellman, E. 1976. Specific ultraviolet effects in plant morphogenesis. Photochem. Photobio. 24:659-660.

71. Wellman, E. 1983. UV radiation in photomorphogenesis. *In* W. Shropshire Jr. and H. Mohr (eds.), Encyclopedia of Plant Physiology, new series, Photomorphogenesis, vol 16B. Springer-Verlag, Berlin, Heidelberg New York, 745p.

72. Wollenweber E., and K. Egger. 1971. Die lipophilen Flavonoide des Knospenol von *Populus nigra*. Phytochem. 10:225-226.

# DNA DAMAGE ACTION SPECTROSCOPY AND DNA REPAIR IN INTACT ORGANISMS: ALFALFA SEEDLINGS

B.M. Sutherland, F.E. Quaite,[1] and J.C. Sutherland

**Biology Department, Brookhaven National Laboratory, Upton, NY 11973 USA**

Key words: alfalfa, dimers, DNA damage, DNA repair, excision, photorepair, UV-A, UV-B

**ABSTRACT**

Although sunlight is essential for plant life, the ultraviolet components in the solar spectrum pose potential threats to DNA-dependent metabolism (e.g., transcription of mRNAs for essential proteins) and to genetic integrity. To evaluate the risks to plant DNA of UV present in solar spectra reaching the earth today and in anticipation of increases in UV-B, we have developed methods for quantitating DNA damages at low, biologically relevant levels (to 1 lesion/ 2 million bases) in nanogram quantities (total requirement 100 ng/determination) of non-radioactive plant DNA.

In our method, plant tissue is embedded in agarose, non-DNA components digested enzymatically, and the DNA treated with an enzyme or other agent which specifically and quantitatively induces a single strand nick at each lesion site, while a companion DNA sample is incubated without the agent. After gel electrophoresis with molecular length standards under denaturing conditions, the DNA is renatured, stained with ethidium bromide and a quantitative electronic image obtained using a charge-coupled device-based camera system. The number of average molecular lengths of the treated and untreated DNAs are calculated, and the frequency of lesions computed.

This approach has allowed us to determine the efficiency of monochromatic radiation in the UV-B and UV-A range in inducing pyrimidine dimers in the DNA of irradiated intact alfalfa seedlings. Although UV-B is indeed quite efficient in damaging DNA in the seedlings, UV-A can also induce dimers in the DNA. Since today's solar spectrum contains far more UV-A than UV-B, the damage from UV-A (calculated from the product of the efficiency of a given wavelength in damaging DNA times the number of photons at that wavelength in the solar spectrum) is a significant portion of the total damage inflicted by sunlight on this plant.

What are the biological consequences of such DNA damage? Since only unrepaired or misrepaired lesions pose biological problems for living systems, it is essential to know the ability of plant systems to cope with damages to their DNA. We are measuring repair of pyrimidine dimers in UV-irradiated seedlings in the dark and in the presence of visible light. Alfalfa seedlings employ both excision and photorepair for removal of pyrimidine dimers; however, the relative importance of these repair paths seems to depend strikingly on the initial damage level to the DNA of the seedling.

[1] Present Address: Biology Department, Argonne National Laboratory, Argonne IL 60439 USA

NATO ASI Series, Vol. I 18
Stratospheric Ozone Depletion/
UV-B Radiation in the Biosphere
Edited by R. H. Biggs and M. E. B. Joyner

## INTRODUCTION

Evaluating the effects of increased UVB (290-320 nm) from stratospheric ozone depletion requires knowing the biological effects of the wavelengths that will be increased as well as the effects of other damaging wavelengths. Since UVB radiation is absorbed much more effectively by DNA than UVA (320-400 nm), UVB has long been considered the major biologically-damaging radiation in sunlight.

More recently, direct and indirect damaging effects of UVA on DNA have been found, but generally at efficiencies per photon of less than 1% that of UVB. Can such low efficiency processes be biologically significant? The answer lies in two factors: first, the number of DNA lesions induced by a specific wavelength of radiation is the **product** of the efficiency with which photons of that wavelength induce the damage **times** the number of photons of that wavelength which are absorbed by the target DNA. Second, in sunlight at the earth's surface, the number of UVA photons is far greater than the number of UVB photons. Thus the total DNA damage inflicted by UVA--the sum over all UVA wavelengths of the **products** of damage efficiency at given wavelength times the number of photons at that wavelength--is much larger compared to UVB-induced damage than would be anticipated based on the efficiencies alone.

## ACTION SPECTROSCOPY OF DNA DAMAGE

We can measure the distribution of UVA and UVB photons in today's solar spectra, and can predict the spectra for various scenarios of differing degrees of ozone depletion. Thus we require "only" the efficiency measurements (action spectra) for DNA lesion induction to calculate the burden of damages inflicted by today's spectrum and by any predicted spectrum of interest. We have said "only" with regard to obtaining action spectra because of the many pitfalls that must be avoided and the immense amount of work required to obtain good, statistically reliable data at a sufficient number of wavelengths for determination of an accurate and useful action spectrum.

What problems make action spectrum measurements so difficult? A complete treatment of action spectroscopy is beyond the scope of this review, and the reader is referred to the chapter by Thomas Coohill in this volume and to John Jagger (3). However, special problems likely to be encountered in action spectrum measurements on intact organisms are particularly pertinent to problems of ozone depletion, and we shall discuss the most important of them.

### ACTION SPECTROSCOPY IN INTACT ORGANISMS

Three critical issues that must be addressed to measure accurate action spectra are 1) selection of suitable sources of monochromatic radiation covering the desired wavelength range with sufficient intensity over a large enough area to irradiate the desired organism(s), 2) design of the experiments, and 3) analysis of data. Each of these points presents pitfalls to the unwary, especially in studying intact organisms.

We must first distinguish "total effect studies," in which target organisms are irradiated with a broad spectrum source approximating a total environmental condition, from analytical action spectra in which the efficiencies of individual monochromatic or very narrow band radiation are determined for a number of wavelengths. In the latter case, the product of the efficiency at different wavelengths and photon intensity can be obtained for any desired broad spectrum source, including sunlight, by convolution of the intensity of the spectrum or "spectral irradiance" and the efficiency of damage at each wavelength in that spectrum.

#### *Radiation*

To determine an analytical action spectrum, we must have a light source that can produce sufficient radiation of uniform intensity over the entire specimen area (dish/plant/animal) at each wavelength of interest to induce a measurable effect in a finite time. For most biological systems, this implies a high intensity lamp (Hg or Hg/Xe), grating or prism monochromator, and photometer calibrated at each desired wavelength to allow quantitation of the incident photon flux. A particular problem in measuring effects at wavelength regions of low efficiency is that the presence of a very small quantity (1% or less) of shorter wavelength, high damaging efficiency radiation can alter significantly the observed effect. The use of short wavelength cutoff filters and bandpass filters at such wavelengths, for example the UVA, can be used to prevent such problems (3-5).

*Interval Between Wavelengths.* On first consideration of carrying out a convolution of DNA damage data with a continuous source such as sunlight, one might think that DNA action spectra should be carried out to similar resolution. However, consideration of the absorption spectra of DNA and other biological macromolecules in solution indicates that this is unnecessary. First, consider the absorption of gas atoms: they show very strong absorption lines over tenths of nanometers. Thus, if one were determining the action spectrum for excitation of such a gas, one would have to measure excitation at very closely spaced wavelength intervals or risk missing large absorptions. However, biological molecules have broad absorption spectra, with smooth envelopes. The absence of fine structure in the UV *absorption* spectra of biological molecules indicates that the action spectra of these molecules will also be a smooth function. [See pp. 15-17 of Jagger (3) for an excellent discussion.]

The properties of biological action spectra that are important to determine are the wavelength maximum of the spectrum and the magnitudes—the absolute magnitudes if possible, the relative values if the absolute magnitudes cannot be determined—of the effect at several wavelengths in the region of interest.

*Light Sources.* High pressure mercury arcs offer several useful lines in the ultraviolet, including 275, 280, and 289 nm in the UVC region, 295, 302, and 313 nm in the UVB, and 334, 365 and 385 nm in the UVA, as well as 405 nm in the short wavelength visible. Mercury-Xenon arcs supplement the interline intervals and allow more flexible choices of wavelength regions. It must be stressed that just because one has set a monochromator to a particular wavelength does not mean that wavelength is the only, or even the principal, biologically effective radiation emerging from the system; scattered or stray light, or light from a strong nearby line may comprise the principal *biologically* effective radiation although its presence is not revealed by the monochromator setting.

## EXPERIMENTAL DESIGN

*Reciprocity.* In action spectroscopy, preliminary experiments are essential to indicate the light flux and, in practical terms, the time required to obtain a measurable biological effect. Suppose that the time required is 20 seconds at 275 nm and 200 minutes at 365 nm (in our example, 275 nm is absorbed much more efficiently than 365 nm). How will we know that the biological damage we measure at 365 nm isn't artificially decreased by repair which occurs during the 200 minutes of irradiation? There are two approaches to solving this problem. First, we can establish whether time and fluence reciprocity holds. That is, will the same number of photons administered over the entire range of times to be used in the experiment give the same result? If so, reciprocity holds, and results obtained using the exposure durations within the range we tested are valid. Time-intensity reciprocity is found to hold in many, but not all, purely photochemical reactions. Figure 1 shows a demonstration of reciprocity for pyrimidine dimer formation in alfalfa seedlings. The solid symbols represent dimer yields in plants irradiated at 275 nm at a rate of 0.3 $mW/cm^2$, while the open symbols show data for plants irradiated at 1.1 $mW/cm^2$, about 3.7 times as high a rate. Clearly, over the time spans required for these irradiations, the exposure-response lines for the two exposure rates are indistinguishable and reciprocity holds for these plants under these experimental conditions (6,7).

In many biological systems, reciprocity does not hold and the second approach is necessary. For example, if one is measuring DNA photoproduct formation in a DNA repair-proficient organism, repair enzymes might remove many lesions during a 200 minute irradiation at 365 nm (a relatively inefficient wavelength in damaging DNA, thus requiring long irradiation times) but very few during a 20 second irradiation at 275 nm, which is quite efficient in lesion induction. We would thus underestimate the efficiency of 365 nm in damaging DNA. In addition, many organisms carry out photorepair, in which a photoreactivating enzyme uses light to reverse UV-induced cyclobutyl pyrimidine dimers in DNA; during the long 365 nm exposure in our example, the organism could photoreactivate many dimers, again leading to an underestimate of the damaging effect of 365 nm radiation.

One solution would be to use equal times of irradiation for all wavelengths; this will should allow equal action of all non-light-dependent processes, such as excision repair. Photorepair might be avoided by irradiating chilled organisms (to decrease the rate of formation of enzyme-substrate complexes). However, it is usually desirable to produce comparable

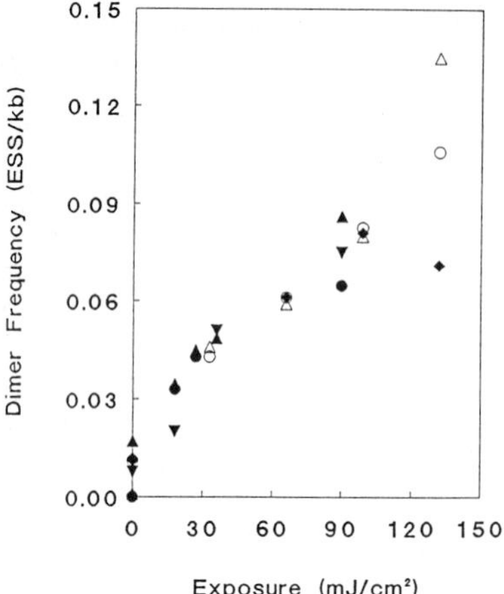


Figure 1. Pyrimidine dimer yields induced in alfalfa seedlings exposed to 275 nm radiation at an exposure rate of 1.1 mW/cm² (open symbols) or 0.3 mW/cm² (closed symbols). Seedlings were harvested, minced, embedded in agarose, and digested with Proteinase K. Companion samples were incubated with or without saturating levels of a pyrimidine-specific endonuclease which makes a single strand nick adjacent to each pyrimidine dimer. The pyrimidine dimer content was determined by an electrophoretic gel assay (1,7). The DNAs were denatured, separated along with DNAs of known size according to single strand molecular length on alkaline agarose gels; the gel was renatured, stained with ethidium bromide and a quantitative electronic image obtained (11). The number average molecular length of each DNA sample was determined and the pyrimidine dimer content determined as previously described (7). The exposure-response functions for the two exposure rates are indistinguishable, indicating time-exposure rate reciprocity holds.

damage levels at all wavelengths studied in an action spectrum, and the very great differences in the efficiencies of different wavelengths in producing damage mean that it is usually difficult to produce a given level of damage in the UVA with the same irradiation times as are possible with irradiation in the UVB. Extending the irradiation times in the UVB to match those in the UVA can result in extremely long experiments. Thus, one may simply have to take into account such limitations in interpreting the data.

*Fluence-response relations*. Once preliminary experiments indicate suitable conditions for obtaining measurable effects, one must determine the effect produced by increasing photon exposures at each wavelength, the "dose-response" or fluence-response lines. For constructing an action spectrum that can be used to predict the level of damage produced by broad spectrum UV radiation, the effect ideally should be a linear function of the exposure of UV at each wavelength. The slopes of these exposure-response lines, in units of effect/photon/area, are then used to construct the action spectrum.

Why is it important to determine fluence-response lines—why not just determine the effect of one exposure at each wavelength, calculate the photon exposure required to induce a constant effect (e.g., 37% killing; 50% substrate conversion) at each of those wavelengths? Just as important as the slope is the *shape* of the line. If the shapes of the lines are different at the different wavelength regions, this indicates that the biological processes involved in the irradiations at the different wavebands were different, and it is not correct to construct a single "action spectrum" for the biological effect across the entire wavelength range.

What if the fluence-response lines are not linear, but show a sigmoid shape? An example is the sigmoid shape of the exposure-response function of transformation of normal, repair-proficient human cells by UV (9,10). At low photon exposures, almost no transformants are observed, presumably due to efficient repair of virtually all the lesions. As the exposure increases, however, transformants are detected in increasing frequency. In this case, the use of slope of the fluence-response line is not valid, but the use of a constant level of biological effect (e.g., transformants per survivor/photon/m$^2$) allows computation of action spectra for human cell neoplastic transformation (9,10).

*Calculation of Action Spectra*

After obtaining linear fluence-response lines at the different wavelengths, one is ready to calculate an action spectrum. Either the slope of the lines (effect/ photon/area) or the reciprocal of the photon flux required to achieve a constant biological effect [1/ (photon/area/effect), which yields the same effect/ photon/area)] is plotted as a function of wavelength. Although measurements of photon exposure are sometimes made with instruments whose output reads in units of energy, it is essential for comparison with absorption spectra to convert these exposures to number of photons using the relationship

$$E = (hc)/\lambda,$$

where h is Planck's constant, c is the speed of light, and $\lambda$ is the wavelength of the incident light. To allow for facile comparison with absorption spectra of suspected target molecules or to allow convolution with broad spectrum sources whose emission at different wavelengths varies over many orders of magnitude, action spectra are most conveniently displayed on a semi-logarithmic scale.

At this point, one must decide whether or not to normalize. Frequently one sees comparisons of action spectra in which all the spectra have been normalized to the same value at a particular wavelength. Although this practice allows comparison of spectra for which obtaining absolute cross-sections is not possible (e.g. tumorigenesis), normalization has many pitfalls. Panel A of Figure 2 shows a shallow container of DNA in solution and Panel A of Figure 3 shows the absolute action spectrum for inducing photoproducts in that DNA(6). Panel B of Figure 2 shows the same DNA, but now a filter has been interposed between the light source and the DNA (but the DNA solution is unchanged). The filter is transparent above 320 nm, but absorbs 99% of all radiation less than 320 nm. In this case, since we measure the light incident on the entire system (in this case the filter), not on the DNA surface, the action spectrum for damaging the DNA is shown in the solid symbols in Figure 3B. The sharp discontinuity in the example results from the sharp change in transmittance of our hypothetical filter.

If these spectra are normalized at a wavelength where there is no shielding by the filter, the relative shapes of the two spectra remain the same and only the absolute values change. However, if one normalizes at a wavelength where there is significant shielding by cell or tissue components, e.g. 300 nm, the resulting action spectra look very different (remember that the DNA had not changed, only the filter was added). Panel C of Figure 3 shows the result of normalizing the spectra in Panel B of Figure 3 at a wavelength where there is significant shielding. One would conclude from these normalized spectra that the effect in the longer wavelength region was abnormally high. In fact, it was invariant. Thus, arbitrary sharp ] can lead to highly misleading comparisons.

A second disadvantage in normalizing action spectra is that quantitative comparisons between different systems, which relate ultimately to differing levels of biological sensitivity, are lost upon normalization. Figure 4 shows the absolute action spectra for inducing pyrimidine dimers DNA in solution (6) and for inducing dimers in the DNA of two intact organisms, human skin *in situ* (2) and intact alfalfa seedlings (6). The absolute action spectra clearly show that DNA in solution is about 100 times more sensitive in the UVB region than DNA in either human skin or in seedlings.

*Action Spectra in Prediction of Effects of Ozone Depletion*

What information does the alfalfa action spectrum provide for evaluating the effects of environmental light on plants? First, the spectrum indicates that radiation throughout the region 270-365 nm induces pyrimidine dimers in alfalfa DNA (6). We could not detect any significant level of dimers over background at either 385 or 405 nm. The absolute efficiency of photons in the UVA region in damaging DNA is approximately $10^{-4}$ that of photons in the UVB region.

How, then, can such a low efficiency process be of biological significance? The bold solid line in Figure 4 shows a spectrum for solar radiation reaching the surface of the earth under today's conditions of stratospheric ozone at a particular latitude, time of year, and time of day (8). The amount of damage inflicted by a broad waveband, such as the UV present in sunlight, is the product of the efficiency of each wavelength for producing damage within that waveband in damaging the biological system times the number of photons at each wavelength within the waveband, summed over all wavelengths in that waveband. Examination of Figure 4 shows that although UVB is quite effective in damaging DNA in

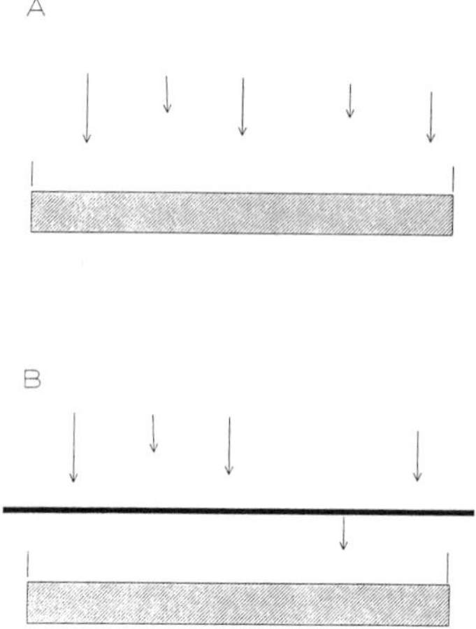


Figure 2. Irradiation of DNA in solution without (Panel A) and with (Panel B) an intervening filter. Panel A shows a solution of DNA (hatched lines) exposed to broad spectrum UV radiation from above. Panel B shows the same DNA solution exposed to the same broad spectrum source, but now with an intervening filter which absorbs 99% of all radiation shorter than 320 nm. This situation models the difference between measurements of damage directly in isolated molecules and in the same molecules in intact organisms, in which the molecules are shielded by cellular and tissue components (e.g., skin, cell walls, cell pigments).

the alfalfa seedlings, there is very little UVB in today's solar spectrum; conversely, although the "per photon" efficiency of UVA radiation is small, sunlight contains very large quantities of UVA radiation. This indicates that both UVA and UVB damage DNA in living organisms.

What will be the impact of ozone depletion on DNA damage to alfalfa? The bold dashed line in Figure 4 shows a spectrum for radiation reaching the earth's surface under conditions of 50% depletion of stratospheric ozone under the same conditions of latitude, etc., as the bold solid line (8). More UVB radiation will reach the earth's surface, and thus the plants will suffer increased DNA damage; however, the quantity of UVA is largely unchanged and the damage from this wavelength region will remain constant. These data indicate that ozone depletion would indeed result in more DNA damage in the seedlings. Since a significant portion of damage at the present time is inflicted by UVA (and that radiation will remain constant), an increase in UVB will not have as large a detrimental effect on living plants as if UVB were the only damaging radiation. In addition, the effect of the increased UVB compared to the constant UVA is modulated by the greater attenuation of UVB by the plant; that is, the cross section for UVB damaging the plant's DNA is reduced compared to the ability of UVB to damage unshielded DNA. In contrast, for most of the UVA the response of DNA in the plant is much closer to that of unshielded DNA.

Many cellular components can be damaged by UV. Why should the level of damage to DNA be of especial concern? DNA damages can block DNA replication and, if replication does proceed in the presence of damage, mutations may be produced. Even if cells are not actively dividing, DNA damage can have serious consequences: DNA lesions can block transcription, or produce decreased quantities of messenger RNAs for critical cellular proteins. Since some important cellular proteins are thought to be damaged by UV, their replacement by [word missing? or is "by" superfluous?] would be essential for cellular function, and UV lesions within their genes would have serious cellular consequences. Finally, even if a messenger RNA coding for an cellular RNA were long lived, some tRNAs are inactivated by UV, and synthesis of their replacements may be essential for continued protein synthesis.

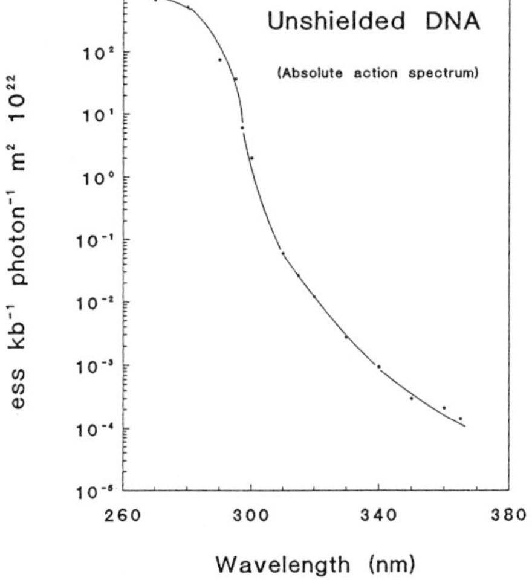


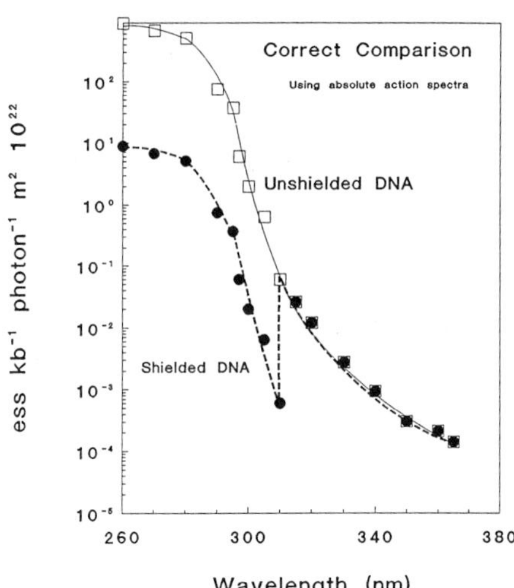


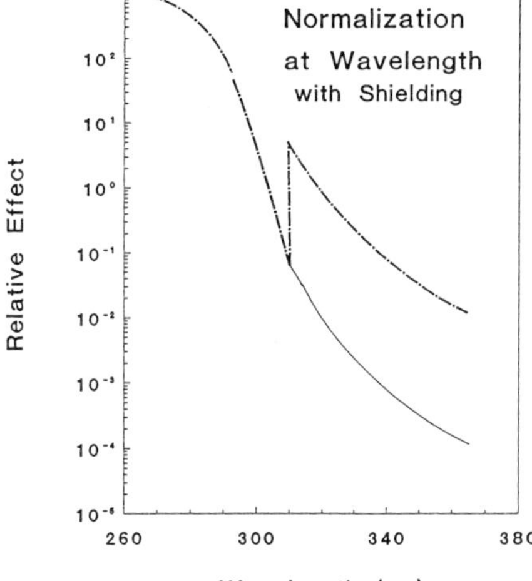


Figure 3. Absolute (non-normalized) and normalized action spectra for damaging DNA irradiated as in Figure 2. Panel A: An absolute action spectrum for damaging DNA irradiated in solution, as shown in Figure 2A. Panel B: Absolute action spectrum calculated for DNA shielded by the filter shown in Figure 2B (●) (6), compared with that for unshielded DNA (□). The filter absorbs 99% of all radiation below 320 nm, thus reducing the action spectrum by two orders of magnitude in the wavelength range 260-320 nm. Panel C: Action spectra for unshielded DNA (smooth curve) and for shielded DNA (dashed curve) normalized at a wavelength less than 320 nm, where the filter shields the DNA. Note that—although the biological system, in this case the DNA, is unchanged—normalization using a wavelength where there is significant shielding producesa misleading appearance of excessively high values in data which are in fact not altered.

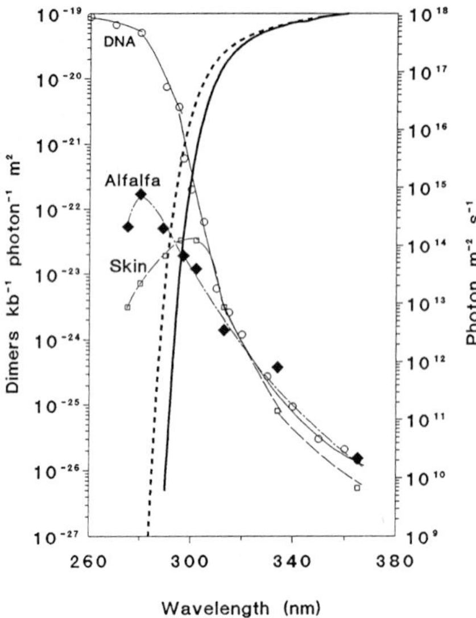


Figure 4. Absolute action spectra for pyrimidine dimer production in unshielded DNA (○) (6), intact alfalfa seedlings (◆) (6), or in skin of healthy human volunteers (□) (left axis) (2). Since these are absolute action spectra, they can be compared directly and quantitatively without normalization. The bold solid line shows solar spectrum for downward global solar flux at the earth's surface for a solar angle of 40° and 0.32 cm stratospheric ozone, compared with that for 0.16 cm ozone (bold dotted line) (8) (right axis).

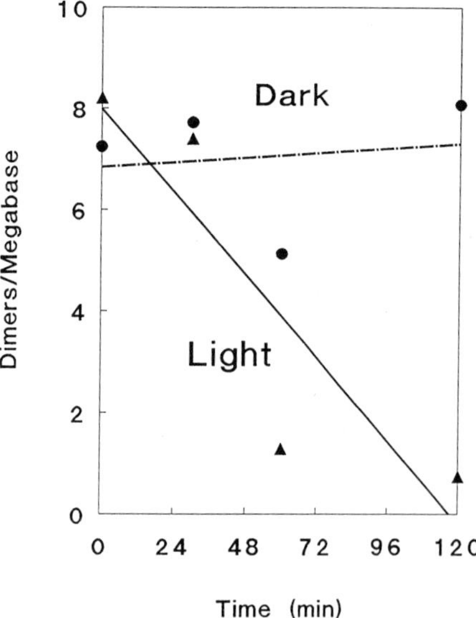


Figure 5. Repair in alfalfa seedlings exposed to 280 nm radiation producing about 8 pyrimidine dimers/ million bases. Seedlings were then kept in the dark (●) or photoreactivated (▲) (exposed to blue photoreactivating light filtered through a yellow filter which removed light below about 405 nm, thus preventing DNA damage induction during photorepair). Pyrimidine dimer contents were determined as in Figure 1; the lines were fit to the mean of three determinations by least square analysis.

## DNA REPAIR IN INTACT ORGANISMS

Organisms have evolved in the presence of solar radiation and are coping adequately with stresses inflicted by environmental UV. Most organisms have mechanisms for removing DNA damages, and the overwhelming majority have multiple, at least partially overlapping, systems for repairing DNA. Two repair paths that are prominent in most organisms are excision and photorepair.

### DNA REPAIR PATHS

Excision is carried out by a multi-enzyme pathway and generally can remove many kinds of lesions, especially bulky ones, from DNA. Sequentially or concomitantly, new synthesis using the undamaged strand as template restores the base sequence and, finally, ligation restores the integrity of the DNA backbone. Excision is thus flexible, carrying out repair of many types of lesions; on the other hand, it requires cellular energy and, during resynthesis, poses the possibility of errors that can lead to mutation.

Photoreactivation is carried out by a single enzyme pathway, repairs cyclobutyl pyrimidine dimers, and uses longer wavelength UV or visible light as the sole required energy source. It is thus less flexible, with only one class of lesions as substrate, but uses light, not cellular energy, as an energy source (and is thus well suited to organisms exposed to sunlight), and is error-free.

*Repair in Alfalfa.* What repair paths does alfalfa use for dealing with UV damages? Figure 5 shows the fate of pyrimidine dimers induced at a low exposure of 280 nm radiation to the seedlings, producing an initial level of about 8 dimers/million bases. (In comparison, 1 J/m$^2$ of 254 nm radiation produces about 6.5 dimers/million bases in unshielded DNA.) After UV exposure, the seedlings were kept in the dark or exposed to blue light filtered by a yellow filter which excluded wavelengths shorter than about 405 nm (thus preventing DNA damage by the photoreactivating light). Seedlings exposed to this level of UV actively carry out photorepair (△), but do not seem to remove dimers by excision (⊙) within the times examined. The relative roles of photorepair and of excision at other exposures, or of damages induced by other waveband sources, remains to be determined.

## CONCLUSIONS

Understanding the effects of UV, and increased levels of UV, on DNA in living organisms requires knowledge of both the frequencies of damages induced by the quantities and quality (wavelength composition) of the damaging radiation, and of the capacity of the organism to carry out efficient and accurate repair. The major levels of uncertainty in understanding the responses of intact organisms, both plant and animal, to UV indicates that we cannot assess accurately the impact of stratospheric ozone depletion without major increases in knowledge of DNA damage and repair.

**Acknowledgments**

Research is supported by a grant from the USDA to JCS and BMS, and by the Office of Health and Environmental Research of the US Department of Energy.

## REFERENCES

1. Freeman, S.E., A.D. Blackett, D.C. Monteleone, R.B. Setlow, B.M. Sutherland, and J.C. Sutherland. 1986. Analyt. Biochem. 158:119-129.

2. Freeman, S.E., H. Hacham, R.W. Gange, D. Maytum, J.C. Sutherland, and B.M. Sutherland. 1989. Proc. Natl. Acad. Sci. USA 86: 5605-5609.

3. Jagger, J. 1967. Introduction to Research in Ultraviolet Photobiology. Biological Techniques Series, A. Hollaender, ed. Prentice-Hall, Englewood Cliffs NJ.

4. Johns, H.E., and A.M. Rauth. 1965. Photochem. Photobiol. 4:693-707.

5. Johns, H.E., and A.M. Rauth. 1965. Photochem. Photobiol. 4:673-692.

6. Quaite, F.E., B.M. Sutherland, and J.C.Sutherland. 1992. Nature 358:576-578.

7. Quaite, F.E., B.M. Sutherland, and J.C. Sutherland. 1992. Appl. Theor. Electroph. 2:171-175.

8. Shettle, E.P., M.L. Nack, and A.E.S. Green, 1975. *In* Impacts of Climatic Change on the Biosphere,

Part 1, Rep. DOT TST-76-55. pp.2:38-2:49, National Tech. Information Service: Springfield VA.

9. Sutherland, B.M., J.S. Cimino, N. Delihas, A.G. Shih, and R.P. Oliver. 1980. Cancer Res. 40:1934-1939.

10. Sutherland, B.M., N.C. Delihas, R.P. Oliver, and J.C. Sutherland, J.C. 1981. Cancer Res. 41:2211-2214.

11. Sutherland, J.C., B. Lin, D.C. Monteleone, J. Mugavero, B.M. Sutherland, and J. Trunk. 1987. Analyt. Biochem. 163:446-457.

# UV-B-INDUCIBLE AND CONSTITUTIVE GENES THAT MEDIATE REPAIR AND TOLERATION OF UV-DAMAGED DNA IN THE PLANT *ARABIDOPSIS THALIANA*

John B. Hays and Qishen Pang

Department of Agricultural Chemistry
Oregon State University
Corvallis OR 97331-7301

Key words: Arabidopsis thaliana, DRT responses, excision repair, photoreactivating enzymes, UV radiation

## ABSTRACT

Plant responses to UV-induced DNA damage are of particular interest because plants are exposed almost continuously to the sunlight on which they depend for energy and developmental signals. The small crucifer *Arabidopsis thaliana* is an ideal model plant for genetic and molecular biological studies of DNA-damage-repair/toleration (DRT) responses. We previously showed that *Arabidopsis* plants primarily used photoreaction to remove cyclobutane pyrimidine dimers (CPDs), the major UV photoproducts in DNA, from their genomes; excision repair of CPDs was about 5% as efficient. *Arabidopsis* photoreactivating enzyme (photolyase) levels are developmentally regulated (more activity in midlife than young plants), and inducible by UV treatment of plants.

To gain information about the range of DRT activities in *Arabidopsis*, and to obtain DRT gene probes for regulatory studies, we recently selected for *Arabidopsis* cDNAs, in an expression library, that partially corrected the phenotypes of *E. coli* mutants lacking repair/toleration activities. Of the six cDNAs isolated, the products of three—*DRT100*, *DRT111* and *DRT112*—appear to catalyze either strand-exchange or resolution steps of homologous recombination; in plants, these activities may mediate recombinational toleration of unrepaired damage. *DRT101* and *DRT102* may encode UV-specific excision repair activities. Four of the DNA sequences—*DRT100, 101, 111* and *112*—predict chloroplast-targeted proteins.

We now find levels of *DRT100* and *DRT101* mRNA to be increased three- to four-fold by UV irradiation of whole plants. *DRT100* and *DRT101* induction patterns differ from one another and from the pattern for the UV-inducible *CHS* gene (encodes the flavonoid-biosynthesis enzyme chalcone synthase): 1) all three genes are induced by UV light, only *DRT100* and *DRT101* by the DNA crosslinking agent Mitomycin C, and only *DRT100* by the alkylating agent methylmethane sulfonate; 2) *CHS* and *DRT101* mRNA levels are maximal immediately after a 2-h UV-B pulse, whereas *DRT100* mRNA peaks 3-h later; 3) a UV-B pulse induces *CHS* mRNA maximally in 7-day-old plants, and much less in older plants, whereas *DRT100* and *DRT101* show higher inducibility at 14 days and 21 days. This last result, and the developmental pattern of photolyase expression, suggest that older *Arabidopsis* plants may rely more on DRT activities, and perhaps less on shielding by flavonoids, than younger ones.

## INTRODUCTION

### CELLULAR RESPONSES TO UV LIGHT AND OTHER GENOTOXIC STRESSORS

Among the genotoxic agents that threaten cellular genomes, sunlight in the UV-B band is currently of particular concern, because ongoing depletion of stratospheric ozone seems certain to significantly increase terrestrial UV-B irradiance, even in temperate latitudes. Many cells use some of the same systems

NATO ASI Series, Vol. I 18
Stratospheric Ozone Depletion/
UV-B Radiation in the Biosphere
Edited by R. H. Biggs and M. E. B. Joyner

to defend themselves against UV photoproducts and against chemical-DNA adducts. They avoid DNA damage by shielding or detoxification mechanisms, remove DNA damage by repair processes, and mitigate the toxicity of unrepaired damage by toleration responses. Cells may adapt to genotoxic stress by increasing expression of gene products that mediate one or more of these three defense responses.

Studies with microbes and cultured mammalian cells continue to advance knowledge of mechanisms of resistance to genotoxic stress. However, elucidation of tissue-specific and age (growth-stage)-specific adaptation requires study of whole differentiated organisms. Plants are differentiated organisms whose obligatory exposure to a potent genotoxic agent, sunlight, has biologically interesting consequences.

PLANT MODELS FOR ADAPTATION TO GENOTOXIC STRESS

Plants need sunlight, for photosynthesis and developmental signals; this constrains the nature of defenses against potentially genotoxic UV-B components. Unlike the distinct germ lines of animals, plant gametes derive from meristematic somatic cells. Thus, DNA-damage-induced somatic mutations (if not haploid-lethal) may give rise to mutant gametes, making avoidance of damage-induced mutation a high priority.

The small crucifer *Arabidopsis thaliana* has become a popular model system, because it lacks many of the experimental disadvantages of other plants, and provides some special advantages [26]. Its small size, large seed set ($10^4$ per plant) and automatic self-fertilization facilitate the production of large "second-generation" (F2) populations of plants homozygous for mutations originally induced by treatment of seeds; thousands of F2 plants can easily be screened for mutations of interest. Because of its short life cycle (6-7 weeks) and its amenability to outcrossing, *Arabidopsis* is a good organism for classical genetics; the efforts of hundreds of workers have produced a genetic map increasingly dense with mutational, RFLP, and RAPD markers [27].

The small, relatively nonrepetitive genome (about ten times the size of the yeast genome) makes feasible such techniques as "chromosome-walking" from known DNA site to mapped genetic locus, and screening of yeast-artificial-chromosome (YAC) libraries for genes of interest. The amenability of *Arabidopsis* to transformation, with *Agrobacterium tumefaciens* T-DNA or via other techniques, and widespread interest in improvement of cell/tissue-culture and transformation techniques, make production of transgenic plants increasingly straightforward.

CELLULAR DNA-DAMAGE-RESISTANCE ACTIVITIES AND THEIR REGULATION

*Nature of DNA-damage-resistance/toleration (DRT) Activities.* Cells remove some kinds of damage by direct-reversal processes. Photoreactivation of the most common UV photoproduct in DNA, cyclobutane pyrimidine dimers (CPDs), requires only UV-A/visible light energy and a single enzyme, photolyase. Some organisms may photo-reactivate the next most common photoproduct, pyrimidine-(6-4′)-pyrimi-dinone photoproducts [(6-4) products], as well [39]. Damaged DNA bases are excised and replaced by undamaged DNA via two types of pathways.

In **base excision repair**, DNA glycosylases targeted to specific altered bases detach them from backbone deoxyriboses; apurinic/pyrimidinic (AP) endonucleases nick the DNA near these abasic sites. In **nucleotide excision repair**, recognition of a broad range of helix-distorting lesions provokes strand incision, apparently by the same sets of repair proteins, e.g. the UvrABC proteins in *E. coli*, the *RAD3* epistasis group in yeast, the set of gene products defined in part by human *Xeroderma pigmentosum* mutations.

The remarkable conservation of mechanisms of DNA repair suggests strong evolutionary pressure. Glycosylase/AP-endonuclease activities specific for CPDs have been described for bacteriophage T4 [35], the bacterium *M. luteus* [40], and yeast [18]. Both CPDs and (6-4) products, and other bulky lesions, are recognized by most nucleotide-excision-repair systems.

A primary cytotoxic consequence of unrepaired DNA damage is thought to be blocked DNA replication. Cells can bypass these blocks, and/or "fill" the daughter-strand gaps that result from reinitiation of replication downstream, by homologous recombination (sister-chromatid exchange). In *E. coli*, strand-exchange steps require RecA protein [8]; efficient resolution of recombination intermediates appears to require RuvC and/or RecG activity [24]. In yeast, *RAD52* epistasis group proteins are involved in recombinational toleration. In mammals, similar processes are poorly characterized. Replication-blocking DNA lesions may also be bypassed by (necessarily error-prone) translesion synthesis. This involves DNA-damage-activated RecA and UmuCD activities in *E. coli*, and apparently *RAD6*-group activities in yeast.

*Regulation of DRT Activities*. Genotoxic stressors are known to increase expression of activities responsible for shielding/detoxification, DNA-damage-repair/toleration, and regulation of cell growth. The mix of DNA-damage-induced activities appear to vary considerably among different organisms.

The *E. coli* SOS system - the paradigm of DNA-damage-induced responses - comprises over twenty genes whose transcription is more-or-less coordinately stimulated by a broad variety of DNA-damaging and replication-blocking treatments (reviewed in [41]). SOS responses include nucleotide-excision-repair proteins (UvrA,B), activities involved in recombinational (RecA, RecN, RuvA,B) or mutagenic (UmuC,D, RecA) toleration, and some associated with growth delay (SulA). Most respond to a single master regulatory switch: cleavage of the SOS LexA repressor mediated by DNA-damage-activated RecA.

There may be 80 or more genes induced by DNA damage in the yeast *S. cerevisiae* [37]. A lesser fraction of these seem to be *DRT* genes than in *E. coli*, but inducible yeast activities include photoreactivation (*PHR1*) and nucleotide-excision-repair (*RAD2, RAD7, RAD23*), and both recombinational (*RAD54*) and mutagenic (*RAD6, RAD18*) toleration [38,18, 9,25]. There is no evidence for a LexA/RecA-like master switch. Some genes are both induced by DNA damage and regulated during the cell cycle [33].

Over 70 DNA-damage-induced mammalian genes have been identified (reviewed in [17,20]). Surprisingly, only one—a protein that binds more tightly to UV-irradiated than unirradiated DNA [1]—appears likely to be a nucleotide-excision-repair activity; there is evidence for induction of DNA polymerase $\beta$ and DNA ligase. A large group of genes, including some encoding protein kinases involved in signal transduction, are induced by growth arrest as well as DNA damage.

*Regulation of Plant Photoshielding Activities*. In response to UVB light, fungal elicitors, leaf wounding, and certain other stressors, plants increase biosynthesis of phenylpropanoids—UV-light-absorbing compounds derived from phenylalanine [7,2,23]. Induced transcription of genes encoding enzymes mediating early steps in the general phenylpropanoid pathway [e.g., phenylammonia lyase (*PAL*) and 4-coumerate:Co A ligase (*4Cl*)] and steps in the branch leading to UV-B-absorbing flavonoids and iso-flavonoids [e.g., chalcone synthase (*CHS*) and chalcone isomerase (*CHI*)] has been extensively studied. Young *Arabidopsis* seedlings treated with UV-containing light show dramatic increases in *PAL, CHS*, and *CHI* mRNA [21]. Last and coworkers found *Arabidopsis chs* and *chi* mutants to be respectively moderately and severely UV-sensitive [22].

## RESULTS

### *ARABIDOPSIS DRT* ACTIVITIES AND THEIR REGULATION

Activities and genes involved in DNA-damage-repair/toleration have in the past been identified by intact-organism physiology, crude-extract or purified-protein biochemistry, and isolation of proteins/genes that correct the phenotypes of damage-sensitive mutants. We began by measuring photoreactivation and dark (excision) repair of CPDs in whole plants. To obtain *Arabidopsis DRT* cDNAs, for use as regulatory probes and sources of Drt proteins, we employed a shortcut: selection for cDNAs that correct microbial DNA-damage-sensitivity mutations.

### PHOTOREACTIVATION AND EXCISION REPAIR OF CPDs

We adapted a UV-endonuclease-nicking technique to assay of CPDs in UV-irradiated plants, and showed that photoreactivation of these photoproducts was over 20 times as rapid as their excision in the dark [28] (Figure 1A). (Britt and coworkers recently showed that removal (presumably excision) of (6-4) products is considerably faster [3].) We found that photolyase activity (assayed in extracts) increased during early plant growth, and could be significantly induced by UV-B light [28] (Figure 1B).

### SELECTION OF *ARABIDOPSIS DRT* cDNAs.

We sought cDNAs to use as probes for studies of regulation of *DRT* genes in plants, with the aim of then isolating from genomic libraries the promoter regions of *DRT* genes that proved to be induced by DNA damage, for further analysis. We also expected to use the respective cDNAs to overproduce Drt proteins for

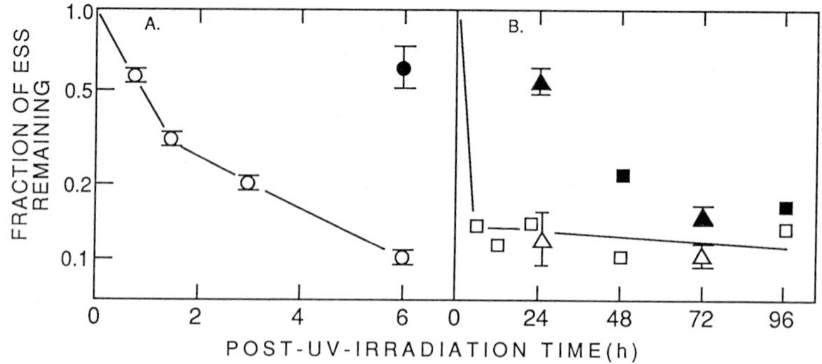


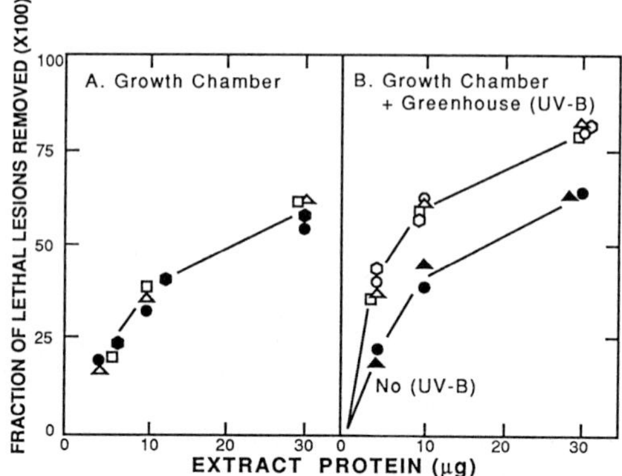


Figure 1. Photoreactivation of cyclobutane pyrimidine dimers (CPDs) in *Arabidopsis*. Upper panels. Removal of CPDs *in vivo*. Growth of plants for ten days at 22°C, radiolabeling ($^3H$) of plant DNA, irradiation with 1000 J/m$^2$ at 254 nm, incubation under white light (open symbols) or in the dark (filled symbols), extraction of plant DNA, and measurement of number of sites sensitive to the CPD-specific phage T4 UV endonuclease (ESS) were as described [28]. 1.0 corresponds to about one ESS per 6 x $10^5$ nucleotides. A: Continuous illumination. B: Extended photoperiodic illumination. Lower panels. Photoreactivating enzyme (photolyase) activity in extracts from unirradiated and UVB-irradiated plants. A: Growth of plants in growth chamber under white light with various Mylar shielding protocols or with no shielding except growth container (Δ), preparation of plant leaf extracts, and measurement of photolyase activity (transformation assay of removal of lethal lesions (CPDs) from UV-irradiated plasmids and protein concentration were as described [28]. After nine days in growth chamber plants were shifted to a greenhouse and irradiated for 6 h (280 J/m$^2$) with a UV-B lamp, filtered through cellulose acetate (open symbols) or Mylar (filled symbols). Reprinted from Pang and Hays (1991), Plant Physiol. 95:536-543.

subsequent biochemical characterization. We selected directly for *DRT* cDNAs in an expression library, because the alternative - identification of plant *drt* mutations, mapping mutations to the vicinity of known markers, chromosome walking to loci that complement the mutations, isolation of the corresponding cDNAs - can take two years or more. We employed two kinds of *E. coli* DNA-damage-sensitive mutants for the initial selections: $Phr^-UvrC^-RecA^-$ mutants lack all defenses against photoproducts in DNA, hence are exquisitively sensitive to UV light (and DNA-damaging chemicals); $RuvC^-RecG^-$ mutants, apparently unable to resolve recombination intermediates (Holliday junctions), lack recombinational toleration of a range of DNA lesions [24]. The *Arabidopsis* cDNA library was prepared by Elledge, Davis and colleagues, using their λ YES vector [13].

When λ YES infects bacteria (homoimmune λ lysogens) expressing phage P1 Cre activity, a region of phage DNA flanked by P1 *lox* sites is excised by Cre-*lox* site-specific recombination. The excised circles encode elements that provide for selection and stable maintenance as yeast or *E. coli* plasmids. Depending on the orientation of the cDNA inserts, they are transcribed by controllable promoters - *E. coli* $P_{lac}$ or yeast $P_{GAL1}$.

We established *Arabidopsis* cDNA plasmids in $10^{11}$ $Phr^-UvrC^-RecA^-[\lambda(ind^-)\text{-}Cre]$ bacteria and irradiated to $10^{-7}$ survival. Screening 840 survivors yielded six (four unique) plant cDNAs - *DRT100, DRT101, DRT102, DRT103*; they increased bacterial UV resistance 10- to 60-fold [32]. Of $10^{11}$ $RuvC^-RecG^-$ mutants containing cDNA plasmids, 25 survived on methylmethanesulfonate (MMS) plates (toxic to all of $10^{11}$ plasmidless controls); the 25 survivors yielded four (two unique) cDNAs, *DRT111* and *DRT112* [30]. It is noteworthy that despite the low apparent phenotype-correction efficiencies of these cDNAs, the powerful selection procedures yielded them readily. From the apparent correction factors and the representation of the cDNAs among the survivors, we estimated the original abundances of *DRT100-103* to be roughly 0.01% [32] - about that expected for average representatives of the 15,000 or so genes expected for plants [16]. [From nuclear runoff experiments we estimate transcription rates of the six *DRT* genes to be 15 to 60% of the rate for the *Arabidopsis* actin gene (Table 2 in discussion).] Correction of the mutant phenotypes by the plant cDNAs falls farther and farther short of full restoration of wild-type phenotypes the greater the demand for activity; apparent correction efficiencies decrease with increasing UV dose, for example [31,32,30] (Figures 2,3). This may reflect poor expression in *E. coli*; although transcribed via the *E. coli* $P_{lac}$ promoter, these cDNAs generally lack in-frame bacterial translation signals. Ultimately, validation of the DRT character of these cDNAs requires biochemical characterization of the respective proteins and analysis of their functions *in planta*. However, by more extensively characterizing their activities in *E. coli*, and analyzing their expression in plants treated with DNA-damaging agents (see below), we obtained strong evidence for the repair/toleration character of at least four of these six *DRT* genes. In all cases, correction activity requires transcription of the respective genes in *E. coli* [31,32,30].

*DRT100, DRT111, DRT112*: POSSIBLE PLANT RECOMBINATIONAL TOLERATION PROTEINS

These three activities increase resistance of *E. coli* $Phr^-UvrB^-/C^-RecA^-$ or $RuvC^-RecG^-$ mutants to UV light, mitomycin C, and MMS [31,30]. They might in principle be novel polypeptides that act on a broad variety of DNA lesions, independently of the UvrABC system. It seemed more likely however, that they mediate recombinational toleration of a variety of DNA-replication-blocking lesions. Indeed, we found them able to increase conjugal recombination in bacterial mutants: *DRT100* partially corrected a $RecA^-$ recombination deficiency [31] (Table 1); *DRT111* and *DRT112* corrected putative resolvase deficiencies remarkably well: they restored recombination in $RuvC^-$ bacteria 68% and 29% of wild-type frequencies, respectively, and in $RuvC^-RecG^-$ mutants to 11% and 6% [30] (Table 1). These cDNAs may restore conjugal recombination proficiency more efficiently than they restore tolerance of extensive DNA damage because the former process requires only one or two crossover events per cell, whereas high levels of DNA damage cause multiple replication blocks that engender multiple recombination events. Their N-terminal sequences suggest that Drt100, Drt111, and Drt112 proteins are all nuclear-encoded but chloroplast-targeted. Drt100 resembles bacterial RecA proteins only with respect to its predicted molecular weight and its predicted nucleotide binding site (Walker box). Drt111 and Drt112 do not resemble prokaryotic Holliday junction resolvases (but these latter proteins

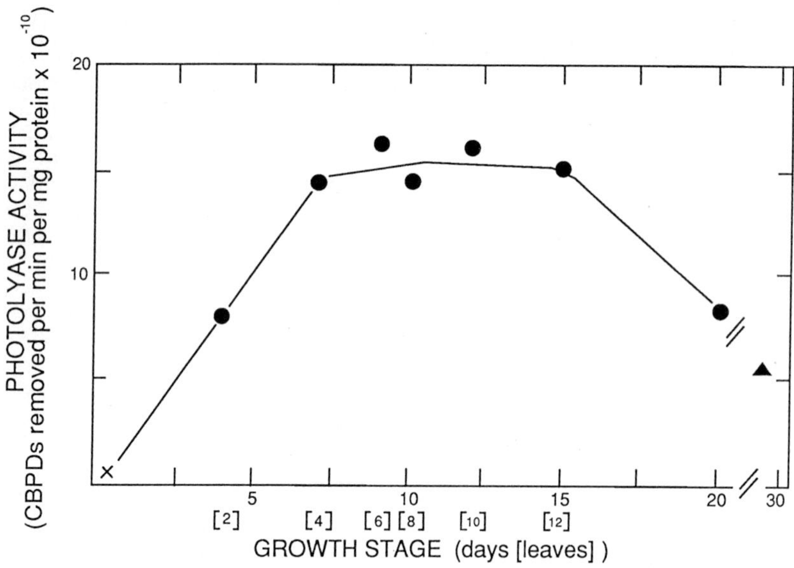


Figure 2. *Arabidopsis* photolyase activity at various growth stages. Preparation of extracts and measurement of specific photoreactivation activity (plasmid transformation assay) were as described [28]. Reprinted from Pang and Hays (1991), Plant Physiol. 95:536-543.

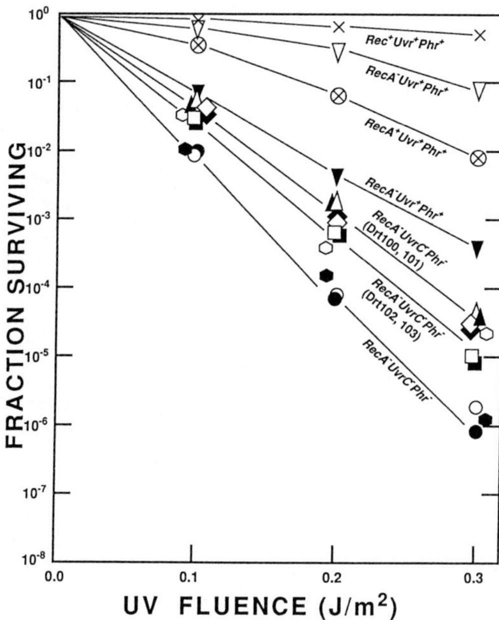


Figure 3. Resistance to UV light of DNA-damage-sensitive *E. coli* RecA⁻UvrC⁻Phr⁻ mutants expressing *Arabidopsis DRT* cDNAs. Transformation of mutant bacteria with plasmids encoding indicated *Arabidopsis* cDNAs under *lac* promoter control, growth of cells and induction of expression of cDNAs with IPTG, UV-irradiation and measurement of surviving fractions, with (open symbols) or without (filled symbols) photoreactivating light were as described [32]. cDNAs were none (○,●), *DRT100* (Δ,Δ), *DRT101* (◇, ◆), *DRT102* (□, ■), *DRT103* (⬠, ⬟). Reprinted from Pang et al. (1993), Plant Mol. Biol. 22:411-426.

Table 1. Partial correction of recombination deficiencies of bacterial mutants by *Arabidopsis* cDNAs.

| Mutant phenotype | Plant cDNA | Relative frequency of conjugal recombinants in Hfr x $F^-$cross (x100) |
|---|---|---|
| $RecA^+$ | none | (100) |
| $RecA^-$ | none | 0.01 |
| $RecA^-$ | *DRT100* | **0.59** |
| $RecA^-$ | *DRT111* | 0.01 |
| $RecA^-$ | *DRT112* | 0.01 |
| $RuvC^+$ | none | (100) |
| $RuvC^-$ | none | 21 |
| $RuvC^-$ | *DRT111* | **75** |
| $RuvC^-$ | *DRT112* | **44** |
| $RuvC^+RecG^+$ | none | (100) |
| $RuvC^-RecG^-$ | none | 0.3 |
| $RuvC^-RecG^-$ | *DRT111* | **11.0** |
| $RuvC^-RecG^-$ | *DRT112* | **5.8** |
| | | Relative survival, UV + mito C |
| $Rec^+$ | none | (100) |
| $RecG^-$ | none | 3 |
| $RecG^-$ | *DRT111* | **52** |
| $RecG^-$ | *DRT112* | 4 |

Transformation of *E. coli* mutants with indicated phenotypes, with plasmids expressing indicated *Arabidopsis DRT* cDNAs, and scoring of recombination between chromosomes of plasmid-containing mutant recipients and Hfr donor DNA transferred in during conjugal mating—as the frequency of stably recombinant recipients—was as described [30,31]. Measurement of resistance to UV light (30 J/m$^2$) plus mitomycin C (0.2 μg per ml) was as described [30]. Data from Pang et al. (1993), Nucleic Acids Res. 21:1647-1653 and Pang et al. (1992) Proc. Natl. Acad. Sci. USA were combined here.

do not resemble one another). Jagendorf and coworkers isolated an *Arabidopsis* cDNA encoding a chloroplast-targeted homolog of bacterial RecA proteins (50% identity), but have not demonstrated RecA-like activity in *E. coli* [6].

*DRT101, DRT102*: POSSIBLE UV-SPECIFIC EXCISION-REPAIR PROTEIN

These activities do not increase resistance of *E. coli* mutants to mitomycin C or MMS, or restore recombination proficiency to *recA* mutants [32]. Possible analogues include the CPD-specific DNA-glycosylase/abasic-endonucleases of phage T4, *Micrococcus luteus*, and yeast, and the pyrimidine-hydrate-glycosylases found in many organisms; the few sequences available for these proteins show little extended amino acid homology with Drt101 and Drt102. The apparent (6-4) photoproduct-specificity of the defect in the UV-sensitive *Arabidopsis* mutant isolated by Britt and coworkers [3] suggests that there may be (a) (6-4)-specific endonuclease(s), which could be encoded by *DRT101* and/or *DRT102*. The

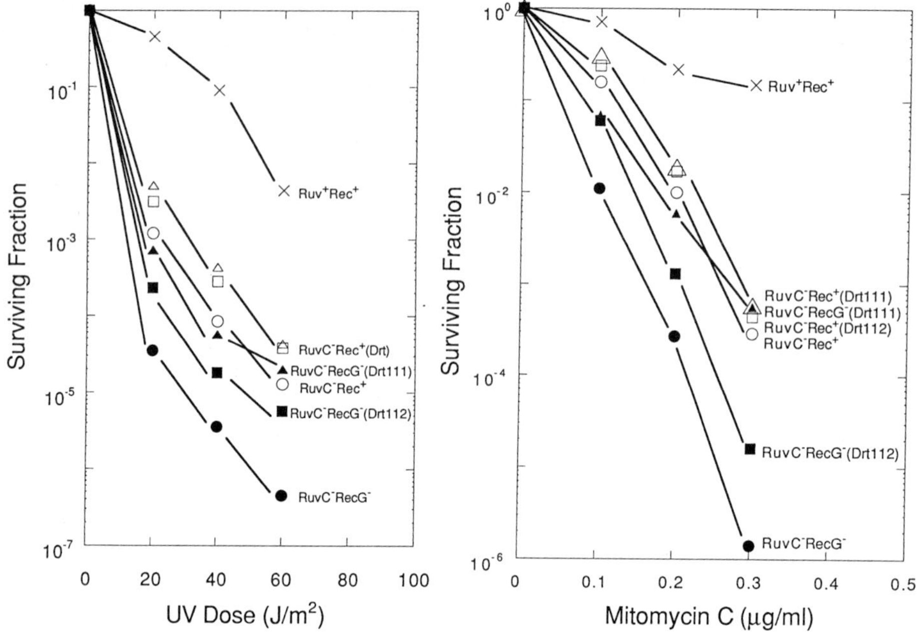


Figure 4. Resistance to UV light and mitomycin C of DNA-damage-sensitive RuvC⁻RecG⁻ mutants expressing *Arabidopsis DRT111* or *DRT112* cDNA. Transformation of mutant bacteria with *DRT*-expressing plasmids and determination of resistance to UV light (left) or mitomycin C (right) were as described [3]. Bacterial phenotypes were: wild-type (X); RuvC⁻RecG⁺ with no (○), *DRT111* (△), or *DRT112* (□) plasmids; RuvC⁻RecG⁻ with no (●), *DRT111* (△), or *DRT112* (■) plasmids. Reprinted from Pang et al. Nucleic Acids Res. 21:1647-1653.

Drt101 N-terminus appears to be a chloroplast-targeting sequence.

ELEVATED TRANSCRIPTION OF *DRT100* AND *DRT101* IN WHOLE TREATED WITH UV LIGHT AND DNA-DAMAGING CHEMICALS

We hybridized poly-(A⁺) RNA from plants with six *DRT* cDNAs, plus *Arabidopsis CHS* (a known UV-induced gene). To provide an internal loading control for these "northern" analyses, we included a "housekeeping" gene probe, *Arabidopsis Aac1* (actin). Both *DRT100* and *DRT101* showed substantial induction by DNA-damaging agents. Several differences in induction pattern suggest that there are interesting differences in regulatory circuitry among *CHS*, *DRT100*, and *DRT101*. First, differential responses to different agents: UV light induced all three, to about the same extent [29] (Figure 4); *CHS* responded to UV irradiation but not to chemical agents, as if UV light per sé were the signal; Mitomycin C induced both *DRT100* and *DRT101* (Figure 4), but MMS induced only *DRT100* (not shown). A "nuclear runoff" experiment showed that at least the major portion of the UV-induced increases in *CHS*, *DRT100*, and *DRT101* mRNA levels were the result of increases in the rates of mRNA synthesis, and that constitutive mRNA levels in untreated plants were 15% to 60% of the levels for the "housekeeping" gene for actin (data not shown; [29]). Second, differences in post-UV kinetics of induction: *CHS* and *DRT* mRNA levels were maximal immediately after a 2-h UV treatment, whereas *DRT* levels peaked later (Figure 5). Third, growth stage affected the responses to UV challenge differently: *CHS* was induced maximally in 1-week-old plants; *DRT100* and *DRT101* showed greater induction in 2- and 3-week-old plants (Figure 6).

## DISCUSSION

Some properties of the six *Arabidopsis DRT* genes described here, and some inferred for *Arabidopsis PHR* [28] and *recA*-homologue [6] genes are summarized in Table 2. Plants appear to use all four DNA-damage-repair/toleration (DRT) processes to resist genotoxic agents such as UVB light. We and others have demonstrated both photoreaction and excision repair of CPDs [28] and repair of (6-4) photoproducts [3] in *Arabidopsis*. The isolation of four *Arabidopsis* cDNAs encoding strand-exchange or recombination-intermediate-resolution activities [31,30] argues strongly for the importance of recombinational toleration of DNA damage, at least in chloroplasts. The ability of DNA-damaging-agents to increase mutation frequencies in various plant systems [34] is indirect but strong evidence for error-prone translesion DNA synthesis. We have shown here that plants adapt to genotoxic stress by increasing transcription of at least several *DRT* genes. The differences in regulatory patterns revealed by our initial studies of induction *DRT100* and *DRT101*, together with the patterns of induction of *CHS* and other phenylpropanoid-biosynthesis genes described by others, suggest the existence of a complex network of adaptive responses.

Irradiation of whole *Arabidopsis* plants with UV-B light, or treatment with DNA-damaging chemicals, substantially increases levels of *DRT100* and *DRT101* mRNA. These mRNAs correspond to two of six *Arabidopsis* cDNAs previously isolated by Pang *et al.* [31,32] on the basis of their ability to partially correct the phenotypes of *E. coli* DNA-damage-sensitive (*uvrB recA phr*) mutants. At least part of the UV-stimulated increases in *DRT100* and *DRT101* reflects increased transcription rates. The constitutive rates of transcription of the six *DRT* mRNAs are neither unusually high nor extremely low. The patterns of induction of expression of *DRT100* and *DRT101*, and of the flavonoid-biosynthesis gene *CHS*, differ among themselves with respect to effects of different DNA damaging agents, post-exposure kinetics, and the growth stage at which inducibility is maximal.

First, different agents induce different sets of genes: UVB light causes levels of *DRT100*, *DRT101*, and *CHS* mRNAs to increase, the DNA-crosslinking chemical, mitomycin C elevates only *DRT100* and *DRT101* mRNAs, and the DNA-alkylating agent MMS affects only *DRT100*. This suggests that *CHS* responds not to UV-induced DNA damage, but to stimulation of (a) receptor(s) for UVB and blue light [15,21]. Alternatively, the UVB effect on *CHS* transcription seen here might be part of a generalized plant stress response. However, stressors such as fungal elicitor [11] or bacterial infection [12] do not affect *Arabidopsis CHS* expression. The marked elevation of *DRT100* and *DRT101* mRNA in mitomycin-C-treated plants suggests that they respond directly to DNA damage, induced by UV-B light or chemical agents, but we cannot rule out a *CHS*-like response to UV light by *DRT100* and *DRT101*.

Since expression of *Arabidopsis DRT101* cDNA in *E. coli* mutants increases their resistance to UV light but not to mitomycin C [32], the marked stimulation by the latter chemical of *DRT101* message levels in *Arabidopsis* was unexpected. This phenomenon may reflect involvement of Drt101 in repair of chemical-DNA adducts in plants (but not in bacteria). However, several DNA-alkylating agents stimulate transcription of the yeast *PHR1* gene, whose product, photolyase, is highly specific for repair of cyclobutane pyrimidine dimers [38]; mitomycin C induction of *DRT101* may be a similar case of so-called gratuitous induction. Induction of *DRT100* by both mitomycin C and MMS, but induction of *DRT101* only by the former chemical, is somewhat surprising, since agents that induce one SOS response in *E. coli* usually induce them all [41]. It seems premature to conclude that there is more than one DNA-damage-signalling pathway for induction of repair/toleration genes in *Arabidopsis*, but the regulatory system appears more complex than that in bacteria.

The contrast between four-fold stimulation of *DRT100* and *DRT101* mRNA levels by UVB light, and ten-fold stimulation by DNA-damaging chemi-

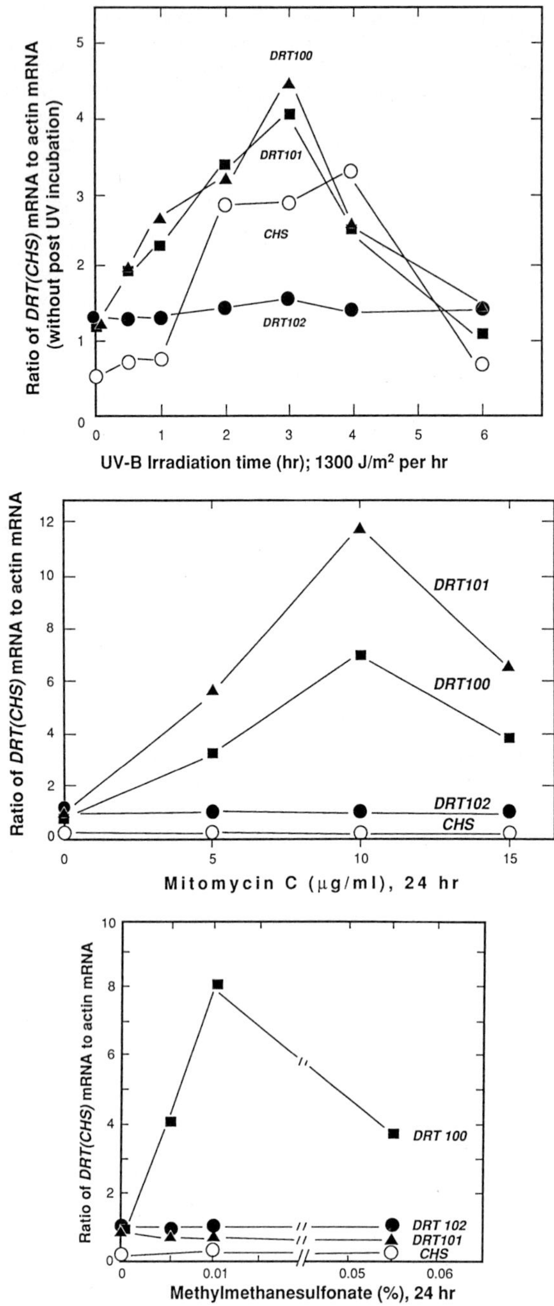

Ratio of DRT(CHS) mRNA to actin mRNA (without post UV incubation)
DRT100
DRT101
CHS
DRT102
UV-B Irradiation time (hr); 1300 J/m² per hr
Ratio of DRT(CHS) mRNA to actin mRNA
DRT101
DRT100
DRT102
CHS
Mitomycin C (μg/ml), 24 hr
Ratio of DRT(CHS) mRNA to actin mRNA
DRT 100
DRT 102
DRT101
CHS
Methylmethanesulfonate (%), 24 hr

Figure 5 (left). Induction of gene expression by treatment of plants with UVB-light, mitomycin C, or methylmethanesulfonate. Treatment of two-week-old *Arabidopsis* seedlings with indicated UV-B light (1300 J/m$^2$ per h) (upper panel) for indicated times, or treatment with indicated concentrations of mitomycin C (middle panel) or methyl-methanesulfonate (MeMes) (lower panel) for 24 h, harvesting and freezing of whole plants, extraction of poly ($A^+$) mRNA, electrophoresis of RNA and transfer to filters, hybridization of filters with $^{32}$P-labeled DNA probes [*Arabidopsis Aac1* (actin) plus *CHS* (chalcone synthase) (○) or *DRT100* (■) or *DRT101* (△) or *DRT102* (●)], autoradiography and densitometry, and successive stripping and rehybridization of filters with new probe mixtures were as described [29]. (Ratios of *DRT/CHS* mRNA to *Aac1* mRNA depend on respective radioactivities of probes and thus are not comparable from experiment to experiment.)

cals, at the respective optimal doses, may reflect the inability of UVB light to penetrate the interior cells, so that whole irradiated plants are actually a mixture of induced and uninduced cells. UVB light could also be less stimulatory because photoproducts are rapidly repaired after the UVB pulse [28,3]. Unlike the UV treatment, exposure to chemicals (and presumably formation of DNA-chemical adducts) continued until the time mRNA was isolated. [Induction of *DRT100* and *DRT101* might have been even higher if exposures were extended beyond 24 h.] However, the fact that extended UVB treatment did not further increase induction of any genes (Figure 4) argues against repair of photoproducts as a major factor limiting the inductive response. It would be of interest to compare maximum UV-B-stimulated and mitomycin-C-stimulated mRNA levels in protoplasts.

A second respect in which *DRT100*, *DRT101*, and *CHS* differ is in the timing of maximum induction by UV-B: immediately after a 2-h pulse in the case of *DRT101* and *CHS*, about 3 h later in the case of *DRT100*. It is thus possible that both *DRT101* and *CHS* are induced relatively rapidly by a UV-light signal (and *DRT101* more slowly by mitomycin-C-engendered DNA damage). *DRT100* and *DRT101* do not show large differences in the rate of mRNA accumulation during mitomycin C exposure, but here induction kinetics may be obscured by the time required for the chemical to diffuse into the plant and reach the nucleus, and by continual formation new chemical-DNA adducts.

A third respect in which these UV-induced genes differ, the growth stage at which they responded maximally to UV-B light, may reflect changing UV-resistance strategies as plants develop. High *CHS* induction seems characteristic of young *Arabidopsis* plants. Kubasek *et al.* [21] found that in seedlings grown under continuous white light or under white light supplemented with UV-B for more than four days after germination, *CHS* mRNA levels were negligible. Here we found that in plants grown photoperiodically in white light without UV, the *CHS* response to a UV-B pulse was strongest in 7-day-old plants, although still significant in 14-day and 21-day

Figure 6. Time course of UVB induction of expression of plant genes. Analysis of relative mRNA levels [ratio of *CHS* (●) or *DRT100* (○) or *DRT101* (□) or *DRT102* (△) mRNA to *Aac1* (actin) mRNA, divided by corresponding ratio for untreated plants (X)] at indicated times after treatment of two-week-old *Arabidopsis* plants with UV-B light for 2 h (total 2600 J/m$^2$) was as described [29]; see legend to Figure 5]. Reprinted from Pang and Hays, manuscript submitted.

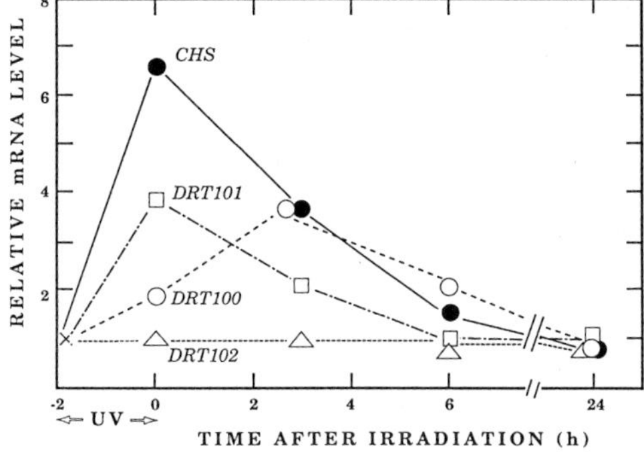

Table 2. Properties of plant DNA-damage-repair/toleration genes from *Arabidopsis thaliana*

| | DRT100[*] | DRT101[*] | DRT102[*] | DRT103[*] | DRT111[*] | DRT112[*] | *(PHR)*[#] | *recA*[+] |
|---|---|---|---|---|---|---|---|---|
| Predicted (unprocessed) protein (amino acid) | 403 | 276 | 230 | >428 | 383 | 167 | ? | 430 |
| Chloroplast targeted? | Yes | Prob | No | No | Prob | Yes | ? | Yes |
| Light-dependent? | No | No | No | Yes | No | No | Yes | No |
| UV resistance in *E. coli*? | Yes | Yes | Yes | Yes | Yes | Yes | Yes | ? |
| Chemical resistance in *E. coli*? | Yes | No | No | No | Yes | Yes | (No) | ? |
| Recombination activity *E. coli*? | Yes | No | No | No | Yes | Yes | (No) | ? |
| Likely *Eco* activity replaced | RecA | UvrABC | UvrABC | Phr? | RuvC | RuvC | (Phr) | RecA |
| Possible activity in *Arabidopsis* | Strand exchange | UV endo-nuclease | UV endo-nuclease | Photolyase ? | Holliday-junction resolvases | | Photolyase | Strand exchange |
| Uninduced transcription rate (relative to actin gene) | 0.28 | 0.15 | 0.60 | 0.20 | 0.44 | 0.43 | ? | ? |
| UV-induced? | Yes | Yes | No | ? | No | No | Yes | Prob |
| MeMeS-induced? | Yes | No | No | ? | ? | ? | ? | Yes |
| Mito-C-induced? | Yes | Yes | No | ? | ? | ? | ? | Yes |

[*]From references (28-32) and unpublished results.

[#]Gene not isolated, but activity inducible (28).

[+]Inducibility shown for pea (4), but not tested for *Arabidopsis recA* (6).

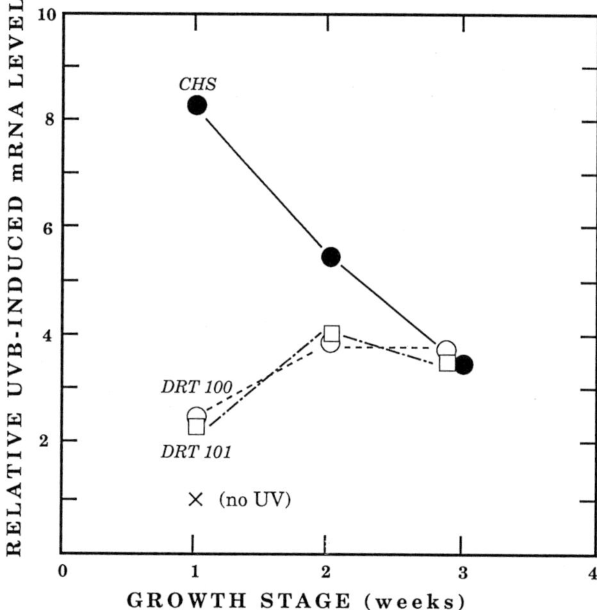


Figure 7. UVB induction of plant gene expression at various growth stages. Growth of plants to indicated stages, treatment with UVB light for 2 h (2600 $J/m^2$), immediate harvesting of plants, and analysis of relative mRNA levels [ratio of *CHS* (●) or *DRT100* (○) or *DRT101* (□) mRNA to *Aac1* (actin) mRNA, divided by corresponding ratio for two-week-old untreated plants (X).] were as described [29]; (see legend to Figure 5). Reprinted from Pang and Hays, manuscript submitted.

plants. Our observation that *CHS* remains partially inducible in older plants is similar to the strong induction of *CHS* transcription, by high-intensity white light, seen in plants grown in a greenhouse for 3 1/2 weeks [14]. Accumulation of flavonoids during continuous white light treatment [21] may inhibit *CHS* transcription by some feedback mechanism. This cannot entirely be explained by attenuation of the inducing signal by intracellular flavonoids, because UVB-supplemented white light stimulates levels of mRNAs transcribed from the general phenylpropanoid-biosynthesis gene *PAL* (phenyl-ammonia-lyase) [21]. The differences between the responses of *CHS* and of *DRT100* and *DRT101* to UVB in older plants suggest that even when flavonoid levels are not high enough to absorb all UV light reaching plant cells, biosynthesis of at least some flavonoids slows, and the plants place more reliance on DNA-damage-repair/toleration responses.

Jagendorf and coworkers detected a RecA-like strand-exchange activity in pea chloroplasts [5], and showed that levels of a chloroplast protein immunologically related to *E. coli* RecA protein were increased as much as three-fold by treatment of pea protoplast with mitomycin C or MMS [4]. These workers have isolated an *Arabidopsis* cDNA which predicts an N-terminal chloroplast transit peptide, followed by an amino-acid sequence highly homologous to those of bacterial RecA proteins [6]. Both *DRT100* and *DRT101* appear to encode chloroplast-targeted proteins [31,32]. It remains to be determined whether the UV-inducible *Arabidopsis* photolyase activity is, similarly to photolyase in *Chlamydomonas* [10] at least partially chloroplast-targeted. At least three, and as many as five, chloroplast-targeted DNA-damage-repair/toleration activities thus appear to be induced by one or more DNA-damaging agents. It will be of interest to determine whether there are additional inducible chloroplast-targeted *DRT* genes, and whether there is a common regulatory circuit, perhaps derived from an SOS system of a primitive bacterial ancestor of chloroplasts.

**Acknowledgments**
Supported by USDA Competitive Grants 90-37280-5597 and 92-37100-7725.

## REFERENCES

1. Abramic, M., A.S. Levine, and M. Protic. 1991. Purification of an ultraviolet-inducible, damage-specific DNA-binding protein from primate cells. J. Biol. Chem. 266:22493-22500.

2. Block, A., J.L. Dangl, K. Hahlbrock, and P. Schulz-Lefert. 1990. Functional borders, genetic fine structure, and distance requirements of cis elements mediating light responsiveness of the parsley chalcone synthase promoter. Proc. Natl. Acad. Sci. USA 87:5387-5391.

3. Britt, A.B., J-J. Chen, and D. Mitchell. 1993. A UV-sensitive mutant of *Arabidopsis thaliana* is defective in the repair of pyrimidine (6-4) pyrimidinone dimers. (Submitted.)

4. Cerutti, H., H.Z. Ibrahim, and A.T. Jagendorf. 1993. Treatment of pea (*Pisum sativum* L.) protoplasts with DNA-damaging agents induces a 39-kilodalton chloroplast protein immunologically related to *Escherichia coli* RecA. Plant Physiol. 102:155-163.

5. Cerutti, H., and A.T. Jagendorf. 1993. DNA strand-transfer activity in pea (*Pisum sativum* L.) chloroplasts. Plant Physiol. 102:145-153.

6. Cerutti, H., M. Osman, P. Grandoni, and A.T. Jagendorf. 1992. A homologue of *Escherichia coli* RecA proteins in plastids of higher plants. Proc. Natl. Acad. Sci. USA 89:8068-8072.

7. Chapell, J., and K. Hahlbrock. 1984. Transcription of plant defense genes in response to UV light or fungal elicitor. Nature 311:76-78.

8. Clark, A.J. 1971. Toward a metabolic interpretation of genetic recombination of *E. coli* and its phages. Ann. Rev. Microbiol. 25:437-464.

9. Cole, G.M., D. Schild, S.T. Lovett, and R.K. Mortimer. 1987. Regulation of *RAD54*- and *RAD52-lacZ* gene fusions in *Saccharomyces cerevisiae* in response to DNA damage. Mol. Cell. Biol. 7:1078-1084.

10. Cox, J.L., and G.D. Small. 1985. Isolation of photoreactivation-deficient mutant *Chlamy-domonas*. Mutat. Res. 146:249-255.

11. Davis, K.R., and F.M. Ausubel. 1989. Characterization of elicitor-induced defense responses in suspension-cultured cells of *Arabidopsis*. Molecular Plant-Microbe Interactions 2:363-368.

12. Dong, X., M. Mindrinos, K.R. Davis, and F.M. Ausubel. 1991. Induction of *Arabidopsis* defense genes by virulent and avirulent *Pseudomonas syringae* strains and by a cloned avirulence gene. The Plant Cell 3:61-72.

13. Elledge, S.J., J.T. Mulligan, S.W. Raner, M. Spottswood, and R.W. Davis. 1991. YES: A multifunctional cDNA expression vector for the isolation of genes by complementation of yeast and *Escherichia coli* mutations. Proc. Natl. Acad. Sci. USA 88:1731-1735.

14. Feinbaum, R.L., and F.M. Ausubel. 1988. Transcriptional regulation of the *Arabidopsis thaliana* chalcone synthase gene. Mol. Cell. Biol. 8:1985-1997.

15. Feinbaum, R.L., G. Storz, G., and F.M. Ausubel. 1991. High intensity and blue light regulated expression of chimeric chalcone synthase genes in transgenic *Arabidopsis thaliana* plants. Mol. Gen. Genet. 226:449-456.

16. Flavell, R. 1980. The molecular characterization and organization of plant chromosomal DNA sequences. Annu. Rev. Plant Physiol. Plant Mol. Biol. 31:569-596.

17. Fornace, A.J. Jr. 1992. Mammalian genes induced by radiation: Activation of genes associated with growth control. Annu. Rev. Genet. 26:507-526.

18. Friedberg, E.C. 1988. Deoxyribonucleic acid repair in the yeast *Saccharomyces cerevisiae*. Microbiol. Rev. 52:70-102.

19. Hamilton, K.K., P.M.H. Kim, and P.W. Doetsch. 1992. A eukaryotic DNA glycosylase/lyase recognizing ultraviolet light-induced pyrimidine dimers. Nature 356:725-728.

20. Herrlich, P., H. Ponta, and H.J. Rahmsdort. 1992. DNA damage-induced gene expression: Signal transduction and relation to growth factor signaling. Rev. Physiol. Biochem. Pharmacol. 119:187-223.

21. Kubasek, W.L., B.W. Shirley, A. McKillop, H.M. Goodman, W. Briggs, and F.M. Ausubel. 1992. Regulation of flavonoid biosynthetic genes in germinating *Arabidopsis* seedlings. The Plant Cell 4:1229-1236.

22. Li, J., T-M. Ou-Lee, R. Raba, R.G. Amundson, and R.L. Last. 1993. Arabidopsis flavonoid mutants are hypersensitive to UV-B irradiation. The Plant Cell 5:171-179.

23. Liang, X., M. Dron, J. Schmid, R.A. Dixon, and C.J. Lamb. 1989. Developmental and environmental regulation of a phenylalanine ammonia-lyase-glucuronidase gene fusion in transgenic tobacco plants. Proc. Natl. Acad. Sciences USA 86:9284-9288.

24. Lloyd, R.G. 1991. Conjugational recombination in resolvase-deficient *ruvC* mutants of *Escherichia coli* K-12 depends on *recG*. J. Bacteriol. 173:5414-5418.

25. Madura, K., S. Prakash, and L. Prakash. 1990. Expression of the *Saccharomyces cerevisiae* DNA repair gene *RAD6* that encodes a ubiquitin conjugating enzyme, increases in response to DNA damage and in meiosis but remains constant during the mitotic cell cycle. Nucleic Acids Res. 18:771-778.

26. Meyerowitz, E.M. 1987. *Arabidopsis thaliana*. Ann. Rev. Genet. 21:93-111.

27. Meyerowitz, E.M. 1989. *Arabidopsis*, a useful weed. Cell 56:263-269.

28. Pang, Q., and J.B. Hays. 1990. UV-B-inducible and temperature-sensitive photoreactivation of cyclobutane pyrimidine dimers in *Arabidopsis thaliana*. Plant Physiol. 95:536-543.

29. Pang, Q., and J.B. Hays. 1993. Stimulation by UV-B light and DNA-damaging chemicals of transcription of *Arabidopsis* DNA-damage-repair/toleration (*DRT*) genes. (Submitted).

30. Pang, Q., and J.B. Hays. 1993. Two plant cDNAs from *Arabidopsis thaliana* that partially restore recombination proficiency and DNA-damage resistance to *E. coli* mutants lacking recombination-intermediate-resolution activities. Nucleic Acids Res. 1647-1653.

31. Pang, Q., J.B. Hays, and I. Rajagopal. 1992. A plant cDNA that partially complements *Escherichia coli recA* mutations predicts a polypeptide not strongly homologous to RecA proteins. Proc. Natl. Acad. Sci. USA 89:8073-8077.

32. Pang, Q., J.B. Hays, I. Rajagopal, and T.S. Schaefer. 1993. Selection of *Arabidopsis* cDNAs that partially correct phenotypes of *Escherichia coli* DNA-damage-sensitive mutants and analysis of two plant cDNAs that appear to express UV-specific dark repair activities. Plant Mol. Biol. 22:411-426.

33. Peterson, T.A., L. Prakash, S. Prakash, M.A. Osley, and S.I. Reed. 1985. Regulation of *CDC9*, the *Saccharomyces cerevisiae* gene that encodes DNA ligase. Mol. Cell. Biol. 5:226-235.

34. Pleura, M.J. 1982. Specific-locus mutation assays in *Zea mays*. Mutat. Res. 99:317-337.

35. Radany, E.H., and E.C. Friedberg. 1980. A pyrimidine dimer-DNA glycosylase activity associated with the *v* gene product of bacteriophage T4. Nature 286:182-184.

36. Riazuddin, S., and L. Grossman. 1977. *Micrococcus luteus* correndonucleases I: Resolution and purification of two endonucleases specific for DNA containing pyrimidine dimers. J. Biol. Chem. 252:6280-6286.

37. Ruby, S.W., and J.S. Szostak. 1985. Specific *Saccharomyces* genes are expressed in response to DNA-damaging agents. Mol. Cell. Bio.5:75-84.

38. Sebastian, J., B. Kraus, and G.B. Sancar. 1990. Expression of the yeast *PHR1* gene is induced by DNA-damaging agents. Mol. Cell. Biol. 10:4630-4637.

39. Todo, T., H. Takemori, H. Ryo, M. Ihara, T. Matsunaga, O. Nikaido, K. Sato, and T. Nomura. 1993. A new photoreactivating enzyme that specifically repairs ultraviolet light-induced (6-4) photoproducts. Nature 361:371-374.

40. Tsujimura, T., V.M. Maher, A.R. Godwin, R.M. Liskay, and J.J. McCormick. 1990. Frequency of intrachromosomal recombination induced by UV radiation in normally repairing and excision repair-deficient human cells. Proc. Natl. Acad. Sci. USA 87:1566-1570.

41. Walker, G.C. 1984. Mutagenesis and inducible responses to deoxyribonucleic acid damage in *Escherichia coli*. Microbiol. Rev. 48:60-93.

# MOLECULAR GENETICS OF NUCLEOTIDE EXCISION REPAIR IN EUKARYOTES

Anne B. Britt

Section of Botany, U.C. Davis
Davis, CA 95616

Key words: *Arabidopsis thaliana*, excision repair, skin cancer, stacking interactions, xeroderma pigmentosum

**ABSTRACT**

The cyclobutyl pyrimidine dimer (CPD) and the pyrimidine (6-4) pyrimidinone dimer make up approximately 75 and 25 percent, respectively, of UV-B induced DNA damage products. In the absence of photoreactivating light pyrimidine dimers are eliminated from the genome by a nucleotide excision repair pathway. Excision repair enzymes recognize a variety of DNA damage products, nicking the damaged strand and removing the damaged oligonucleotide. Because UV-B-induced DNA damage products act as inhibitors of both RNA and DNA synthesis, one would expect that resistance to the growth-inhibitory effects of UV-B might be determined by the rate of repair of UV-induced lesions. Experimental results, however, have not demonstrated a simple positive correlation between the rate of repair of CPDs and the overall level of UV-resistance of various mammalian cell lines. Recent developments suggest that: a) the 6-4 product, rather than the CPD, may be the critical cytotoxic UV-induced DNA damage product, and b) that rapid removal of 6-4s and CPDs from actively transcribed DNA strands, rather than from the genome overall, is required for UV-resistance. The repair of 6-4s also seems to be required for UV-B resistance in plants. We have found that dark repair of 6-4s in the higher plant *Arabidopsis thaliana* is rapid and efficient, while dark repair of CPDs in the genome overall is negligible. We have also found that the growth of a mutant of Arabidopsis which is deficient in 6-4 repair is inhibited by very low levels of UV-B.

## INTRODUCTION

DNA repair occurs via one of two of strategies; the damaged base can be directly repaired (through the actions of pyrimidine dimer photolyase, or $O^6$ methylguanine methyltransferase), or the altered base can be removed from the DNA, and replaced with undamaged bases. This second process, termed excision repair, can be further subdivided into two categories. Nucleotide excision repair involves several steps: the recognition of the damaged base or bases, the generation of nicks in the damaged strand on both sides of the lesion, the removal of the damaged oligonucleotide by a helicase, the synthesis of a new strand using the undamaged strand as a template, and the sealing of the remaining nick by DNA ligase. This process should be distinguished from alternative base excision repair mechanisms involving lesion-specific glycosylases (such as uracil glycosylase), which cut the damaged base directly from the sugar-phosphate backbone, thereby producing a single type of lesion (an abasic site) which is efficiently repaired via the combined actions of an apurinic/apyrimidinic endonuclease, DNA polymerase, and DNA ligase.

While the glycosylases and photolyases recognize only one, or a few, substrates [2,35], the enzymes involved in the damage recognition step of nucleotide excision repair are strikingly nonspecific in the range of damaged bases recognized [15]. Many of the most frequently generated lesions are recognized and repaired by both lesion-specific glycosylases and the general NER enzymes. For example, the cyclobutyl pyrimidine dimer, which makes up approximately 75% of all UV-induced lesions [34,31], has been

NATO ASI Series, Vol. I 18
Stratospheric Ozone Depletion/
UV-B Radiation in the Biosphere
Edited by R. H. Biggs and M. E. B. Joyner

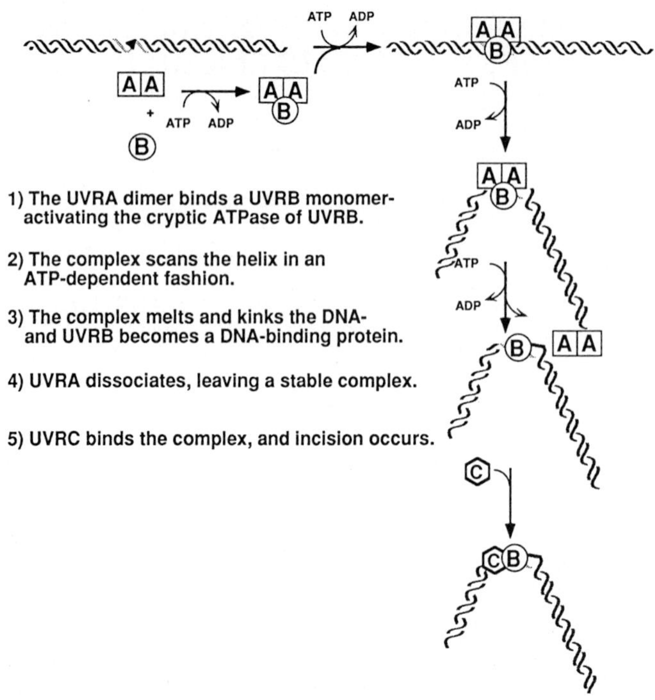
ATP ADP
A A
B
A A
+
B
ATP ADP
A A
B
ATP
ADP
A A
B
1) The UVRA dimer binds a UVRB monomer-activating the cryptic ATPase of UVRB.
2) The complex scans the helix in an ATP-dependent fashion.
3) The complex melts and kinks the DNA-and UVRB becomes a DNA-binding protein.
4) UVRA dissociates, leaving a stable complex.
5) UVRC binds the complex, and incision occurs.
ATP
ADP
B
A A
C
C B

Figure 1. Damage recognition and incision by the *E. coli* UvrABC endonuclease (after [9] ).

demonstrated in many systems to be repaired in both a rapid, light-dependent fashion (via photolyase), and via the slower NER pathway. More recently a photolyase specific for the second most common UV-induced lesion, the pyrimidine [6-4] pyrimidinone photoproduct (the 6-4 product), has been discovered in extracts prepared from drosophila embryos [43]. It is possible that the major present-day biological role of NER is the repair of relatively rare DNA damage products that do not have a "dedicated" (lesion-specific) repair pathway.

This chapter focuses on nucleotide excision repair in eukaryotic systems, including plants, and the significance of NER in UV-resistance. This field has advanced rapidly in recent years, and it is not possible in this format to provide an exhaustive list of new developments. Instead, I will raise what I think are few of the most interesting issues. For more detail on the molecular genetics of NER in yeast and higher eukaryotes, I refer the reader to a recent concise review [13,14].

## NUCLEOTIDE EXCISION REPAIR IN *E. coli*

The genetics of NER have been thoroughly developed in *E. coli*. Efficient excision of UV-induced damage requires the combined actions of four gene products [36] (Figure 1). UvrA and UvrB form a complex which tracks along the DNA, forming a tightly bound, stationary complex at the recognition site. The damaged DNA strand bound by this preincision complex is highly kinked and locally melted. The UvrA protein dissociates from the complex, which is then recognized by the UvrC endonuclease, which generates nicks on both sides of the damaged strand. Turnover of this complex apparently requires the concerted actions of the UvrD helicase and DNA polymerase I [11]. The biochemistry of damage recognition, which is not yet fully understood, is particularly interesting due to the its apparently paradoxical nature. In order to function, enzymes must interact with a substrate that physically matches the enzyme's active site. As it is unlikely that an enzyme could react with a structure defined as "anything abnormal," there must be some chemical moiety present in all lesions which acts as a recognition site for UvrAB. Although alterations to the conformation of sugar-phosphate backbone (bends in the DNA) are often suggested as possible recognition sites, Houten and Snowden [15] argue plausibly that the UvrAB complex recognizes its substrate via stacking interactions between aromatic amino acids in the UvrB protein and the damaged bases and/or undamaged adjacent bases.

## NER IN YEASTS AND MAMMALIAN TISSUE CULTURE

*The Genetics of NER in Eukaryotes is Complex and Incomplete.* Given the remarkable successes of the genetic approach in dissecting the mechanisms and regulation [48] of DNA repair in *E. coli,* several research groups have undertaken a similar study of the genetics of DNA repair in the model eukaryotes *Saccharomyces cerevisiae* and *Schizosaccharomyces pombe* [10,40]. However, in spite of substantial efforts, the genetics of DNA repair in yeast remains incomplete. Each attempt to saturate the system (i.e., to isolate mutations in every gene which is required for DNA repair) reveals additional complementation groups. There are over forty genes in *S. cerevisiae* which can affect DNA repair, and at least eleven of these are required for excision repair. Although several of these genes have been cloned, and certain biochemical activities identified, the biochemical mechanism of the yeast "excinuclease" is not yet understood, and it is difficult to imagine an excision repair process which would require the concerted activity of so many gene products.

The genetics of excision repair has also been developed in mammalian systems, employing cell lines derived from naturally occurring human variants and mutant radiation-sensitive cell lines primarily isolated from immortalized Chinese hamster ovary cells [7]. Complementation groups are determined in these systems by fusion of different cell lines and assay of the repair proficiency of the hybrid line. Because rodent and human lines can undergo fusion to form viable hybrid lines, the rodent mutations can also be identified as homologues of the human genes. In addition, human (and rodent) repair genes can and have been cloned by complementation of repair defects in UV-sensitive cell lines. Under these circumstances, however, that complementation (the restoration of a wild-type phenotype via the provision of a wild-type copy of a gene to a recessive mutant) is extremely difficult to distinguish from suppression (the masking of a mutant phenotype via the alteration of expression of an unrelated gene). Given the generally incomplete nature of the restoration of repair-proficiency in these experiments, the likelihood of suppression effects is quite high.

The genetics of excision repair in the mammalian cell lines seems to be similar to that of yeast. The sequences of the cloned mammalian genes have clear homologies to yeast sequences (reviewed in [14]). The rodent genetics are, like that of yeast, still incomplete, and the human genetics, which has relied mainly on the identification of natural variants, will probably continue to lag behind the rodent systems in the identification of genes involved in repair, as undoubtedly some mutations which are viable in tissue culture will be inviable in an intact organism. In spite of this, it is clear that the yeast and mammalian systems are similar in that there are a large number of genes which are directly or indirectly involved in excision repair. Interestingly, the defect in repair of CPDs can be suppressed in the human repair-deficient lines by transformation with bacterial genes encoding CPD-specific endonucleases [8]. This suggests that the defect in excision repair, in all of these mutants, is at, or prior to, the incision step. The apparent lack of mutations affecting putative repair-specific post-incision helicases, polymerases, or ligases is not especially surprising, considering the frequency at which these enzymes are probably called upon to repair spontaneous lesions such as apurinic sites, which occur at a frequency of over 1000/hr per mammalian cell [39].

How similar are excision repair processes in *E. coli* and eukaryotes? Huang et al [16] have recently demonstrated that the biochemical process of nucleotide excision in humans not unlike that of the *E. coli* holoenzyme. The human endonuclease, like the *E. coli* complex, nicks the damaged strand both 5' and 3' to CPDs, although the human enzyme produces an

Table 1. Excision repair of CPDs is proportional to UV-resistance in xeroderma pigmentosum cell lines. (From [18]).

| Complementation Group | % normal repair synth. | % normal repair CPDs | UV sensitivity |
|---|---|---|---|
| A | 0-5 | 0 | ++++ |
| B | 3-7 | 10 | +++ |
| C | 10-15 | 20 | ++ |
| D | 10-40 | 20 | ++++ |
| E | 40-60 | 50 | + |
| F | 0-10 | 60 | + |
| G | 10-15 | 0 | ++++ |

oligonucleotide of 27 to 29 bases, in contrast to the 12-13 bp *E. coli* product. As stated above, however, the genetics of excision repair in eukaryotes suggests that NER is a more complex process in eukaryotes than in *E. coli*.

*UV-Resistance Is Not Always Proportional to the Rate of CPD Repair.* A simple relationship between UV-resistance and the excision of CPDs was originally suggested by characterization of the rates of repair synthesis and the disappearance of CPDs in UV-sensitive fibroblasts derived from xeroderma pigmentosum (XP) patients [9]. In at least seven of the XP complementation groups, the correlation between the rate of repair of CPDs, the rate of repair-related DNA synthesis, and the UV-sensitivity of the cells was quite good [reviewed in [9] (Table 1). The XP syndrome involves an unusually high rate of skin cancers in exposed areas of the skin and progressive neurological degeneration. The high incidence of UV-induced cancers, combined with a measurable defect in the repair of a UV-induced DNA damage product, made XP a paradigm for the causal relationship between mutagenesis and cancer and the importance of excision repair as an anti-cancer mechanism [6]. The assay of repair rates for CPDs in rodent cells, however, resulted in an apparent discrepancy in the relationship between CPD repair and UV-resistance. While rodent cells are similar to human cells in their resistance to the cytotoxic effects of UV light, they lag far behind human cells in their repair of CPDs (Table 2 [44,50]). This discrepancy was striking enough to be generally termed the "rodent paradox." Similarly, two independently-derived UV-resistant revertants of an XP-A line failed to display a restoration of their ability to repair CPDs, in spite of their nearly wild-type levels of UV-resistance [5]. Further analysis of one of these revertants (the XP129 revertant of line XP12RO) revealed that it had nearly normal levels of UV-induced repair replication, sister chromatid exchange, and mutagenesis [4]. These results suggested that the repair of cyclobutyl dimers is not an important component of UV-resistance in mammalian cells. Thus while the XP cell lines display a defect in the repair of CPDs, it may actually be the failure of the excision complex to repair a less common UV-induced lesion that results in cytotoxicity and mutagenesis.

The use of radioimmunoassays to follow the repair of the 6-4 photoproduct revealed that the rate of repair of this lesion in wild-type and UV-sensitive mutant cell lines shows a better correlation with the cell's sensitivity to the cytotoxic effects of UV (Table 2) [29]. Rodent cells, while inefficient in the repair of CPDs, repair 6-4 photoproducts at nearly the same rate as human cells [28]. The repair of 6-4 photoproducts was also assayed in the above mentioned the XP-A revertant line and found to be close to that of normal cells (Table 2). These results suggest that cell's ability to repair 6-4 products (and perhaps other minor UV-induced photoproducts) may be a major determinant of UV-resistance.

Table 2. UV resistance is related to both overall 6-4 repair capacity and the rate of repair of CPDs in actively transcribed strands. (Data from [22,26,29]).

| Cell type | % Normal UV-resistance* | % normal repair CPDs | %normal repair 6-4s in 6 hrs | % repair of CPDs in *DHFR*† in 24 hrs |
|---|---|---|---|---|
| Human | 100 | 100 | 100 | 97 |
| CHO | 100 | 15 | 90 | 89 |
| XP-A | <15 | <5 | <10 | 6 |
| XP-A revertant | 100 | <5 | 80 | 71 |

* Normal rodent and human cells have 85% survival at 3 J/m$^2$ 254 nm radiation
† Transcribed strand only

*Repair of Certain Critical Regions May Be More Important to UV-Resistance than Repair of the Genome Overall.* The repair rates discussed above reflect the overall rate of repair of DNA damage products in total DNA of the cell. Both CPDs and the 6-4 product can act as blocks to transcription, as well as replication [30,33], and the resulting inhibitory effect of persisting damage on transcription may be an important component of UV-cytotoxicity. For this reason, an alternative explanation for the "rodent paradox" might be that rodents repair DNA damage within transcriptional units at a high rate, and that this site-specific repair provides the cells with a high level of UV-resistance. This hypothesis has been tested using a Southern blot-like assay for the frequency of cyclobutyl dimers within actively transcribed regions [1]. Rodent cells were found to repair CPDs within transcribed regions at a higher rate than in the overall genome, one comparable to that of human cells. Further research into sequence-specific repair has revealed: a) that CPDs in actively transcribed genes are repaired more rapidly than those in inactive genes [23], b) that this effect is generally limited to the transcribed strand of the DNA [26], c) that preferential repair of 6-4 photoproducts also occurs [42] and d) that this phenomena occurs not only in rodent cells, but also in human cells [25], yeast [41], and *E. coli* [27,37]. It is likely that the rapid repair of transcribed strands, perhaps together with the efficient repair of 6-4 photoproducts, may explain the UV-resistant phenotype of rodent cells. Similarly, the UV-resistant XP-A revertant line, described above as proficient in 6-4 photoproduct repair but deficient in CPD repair, has been shown to have wild-type levels of CPD repair in the transcribed strand of the dihydrofolate reductase gene [22], making the revertant cell line a perfect analog of rodent repair.

## DO ALL DNA REPAIR MUTATIONS OCCUR IN DNA REPAIR GENES?

The cloning and sequencing of higher eukaryotic genes involved in DNA repair has provided further evidence for a connection between transcription and DNA repair [3]. For example, the human *ERCC-3* gene, which complements the excision repair defect of XP-B cells, is homologous to the yeast *SSL2* (Suppression of Stem-Loop) gene. The *SSL2-1* mutation, which has no effect on UV-sensitivity, was originally isolated as a second-site suppressor mutation of a defect in transcription of the *HIS4* gene. A generation of a mutation that mimics the XP-B defect in *ERCC-3* results in a UV-sensitive phenotype [12]. Functional disruption of *SSL2* results in lethality. Thus the yeast *ERCC-3* homolog seems to be required for transcription as well as for DNA repair. Similarly, the *ERCC-3* gene product has been shown to be identical to an 89 kilodalton subunit of the human BTF2 (TFIIH) basic transcription factor, which participates both in the formation of the transcriptional initiation complex and in the elongation complex [49].

Thus the human *ERCC-3* gene product, like the product of the yeast *SSL2*gene, plays a role in both transcription and repair.

The XP-B mutation represents a rare disorder (observed in only one individual) that produces the symptoms of both Cockayne's syndrome and xeroderma pigmentosum. Individuals with Cockayne's syndrome (complementation groups A and B) are characterized by extreme photosensitivity (though, interestingly, no predisposition to cancer) and a broad spectrum of developmental abnormalities. Cell lines from individuals with "pure" Cockayne's syndrome are defective solely in site-specific repair, while overall repair is normal [46]. Because the *ERCC-3* gene product is a subunit of a transcription factor which is bound to the polymerase, it is tempting to suggest that this protein is involved in the physical coupling of the polymerase and the repair enzymes. However, the XP-B line is defective in *both* overall and sequence-specific repair (cited in [14]). Thus the *ERCC-3* gene product is also, somehow, involved in overall repair. Two hypotheses have been proposed to explain the role of proteins like the *ERCC-3 gene* product in transcription, strand-specific repair, and overall repair: 1) the protein may function directly in all three processes, probably as a helicase, or 2) the NER-defective phenotype may result from defective transcription of genes involved in NER. Whether *ERCC-3* plays an active role in the biochemistry of damage recognition and excision remains to be determined.

## GENETICS OF EXCISION REPAIR IN HIGHER PLANTS

The use of higher plants as a model system has yielded many insights into basic genetic processes, including the particulate nature of inheritance, the modification of genes via mutation, and the mutagenic effects of transposable genetic elements. Given these successes, it is surprising that our knowledge of the molecular mechanisms of DNA repair and mutagenesis in plants lags so far behind our understanding of these processes in microbial and higher animal systems. The developmental strategy of plants differs from that of animals in a way that enhances the beneficial effects of random mutagenesis and diminishes its harmful effects [19,47]. For this reason the DNA repair strategies of plants may differ from those of animals.

Although progress has been made in the genetics and biochemistry of repair in the single celled green alga Chlamydomonas [38], studies of DNA repair in higher plants have largely been limited to efforts to demonstrate the presence of repair activities (reviewed in [45]). Many early studies presented negative results that led to the suggestion that DNA repair does not occur in plants. Recent improvements in the sensitivity of techniques for the detection of DNA damage products have resulted in the demonstration of both photorepair and excision repair of UV-induced photoproducts in phytoplankton [18] and in several higher plant systems (reviewed in [24]). The recent observation of cyclobutyl pyrimidine dimer photolyase activity in the model higher plant *Arabidopsis thaliana* [32] has stimulated research into the molecular and genetic basis of UV resistance and DNA repair in this model system, which is further described in an article by John Hays in this volume.

We have initiated a search for mutants of Arabidopsis which are specifically defective in excision repair. Other laboratories, notably that of Dr. David Mount (of the University of Arizona) and Dr. Robert Last (of the Boyce-Thompson Institute at Cornell University) have also identified UV-sensitive mutants of Arabidopsis. To enhance the UV-transparency of our plants, all work was performed with a *tt5* (transparent testa [20,21]) line that is defective in the production of UV-absorbing flavonoid pigments. M1 seeds were mutagenized with EMS, grown to maturity, allowed to self-pollinate, and harvested in bulk. The resulting M2 population contains seeds that are heterozygous or homozygous for new mutations. Seeds were harvested from the individual M2 plants, and a sample of the progeny of each plant was tested for UV-sensitivity using a root-bending assay. Approximately 20 seeds from each M2 family were placed on an agar plate and incubated under low light at 22°C. Because the plate is incubated on edge (vertically), the roots of the seedlings grow downward across the surface of the agar. After three days, half of each row of newly germinated seedlings was exposed to 0.5 kJ/m$^2$ of UV-B from a UV transilluminator (UVP, San Gabriel, CA). After irradiation, the plates were rotated by 90° and allowed to grow overnight in complete darkness. Because the plate was rotated, any new growth (post-irradiation) was at right angles to the old growth, and so was easily scored. Families that continued to grow on the unirradiated side of the plate, but failed, uniformly, to

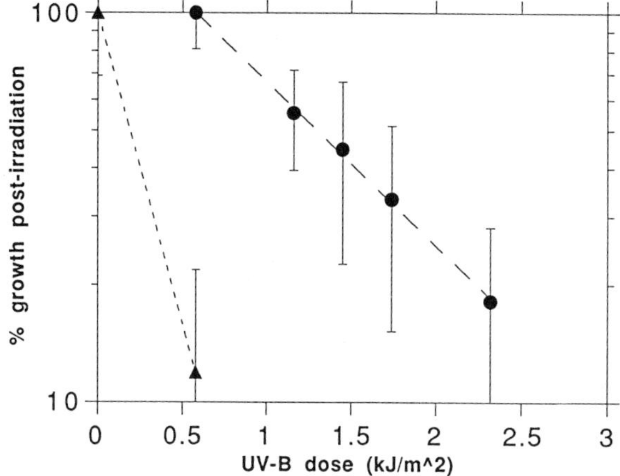


Figure 2. Inhibition of root growth in the *uvr1* strain and its progenitor. New root growth (at right angles to growth pre-irradiation) was measured three days after irradiation at the UV-B doses indicated. After irradiation, seedlings were grown at 22°C in the dark. Growth of the Arabidopsis seedling, under these conditions, ceases by the sixth day after imbibition. The results represent an average of measurements made on between 10 and 31 seedlings. Error bars are one standard deviation.

grow on the irradiated side were scored as UV sensitive mutants. Families that uniformly expressed a UV-sensitive phenotype were further propagated and backcrossed to their progenitor.

Two isolates of a single UV-sensitive mutant were detected in a screen of over 2,000 M2 families. This mutant, designated *uvr1*, was further tested for sensitivity to the growth inhibiting effects of UV light. The mutant was approximately 6 fold more sensitive than its progenitor to the inhibitory effects of UV-B on elongation of the immature root (Figure 2). The *uvr1* mutation is recessive, and segregates as a single locus. Not only is the growth of the root tip sensitive to UV-B irradiation, but the aerial tissues of the plant also display a UV-sensitive phenotype. "Cool white" fluorescent lamps emit a minor fraction of their power (approximately 0.04 W/m$^2$ at the level of the soil) as a small peak at 313 nm ([17], pers. obs.). When grown under these conditions, the mutant plants were smaller than their wild-type progenitor. The harvested stems of the mutant line were blackened on their illuminated surface, in distinct contrast to the bright yellow color of their *tt5* progenitor. Most notably, the leaves of the mutant gradually turned brown and whithered, and the edges curled upward and inward to form a concave bowl. This curling indicates that cell growth is inhibited on the upper surface of the leaf. In dramatic contrast, the phenotype of the *uvr1* mutant grown under cool white lamps filtered through Mylar, which eliminates the UV-B component of the lamp's output, was very similar to that its the progenitor.

To test the possibility that the *uvr1* mutant is sensitive to UV simply because it is more transparent to UV than its progenitor line, we assayed the induction of cyclobutyl dimers, the major UV-induced DNA photoproduct, in mutant and progenitor 5-day-old seedlings. Immediately after irradiation, the concentration of cyclobutyl dimers was assayed by comparing the weight-average molecular weights of single stranded DNA (assayed by alkaline sucrose gradient sedimentation) in samples treated with a cyclobutane dimer specific endonuclease, T4 UV-endonuclease V (gift of Anne Ganesan), with that of the same DNA sample which had not been subjected to digestion. CPDs were induced at the same rate in the mutant and progenitor lines (Figure 3). Hence the mutant and progenitor seedlings were identical in their transparency to UV-B.

To test the possibility that the mutant's UV-sensitivity was due to a defect in DNA repair, we measured the rate of loss of the two major UV-induced photoproducts: the cyclobutyl pyrimidine dimer (CPD) and the pyrimidine(6-4)pyrimidinone dimer (6-4 photoproduct). Seeds of the *uvr1* and

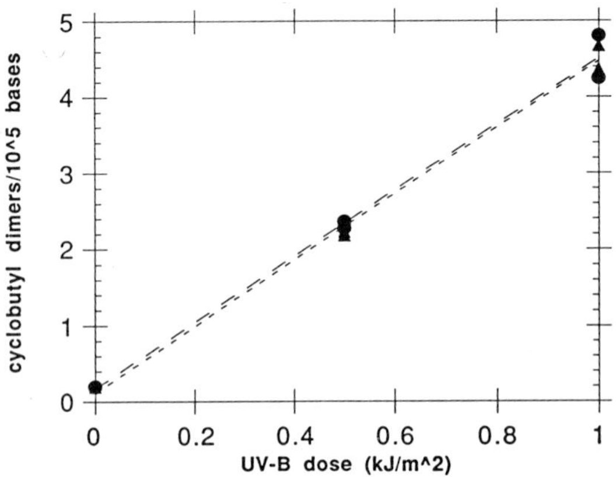


Figure 3. Cyclobutyl dimers are induced at equal rates in the *uvr1* mutant (Δ) and its progenitor (o). Seedlings were grown on vertically oriented nutrient agar plates plus 0.5% sucrose. Approximately 500 to 1000 seeds are sown per 100x100 mm plate. To label the DNA, tritiated thymidine (1 Ci/ml) was also included in the medium.

progenitor lines were sown on agar plates and incubated on edge under continuous low light for five days. At the end of this period, the seedlings had germinated and elongated, growing across the surface of the agar. The five-day-old seedlings were then irradiated with UV-B light, receiving a dose of 1.25 kJ/m$^2$ over a period of 40 seconds. The seedlings were then either immediately harvested (in the dark) for DNA extraction, further incubated in complete darkness for dark repair assays, or incubated under cool white lamps for a period of 1 to 20 hours for photoreactivation studies. In collaboration with Dr. David Mitchell (MD Anderson Cancer Research Center, Smithville, Texas) the concentrations of CPDs and were assayed via a lesion-specific radioimmunoassay [28].

As illustrated in Figure 4A, the wild-type strain repairs CPDs very inefficiently, if at all, in the dark, while photoreactivation of CPDs is fairly rapid. The mutant is indistinguishable from its progenitor in both of these processes. The ability of the wild-type Arabidopsis line to continue to grow normally (Figures 2 and 3) in spite of the persistence of cyclobutyl dimers at a frequency of approximately one dimer per 42,000 bases suggests that Arabidopsis has some capacity for dimer tolerance, which may be mediated by a dimer bypass mechanism, recombinational "repair," or perhaps selective, rapid repair of a small subclass of critical sites, such as actively transcribed genes. Whether preferential repair of transcribed strands occurs in plants is technically difficult to determine, as plants have unusually small (generally under 2 kb) transcriptional units. In order to see an average of one or more dimers per gene very high frequencies of dimers must be generated (in fact, the very small average intron size in plants may have resulted from constant selection for UV-resistance). Thus the assay of site-specific repair of biologically relevant levels of UV-induced damage will require unusual levels of sensitivity. Betsy Sutherland describes in this volume a novel method for the site-specific assay of repair which may be applicable to this problem.

To determine whether the UV-sensitivity of the mutant strain might result from a deficiency in the repair of the 6-4 product, we assayed the rate of repair of the 6-4 photoproduct in wild-type and *uvr1* Arabidopsis lines (Figure 4B). Dark repair of the 6-4 photoproduct occurs much faster than dark repair of CPDs in wild-type Arabidopsis, with approximately 50% of the initial lesions eliminated within the first two hours post-irradiation. Arabidopsis therefore resembles many biological systems, including fish, frog, and rodent cells [reviewed in 29)], in that the overall rate of dark repair of CPDs is low or undetectable, while 6-4 products are rapidly removed.

In contrast to the wild-type strain, the *uvr1* mutant is defective in the repair of the 6-4 product. As

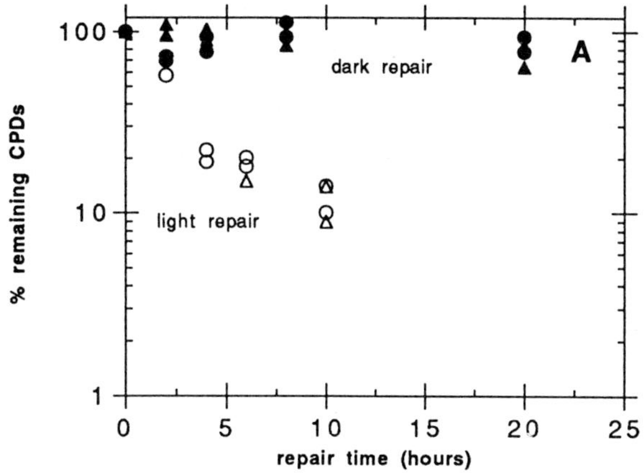


Figure 4. A: Repair of CPDs is similar in both the *uvr1* (Δ) and progenitor (o) strains. Solid symbols indicate dark repair. Seedlings were grown on agar plates and irradiated to a final UV-B dose of 1.32 J/m² as described in Figure 2. The *uvr1* seedlings are derived from a backcrossed line. Photoreactivation was carried out under a bank of GE Cool White fluorescent lamps, filtered through the polystyrene lid of the petri dish. The UV-A fluence at the surface of the agar was approximately 0.08 W/m² (measured using a UV-A radiometer (UVP, San Gabriel). B: The *uvr1* mutant is defective in repair of the 6-4 product. After irradiation, seedlings were incubated at 22°C in the dark. The data represents an average of two to eight experimental values. Error bars are one standard deviation.

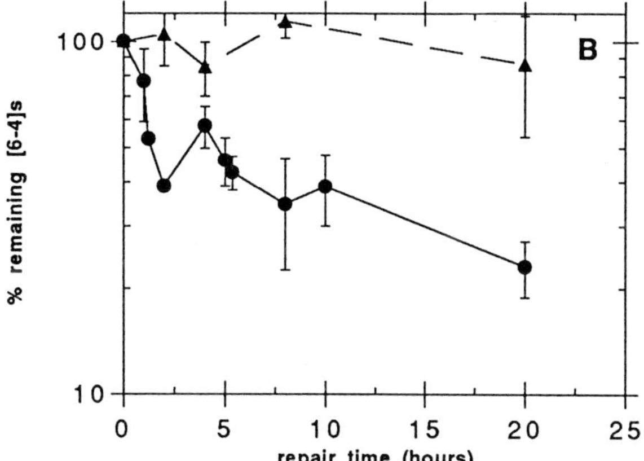


illustrated in Figure 4B, little or no repair of 6-4s occurs in the *uvr1* line. The sensitivity of this strain to low, chronic levels of UV-B suggests that repair of this lesion is essential for resistance to environmental UV-B. In addition, it is possible that the *UVR1* gene is required for the repair of certain minor UV-photoproducts, such as pyrimidine hydrates, and that the persistence of these lesions may contribute significantly to the UV-sensitive phenotype of the *uvr1* mutant. As discussed above, it may also be possible that Arabidopsis, like the mammalian systems, is capable of rapid removal of CPDs from a minor but critical subclass of sites, and that the *UVR1* gene is also required for this process.

It is conceivable that the *uvr1* mutant is UV-sensitive for some reason other than a defect in DNA repair, and that its failure to repair 6-4 products after UV-irradiation is, like its failure to grow after UV-irradiation, the result, rather than the cause, of its UV-sensitivity. The fact that the mutant efficiently photorepairs CPDs at the wild-type rate, however, makes this possibility unlikely, and strongly suggests that the *uvr1* mutation is in a gene specifically required for DNA repair.

Recent awareness of the increasing degradation of the UV-B absorbing ozone layer has focused attention on the mechanisms of UV-B resistance in plants. UV-B intensity also varies widely with season, latitude, and altitude, and so may influence the range of environments at which certain plant species can survive. The continued isolation and analysis of DNA repair mutants in higher plants is expected to make a substantial contribution to our understanding of the role of DNA repair both in resistance to environmental DNA damaging agents and in the evolution of the plant genome.

**Acknowledgments**
This work was upported by NSF Grant #DMB-90-19159.

## REFERENCES

1. Bohr, V.A., et al. 1985. DNA repair in an active gene: Removal of pyrimidine dimers from the DHFR gene of CHO cells is much more efficient than in the genome overall. Cell 40:359-369.

2. Brash, D.E., et al. 1985. *E. coli* DNA photolyase reverses cyclobutane pyrimidine dimers but not pyrimidine-pyrimidone (6-4) photoproducts. J. Biol. Chem. 260:11438-11441.

3. Buratowski, S. 1993. DNA repair and transcription: The helicase connection. Science 260:37-38.

4. Cleaver, J.E., et al. 1987. Unique DNA repair properties of a xeroderma pigmentosum revertant. Molec. Cell. Biol. 7:3353-3357.

5. Cleaver, J.E. 1989. DNA damage and repair in normal, xeroderma pigmentosum and XP revertant cells analyzed by gel electrophoresis: Excision of cyclobutane dimers from the whole genome is not necessary for cell survival. Carcinogenesis 10:1691-1696.

6. Cleaver, J.E. 1990. Do we know the cause of xeroderma pigmentosum? Carcinogenesis 11:875-882.

7. Collins, A.R. 1992. Mutant rodent cell lines sensitive to ultraviolet light, ionizing radiation and cross-linking agents--a comprehensive survey of genetic and biochemical characteristics. Mutat. Res. 293:99-118.

8. de Jonge, A.J.R., et al. 1985. Microinjection of *M. luteus* UV-endonuclease restores UV-induced unscheduled DNA synthesis in cells of 9 xeroderma pigmentosum complementation groups. Mutat. Res. 150:99-105.

9. Friedberg, E.C. 1985. DNA repair. San Francisco: W.H.Freeman and Co.

10. Friedberg, E.C. 1991. Yeast genes involved in DNA repair processes: New looks on old faces. Molec. Microb. 5:2303-2310.

11. Grossman, L., and A.T. Yeung. 1990. The UvrABC endonuclease of *E. coli*. Photochem. Photobio. 51:749-755.

12. Gulyas, K.D., and T.F. Donahue. 1992. SSL2, a suppressor of a stem-loop mutation in the HIS4 leader encodes the yeast homolog of human *ERCC-3*. Cell 69:1031-1042.

13. Hoeijmakers, J.H.J. 1993. Nucleotide excision repair I: From *E. coli* to yeast. TIG 9:173-177.

14. Hoeijmakers, J.H.J. 1993. Nucleotide exzcision repair II: From yeast to mammals. TIG 9:211-217.

15. Houten, B.V., and A. Snowden. 1993. Mechanism of action of the *E. coli* UvrABC nuclease: Clues to the damage recognition problem. BioEssays 15:51-59.

16. Huang, J.-C., et al. 1992. Human nucleotide excision nuclease removes thymine dimers from DNA by incising the 22nd phosphodiester bond 5' and the 6th phosphodiester bond 3' to the photodimer. PNAS 89(April):3664-3668.

17. ISCO Tables. 1970. Lincoln, Nebraska: Instrumentation Specialties Company.

18. Karentz, D., J.E. Cleaver, and D.L. Mitchell. 1991. Cell survival characteristics and molecular responses of antarctic phytoplankton to ultraviolet-B radiation. J. Phycol. 27:326-341.

19. Klekowski, E.J. 1988. Mutation, Developmental Selection, and Plant Evolution. New York: Columbia University Press. p.373.

20. Koornneff, M. 1990. Mutations affecting the testa color in *Arabidopsis*. Arab. Inf. Service. 27:1-4.

21. Li, J., et al. 1993. *Arabidopsis* flavonoid mutants are hypersensitive to UV-B radiation. Plant Cell 5:171-179.

22. Lommel, L., and P.C. Hanawalt. 1993. Increased UV-resistance of a xeroderma pigmentosum revertant is correlated with selective repair of the transcribed strand of an expressed gene. Molec. Cell. Biol. 13:970-976.

23. Madhani, H.D., V.A. Bohr, and P.C. Hanwalt. 1986. Differential DNA repair in a transcriptionally active and inactive protooncogene: c-abl and c-mos. Cell 45:417-422.

24. McLennan, A.G. 1987. The repair of ultraviolet light-induced DNA damage in plant cells. Mutat. Res. 181:1-7.

25. Mellon, I., et al. 1986. Preferential DNA repair of an active gene in human cells. Proc. Natl. Acad. Sci. USA 83:8878-8882.

26. Mellon, I., G. Spivak, and P.C. Hanawalt. 1987. Selective removal of transcription-blocking DNA damage from the transcribed strand of the mammalian DHFR gene. Cell 51:241-249.

27. Mellon, I., and P. Hanawalt. 1989. Induction of the *E. coli* lactose operon selectively increases repair of its transcribed DNA strand. Nature 342(6245):95-98.

28. Mitchell, D.L., C.A. Haipek, and J.M. Clarkson. 1985. (6-4) Photoproducts are removed from the DNA of UV-irradiated mammalian cells more efficiently than cyclobutane pyrimidine dimers. Mutat. Res. 143:109-112.

29. Mitchell, D.L., and R.S. Nairn. 1989. The biology of the (6-4) photoproduct. Photochem. Photobiol. 49:805-819.

30. Mitchell, D.L., J.E. Vaughan, and R.S. Nairn. 1989. Inhibition of transient gene expression in Chinese hamster ovary cells by cyclobutane dimers and (6-4) photoproducts in transfected ultraviolet-irradiated plasmid DNA. Plasmid 21:21-30.

31. Mitchell, D.L., D.E. Brash, and R.S. Nairn. 1990. Rapid repair kinetics of pyrimidine(6-4)pyrimidone photoproducts in human cells are due to excision rather than conformational change. Nucl. Acids Res. 18:963-971.

32. Pang, Q., and J.B. Hays. 1991. UV-B-inducible and temperature-sensitive photoreactivation of cyclobutane pyrimidine dimers in *Arabidopsis thaliana*. Plant Physiol. 95:536-543.

33. Protic-Sabljic, M., and K.H. Kraemer. 1986. One pyrimidine dimer inactivates expression of a transfected gene in xeroderma pigmentosum cells. PNAS 82:6622-6626.

34. Rosenstein, B.S., and D.L. Mitchell. 1987. Action spectra for the induction of pyrimidine(6-4)pyrimidone photoproducts and cyclobutane pyrimidine dimers in normal human skin fibroblasts. Photchem. Photobiol. 45:775-780.

35. Sakumi, K., and M. Sekiguchi. 1990. Structures and functions of DNA glycosylases. Mutat. Res. 236:161-172.

36. Sancar, A., and J.E. Hearst. 1993. Molecular Matchmakers. Science 259:1415-1420.

37. Selby, C.P., and A. Sancar. 1993. Molecular mechanism of transcription-repair coupling. Science 260(2 April):53-58.

38. Small, G.D. 1987. Repair systems for nuclear and chloroplast DNA in *Chlamydomonas reinhardtii*. Mutation Res. 181:31-35.

39. Smith, K.C. 1992. Spontaneous mutagenesis: Experimental, genetic, and other factors. Mutat. Res. 277:139-162.

40. Subramani, S. 1991. Radiation resistance in *Schizosaccharomyces pombe*. Molec. Microb. 5:2311-2314.

41. Sweder, K., and P. Hanawalt. 1992. Preferential repair of cyclobutane pyrimidine dimers in the transcribed strand of a gene in yeast chromosomes and plasmids is dependent on transcription. PNAS 89:10696-10700.

42. Thomas, D.C., et al. 1989. Preferential repair of (6-4) photoproducts in the dihydrofolate reductase gene of Chinese hamster ovary cells. J. Biol. Chem. 264:18005-18010.

43. Todo, T., et al. 1993. A new photoreactivating enzyme that specifically repairs ultraviolet light-induced (6-4) photoproducts. Nature 361:371-374.

44. van Zeeland, A.A., C.A. Smith, and P.C. Hanawalt. 1981. Sensitive determination of pyrimidine dimers in DNA of UV-irradiated mammalian cells. Introduction of T4 endonuclease V into frozen and

thawed cells. Mutat. Res. 82:173-189.

45. Veleminsky, J., and K. Angelis. 1990. DNA repair in higher plants. *In* R. Mortimer, L. Mendelsohn, and R. Albertini (eds.), Mutation and the Environment, Wiley-Liss: N.Y. pp.195-203.

46. Venema, J., et al. 1990. The genetic defect in Cockayne syndrome is associated with a defect in repair of UV-induced DNA damage in transcriptionally active DNA. PNAS 87:4707-4711.

47. Walbot, V. 1985. On the life strategies of plants and animals. TIG 1(6):165-169.

48. Walker, G.C. 1985. Inducible DNA repair systems. Ann. Rev. Biochem. 54:425-457.

49. Zawel, L., and D. Reinberg. 1992. Advances in RNA polymerase II transcription. Curr. Biol. 4:488-495.

50. Zelle, B., et al. 1980. The influence of the wavelength of ultraviolet radiation on the survival, mutation induction, and DNA repair in irradiated Chinese hamster cells. Mutat. Res. 72:491-509.

# IS UV-B A HAZARD TO SOYBEAN PHOTOSYNTHESIS AND YIELD? RESULTS OF AN OZONE-UV-B INTERACTION STUDY AND MODEL PREDICTIONS

Edwin L. Fiscus,[1] Joseph E. Miller,[1] and Fitzgerald L. Booker[2]

[1]USDA/ARS Air Quality Research Unit and Crop Science Department
[2]Botany Department, North Carolina State University
1509 Varsity Drive, Raleigh NC 27606

Key words: ozone, photosynthesis, soybean, UV radiation

## ABSTRACT

Current levels of tropospheric ozone suppress photosynthesis and yield in soybean. Also, it has been suggested that increased ground-level UV-B as a result of stratospheric ozone depletion may have additional deleterious effects. A three-year field study, conducted in open-top chambers, was undertaken to uncover possible interactions between these putative stressors. Ozone treatments resulted in the expected and well-documented reductions in photosynthesis, yield and acceleration of senescence. However, UV-B treatments not only failed to induce any significant interactions, but did not induce any significant reductions in photosynthesis or yield, even at levels simulating a 35% column ozone depletion. Reconciliation of our data with other predictions of physiological dysfunction and crop losses due to increased UV-B was attempted by examining the models used to predict ground-level UV-B, ground truthing, and critically reviewing the literature. Comparison of ground-based measurements at our location with the Green et al. (9) model, frequently used to predict ground-level UV-B, showed consistent over-predictions of clear-sky UV-B of 32% on an annual basis. We believe that similar over-predictions have led some researchers to underestimate the actual dosages used, with the result that the effects reported would normally only occur at much higher UV-B levels and are much greater than would occur at the reported dosages. Lack of ground-level UV-B monitoring in many experiments has obscured and perpetuated this problem. Also, there generally has been no adjustment of enhancement levels for either season or weather conditions, except where modulated systems have been used, so that effects are additionally exaggerated for these reasons. Interpretation of experimental results is confounded by these four factors (model over-prediction, seasonal changes, weather changes, and failure to monitor UV-B ) and made much more difficult when UV-B enhancement experiments are conducted under greenhouse growth conditions. Additional illustrative calculations for greenhouse conditions are included for consideration. Examination of the literature in light of these findings indicates there is little evidence that increased ground-level UV-B, well in excess of current predictions for the next century, will pose any hazard to soybean growth and productivity.

## INTRODUCTION

Recently Stolarski et al. [17] demonstrated statistically significant downward trends in stratospheric ozone over much of the northern hemisphere. The year-round trend was estimated as -1.8% per decade and, although much higher (-2.7%) in the winter, the trends were about -1.0 to -1.3% per decade over most of the crop-growing season at mid-northern latitudes. As a result of these decreases there is a potential for increased ultraviolet-B (UV-B) radiation (280-320nm) at ground level which may pose a hazard to natural aquatic and terrestrial ecosystems as well as suppress

NATO ASI Series, Vol. I 18
Stratospheric Ozone Depletion/
UV-B Radiation in the Biosphere
Edited by R. H. Biggs and M. E. B. Joyner

worldwide crop production. On the other hand, tropospheric $O_3$ currently is recognized as the single most phytotoxic regional air pollutant and already may be responsible for a 15% decrease in production of some crops [10]. It is important to determine if increased UV-B and tropospheric $O_3$ acting in combination pose an additional threat to the normal development and yield of crops. Therefore, the objective of the experiments reported here was to determine what, if any, interactions might occur between pollutant ozone and UV-B with regard to photosynthesis and yield and to determine their extent and identify their cause.

Soybean was chosen because it is the most extensively studied of the crop plants in terms of UV-B response in field situations and we also have a large data base on the response of this species to tropospheric ozone. For this experiment we used three soybean cultivars. One of the cultivars (Essex) has been reported to be susceptible to UV-B damage [12,21] and was found to be normally sensitive to ozone damage.

## MATERIALS AND METHODS

Although details of the materials and methods for this experiment can be found in Miller et al. [13], salient features of that study for the 1990 season will be repeated here for the convenience of the reader.

Three soybean cultivars [*Glycine max* (L.) Merr. cvs. Essex, Coker 6955, and S 53-34] were grown in 15 l pots in a 2:1:1 (by volume) mixture of soil, sand, and Metro-Mix 220 (W.R. Grace Co., Cambridge, MA)[a] in open-top field chambers at Raleigh, NC, USA. Plants were watered daily and fertilized biweekly with "Peters Blossom Booster" (10-30-20 N-P-K) (Grace-Sierra Horticultural Products Co., Milpitas, CA)[a] and three times during the season with "Peters STEM" soluble trace elements and micronutrient mix (Grace-Sierra Horticultural Products Co., Milpitas, CA).[a] Ozone treatments of charcoal-filtered air (CF) and ozone additions to a final concentration of 1.7x ambient were administered over a 132-d period starting at germination and continuing to harvest. The seasonal mean of the 12-h daily mean $[O_3]$ in the CF and supplemental $O_3$ treatments were 24 and 83 nL $L^{-1}$, respectively.

Low, medium, and high UV-B supplements were provided by banks of fluorescent lamps (model UVB-313, Q-Panel Co., Cleveland, OH)[a] suspended in the open-top chambers as previously described. Lamps in the low treatment were wrapped in polyester film to filter out radiation less than 315 nm; therefore, this treatment also served as the control for UV-A (320-400nm) and visible radiation emitted from the lamps. Lamp irradiance in the medium and high UV-B treatments was filtered with cellulose diacetate (0.13 mm thickness) to remove radiation below 290 nm. Lamp banks were kept at a constant height of 0.4 m above the canopy, and the irradiance was varied by fluorescent dimmer controls on the lamp ballasts.

Broadband erythemal meters (model 2D, Solar Light Co., Philadelphia, PA)[a] with a spectral response very similar to that of the Robertson-Berger (R-B) meter were used to set the lamp bank irradiance levels each day and a Robertson-Berger meter was used to monitor solar UV continuously. Ambient and chamber irradiances were also measured with a UV-visible spectroradiometer (model 742, Optronics Laboratories, Inc., Orlando, FL)[a] equipped with a 3.7-m-long quartz fiber optics cable and Teflon diffuser head. The spectroradiometer was calibrated with an NIST-traceable 200W tungsten-halogen lamp standard of spectral irradiance (model 220A, Optronics Laboratories, Inc., Orlando, FL)[a] driven by a current regulated power source (model 65, Optronics Laboratories, Inc., Orlando, FL).[a] Wavelength calibration was checked periodically by comparison with Hg emission lines from a UVB-313 lamp. The broadband erythemal meters were calibrated against the spectroradiometer as previously described [1], and biologically effective UV-B irradiance (UV-$B_{BE}$) was calculated by applying Caldwell's generalized plant action spectrum [5], normalized to 300 nm (PAS300), to the spectroradiometer scans. In 1992 an additional broadband erythemal instrument (Model UVB-1 UV Pyranometer; Yankee Environmental Systems, Montague, MA)[a] was used for monitoring solar UV-B.

Stratospheric ozone losses corresponding to the supplemental UV-$B_{BE}$ radiation levels were calculated from the radiative transfer model derived from Green [9] by Björn and Murphy [3]. The model was extensively modified[b] for ease of use and flexibility, and additional functions were added for PAS300

[a]Mention of a company or product does not imply an endorsement by the United States Department of Agriculture.

weighting and the Robertson-Berger meter spectral sensitivity curve. Inputs and outputs of the model were confirmed by ground-based measurements. The model predicted a total daily UV-$B_{BE}$ irradiance for 21 June of 5.45 kJ $m^{-2}$ with an aerosol coefficient = 1. In 1990 an average daily solar UV-$B_{BE}$ of 4.88 ± 0.12 (s.d.) kJ $m^{-2}$ measured by the Robertson-Berger meter for the 30 d surrounding 21 June indicated acceptable agreement with the model. Also on 21 June, the daily supplemental UV-$B_{BE}$ irradiance in the three treatments was set to 0, 4.43, and 8.13 kJ $m^{-2}$ for the low (control), medium, and high UV-B treatments. Due to an average UV-B shading effect of the open-top chambers of 24%, the total daily UV-$B_{BE}$ irradiance for the three treatments was 4.14, 8.57, and 12.27 kJ $m^{-2}$. The treatments thus corresponded to an increase in the $O_3$ column thickness of 15% and decreases of 20%, and 35% respectively.

Supplemental UV-B treatments were administered as a constant daily addition with the lamp output levels being set each day, from sowing to harvest, while the ozone was being dispensed. To compensate for seasonal changes in photoperiod and solar UV-B irradiance, the supplemental irradiance levels were adjusted biweekly; thus, relatively constant ozone column depletion simulations were maintained [1]. For example, on 1 September the solar clear sky irradiance was calculated as 4.66 kJ $m^{-2}$. On that basis, the lamps in the high UV treatment would be set to deliver an additional 5.27 kJ $m^{-2}$ for a total exposure of 9.93 kJ $m^{-2}$, still corresponding to a column ozone depletion of 35% even though there was 19% less UV-$B_{BE}$ delivered to the plants than on 21 June. At season's end, the actual exposure figures were refined according to the ground-based measurements provided by the R-B meter. Over the entire 1990 experimental period, therefore, the mean daily UV-$B_{BE}$ irradiances were 3.02, 6.24 and 8.98 kJ $m^{-2}$. On overcast days the treatments were discontinued if the UV-$B_{BE}$ irradiance fell below 20% of the maximum calculated for the Raleigh, NC, location on 21 June for more than 30 minutes. Solar UV-$B_{BE}$ was evaluated every 2 h afterward, and treatments were resumed if the overcast cleared.

Net carbon exchange rate (NCER), leaf conductance, and transpiration were made with an LI-6000 portable photosynthesis system (Li-Cor, Inc, Lincoln, NE)[a] and chlorophyll fluorescence induction was measured at room temperature using a pulse amplitude modulated chlorophyll fluorometer (PAM-101-103, H. Walz, Effeltrich, Germany)[a]. The efficiency of excitation energy capture by open PSII reaction centers in dark-adapted leaves was determined by $F_v/F_M$, where $F_v = F_M - F_O$ [11]. The photochemical fluorescence-quenching coefficients ($q_P$) were determined as described by Horton [11].

The Ozone/UV-B interaction portion of the study was conducted for two years (1989 and 1990), while the UV-B portion was continued for a third year (1991). UV-B enhancement details for 1990 are given above while the details for all three years may be found in Miller et al. [13].

All parameters were compared by analysis of variance and differences deemed not significant if $P>0.05$. Further statistical details are also in Miller et al. [13].

## RESULTS AND DISCUSSION

The UV-$B_{BE}$ enhancement treatments for 1990 are shown in Figure 1. The data points and lines are the total measured and calculated doses at the plant canopy level and are the sum of solar and supplemental UV-$B_{BE}$. Seasonal means for 1989 and 1991 are available in Miller et al. [13].

Detailed analysis of the growth, photosynthesis, and yield of these experiments [13] shows the expected declines as well as the accelerated senescence of the plants typical of ozone exposure. Miller et al. [13] could find no UV-B effects on net carbon exchange rate (NCER), stomatal conductance, or transpiration in Essex soybean, and so the UV-B treatments were combined for further analysis. To illustrate the lack of UV-B effect on photosynthesis, we present the NCER data broken down into UV-B treatments in Figure 2. Clearly the ozone treatments accelerated the decline from peak photosynthetic competence without any discernable significant or lasting effects or interactions of the UV-B treatments. Similar segregation of data by ozone, but not UV-B, was apparent in other parameters as well. In particular

---

[b] Copies of the executable code which will run in MS-DOS without an interpreter or compiler may be obtained on request from the senior author. The programs are available on a single 3.5" DD or HD diskette, a single HD 5.25" diskette or two DD 5.25" diskettes. If possible, a return mailer and blank diskettes should accompany the request.

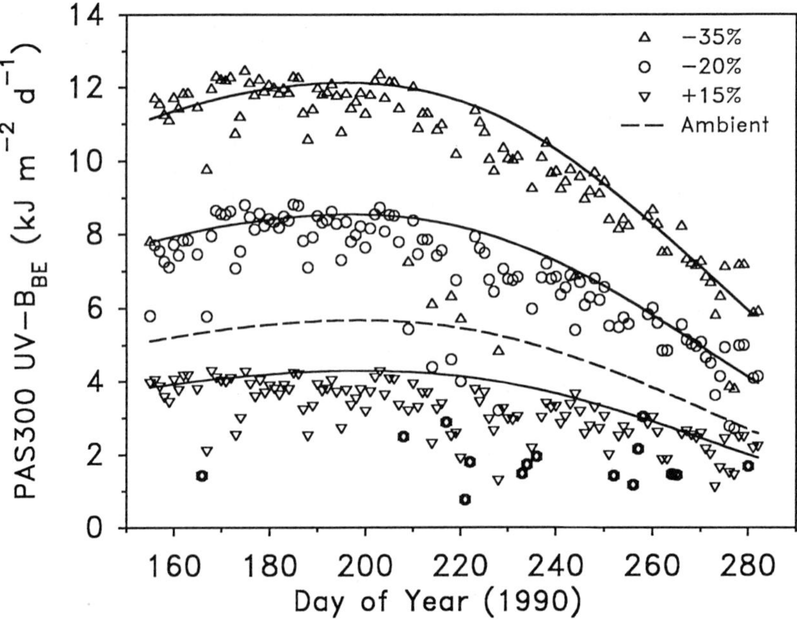


Figure 1. UV-B treatment levels for the 1990 experiment. Lines were calculated from Björn and Murphy (1985). Numbers on figure indicate the simulated column ozone change in the different treatments. Ambient UV-$B_{BE}$ data are not shown to reduce clutter.

there was no UV-B suppression of growth or yield for any of the three cultivars used for any of the three years of the study. In addition, in 1991 analysis failed to show any UV-B effects on the variable component of chlorophyll fluorescence induction or $q_P$ at either 26 or 69 days after planting (Table 1). There was, however, a statistically significant increase in NCER during the period of 69-71 days after planting. The reasons for this difference are not clear; however, it is difficult to attach much physiological significance to such small differences. Other studies failed to find any UV-B effects on peroxidase activity or chlorophyll levels [2]. The only UV-B effect apparently consistent with many other studies was the increase in UV absorbance of leaf extracts indicative of increased production of flavonoids and other phenolic compounds.

Given the current apparent perception that crops are at risk [22,24] from increased UV-B radiation, it is important to reconcile our results with those from other laboratories. Comparisons between laboratories, especially for field and greenhouse studies, are not trivial exercises since reporting of growth conditions, UV-B dose, and dose monitoring is frequently confusing. Therefore, the remainder of this paper will be devoted to a discussion of dosing procedures and calculations, which we believe have lead to large underestimates of actual UV-B treatment levels and consequent large overestimates of potential UV-B effects.

There are four reasons why UV-B dosages could be underestimated:

1) use of a radiative transfer model that overestimates ground level UV-B in certain circumstances;

2) lack of daily monitoring of ambient ground level UV-B during the course of an experiment;

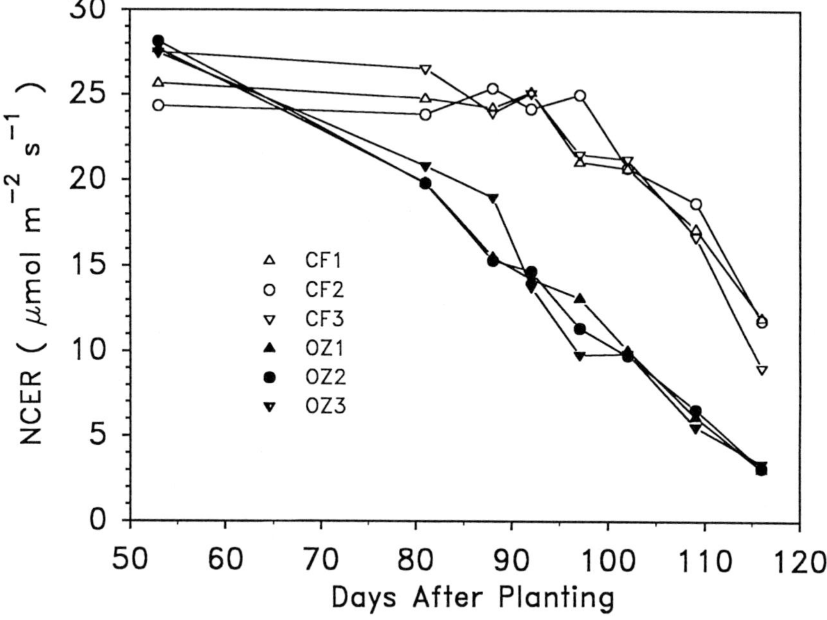


Figure 2. Net carbon exchange rate for the 1990 experiment.

3) lack of treatment level adjustments according to weather conditions; and

4) lack of treatment level adjustments to account for seasonal changes.

In addition there has been infrequent reporting of daily photosynthetically active radiation (PAR). These factors are not completely independent of each other so that inadequate monitoring of ground-level UV-B can worsen and perpetuate the errors based on a poor model prediction. Daily PAR values also will be related to weather and season. To illustrate the difficulties posed by these four factors, we shall use data from our field site at Raleigh, NC, USA for 1992. The measurements and model predictions will require only small adjustments to apply generally to mid-latitudes (30-40°N) and low elevations in the eastern United States. Before proceeding, however, we stipulate, for purposes of discussion later in this paper, our acceptance that as a worst case scenario the current year-round trend of stratospheric ozone depletion of 1.8% per decade [17] will apply to the normal cropping seasons and continue unchanged for the next 100 years.

It is well established that UV-B damage to plants is aggravated by low levels of PAR [6,14,19,23,25]. However, methods of reporting PAR during experiments tend to be cursory and difficult to interpret in relation to UV-B experiments. Although reporting daily midday PAR maxima during the experiment can give a general impression of weather conditions, we feel it would be far preferable to report actual daily integral data throughout the experiment as well as the daily UV-$B_{BE}$ integrals and perhaps the ratio of these two.

Figure 3 clearly illustrates the frequency of days at our location when PAR is below its clear sky value.

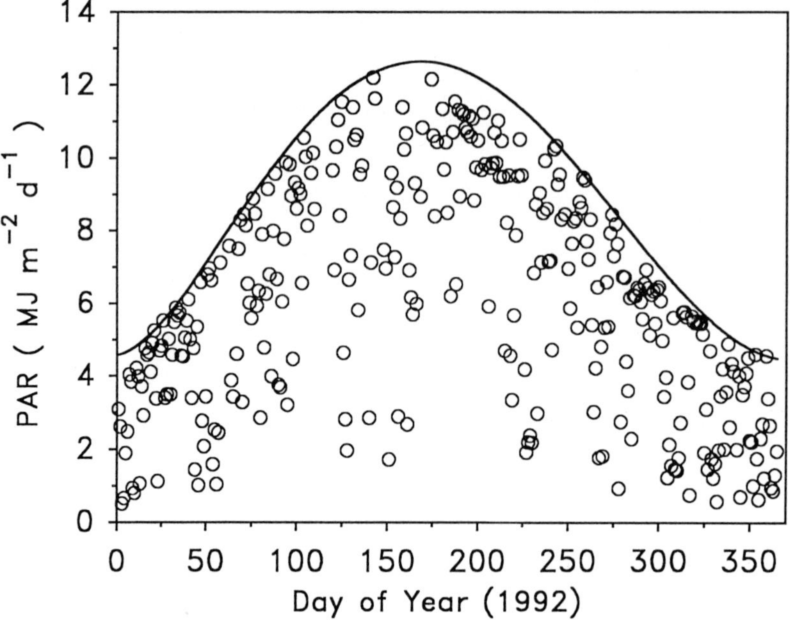


Figure 3. Photosynthetically active radiation falling on a horizontal sensor in 1992. The boundary line shown is an arbitrary function used only for estimating maximum PAR values through the year and is not based on an analytical model.

Analysis of these data indicates that over the entire year, mean PAR averaged only 69% of maximum. Figure 4 illustrates the same point for PAS300 UV-$B_{BE}$ for which the annual mean was only 70% of the clear sky maximum. Also shown in Figure 4 is the excellent agreement between ground-based measurements and the data boundary line calculated from the Björn and Murphy [3] model that we have used throughout these studies.

Further analysis of the data in these two figures shows that the minimum annual ratio of clear sky PAR/UV-$B_{BE}$ (units of each are J $m^{-2}$ $d^{-1}$) is about 1700 and falls to 1150 for a 20% column ozone depletion. Further, the average ratio for days 100 to 300, which under ambient conditions is normally 2250, would be 1490 for a 20% reduction of the ozone column. This ratio varied cyclically throughout the year and rose to an annual maximum of about 7800 during the winter months. It seems only prudent to maintain realistic ratios of PAR and UV-$B_{BE}$ radiation during enhancement experiments to ensure that capacity for photorepair is adequate and that other physiological processes affected by PAR and UV-B are in a natural balance as well as to stimulate synthesis of UV screening compounds [4,14,15].

The choice of radiative transfer model to use is also a matter of great importance as can be seen from Figures 4 and 5. As mentioned earlier, the Björn and Murphy [3] model adequately describes the en-velope of ambient UV-B measurements made at our field site throughout the year when ozone column data from the Nimbus 7 Total Ozone Mapping Spectrometer (TOMS) is used as input (Figure 4). By itself, the model will calculate an ozone column but the algorithm tends to overestimate the average column thickness except for about a two month period in the fall (Figure 5). It would appear from Figure 4 that, even with TOMS inputs, the model overestimates the UV-

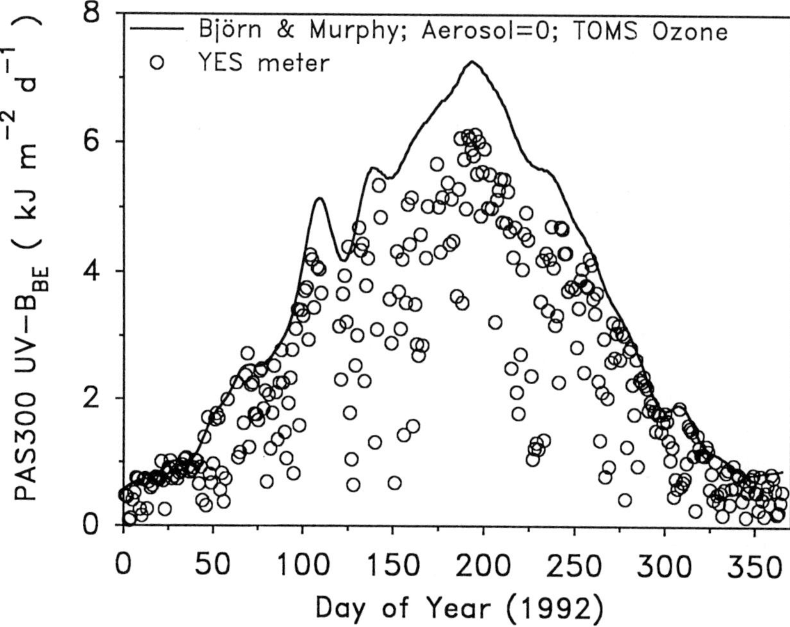


Figure 4. Comparison of ground-based measurements with a YES erythemal sensing instrument. The erythemal data were converted to PAS300 units by using the ratio PAS300/Erythemal irradiance determined from the Björn and Murphy (1985) model for each day of the year. The line was calculated from the model with the following Inputs: atmospheric pressure = 1000 mb; relative humidity = 0.5; aerosol coefficient = 0; environment = rural green farmland; TOMS ozone; and location = 35.75°N, 78.67°W.

$B_{BE}$ in the mid-summer period. That discrepancy can be explained by remembering that the boundary line describes clear sky conditions and the actual summertime radiation is more often attenuated by aerosols, spotty clouds, and haze than it is in winter. Inclusion of an arbitrary annual aerosol function, which rises from 0 in the winter to a maximum of 2 in mid-summer along a cosine curve, drops the boundary peak in Figure 4 onto the top of the data.

In the past, the Green et al. [7,8] model was used to calculate ground-level UV-$B_{BE}$ for the purpose of setting supplemental levels. In Figure 5 we compare the output from both the Green et al. [8] and the later Björn and Murphy [3] models. Clearly evident in this figure is that the Green et al. model overestimates UV-$B_{BE}$ at the ground by a substantial amount throughout the year whether a mean annual value for the ozone column or daily TOMS data are used. When TOMS data are used for both models, the Green et al. model overestimates UV-$B_{BE}$ by an average, over the year, of 32% when compared to the Björn and Murphy model. Therefore, supplements calculated from the Green et al. model will be understated in terms of column ozone depletion. A supplement calculated to simulate a 20% reduction in column ozone by the Green et al. model will actually deliver radiation consistent with a 25% reduction under the best (clear sky) circumstances (Table 1). It is important to note that this 5% discrepancy in column ozone represents a 25% calculational error and a 38% difference in the level of the UV-$B_{BE}$ supplement delivered to the canopy.

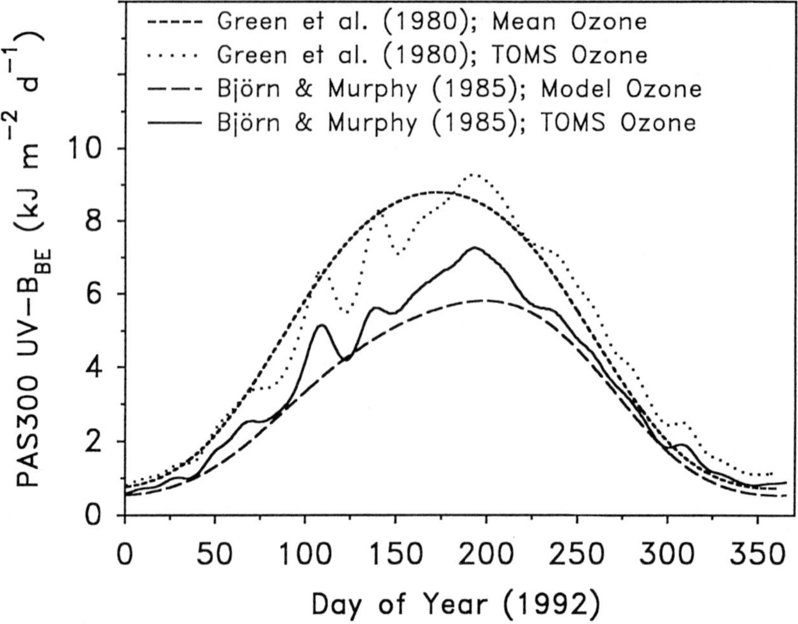


Figure 5. Comparison of the Green et al. (1980) and the Björn and Murphy (1985) models, both with and without TOMS data. Inputs for both models were the same as for Figure 4. Additional inputs for the Green et al. model are elevation = 114m and for the curve not using TOMS data, an annual average of 0.315 cm atm was entered. In the absence of TOMS data, the Björn and Murphy model generates its own column ozone data, shown as the long dashed line.

Calculations presented in Table 2 show how poor model estimates are compounded by not adjusting for clouds or aerosols. These calculations are based on Figure 3 and the assumption that, when PAR is reduced by clouds or other reasons, UV-$B_{BE}$ is also attenuated by the same fraction. While this assumption is probably not exact for any particular day, earlier calculations in this paper show that it is very close on an average basis. That is, while average PAR is 69% of the envelope value, average UV-$B_{BE}$ was 70% of the envelope. In any case, the assumption will satisfy the purpose of illustrating the problems of not adjusting supplements according to weather.

Earlier we showed that a Green et al. (8) supplement designed to simulate a 20% ozone column reduction actually simulated a 25% reduction under clear sky conditions. Now, if PAR only falls 10% (to 90% of $PAR_{MAX}$), as happens on 3 out of every 4 days, that same supplement represents a 27% depletion if calculated as follows: ambient UV-$B_{BE}$ was discounted for clouds by the same fraction as PAR; the ozone column necessary (OCN) to deliver that discounted level under clear skies was calculated; the fixed supplement was added to the reduced ambient value to obtain the total dose (TD); and, finally, the reduction in OCN necessary to pass the TD was calculated. Thus, although the TD seen at canopy height is less under cloudy conditions, the ratio of PAR/UV-$B_{BE}$ is decreased dramatically.

On an average day, with 69% of $PAR_{MAX}$, the calculated supplement actually represents a 30% ozone depletion. On as many as one day in four that

Table 1. Chlorophyll fluorescence ratios of leaf disks and NCER of soybean leaves treated from emergence to maturity with three levels of supplemental UV-B radiation. Values are means ± se; n=18; **P<0.01.

| Treatment | Days after Planting | $F_V/F_M$ | $q_P$ | NCER ($mol\ m^{-2}\ s^{-1}$) |
|---|---|---|---|---|
| Control | 26 | 0.83 ± 0.01 | 0.71± 0.3 | - |
| Medium UV-B | 26 | 0.83 ± 0.01 | 0.71± 0.3 | - |
| High UV-B | 26 | 0.83 ± 0.01 | 0.67± 0.3 | - |
| Control | 69-71 | 0.85 ± 0.01 | 0.74 ± 0.01 | 21.1 ± 0.3 |
| Medium UV-B | 69-71 | 0.85 ± 0.01 | 0.72 ± 0.02 | 22.7 ± 0.2** |
| High UV-B | 69-71 | 0.86 ± 0.01 | 0.72 ± 0.01 | 23.1 ± 0.3** |

Table 2. Influence of cloud cover and model-based overestimates of ground-level $UV\text{-}B_{BE}$ on a simulated 20% ozone column depletion for Raleigh, NC, 1992. The supplement necessary to simulate the depletion was calculated for clear skies for the summer solstice using the Green et al. (1980) model and an aerosol coefficient=0. The effects of cloud cover were estimated assuming that $UV\text{-}B_{BE}$ is reduced by the same fraction as PAR for any particular day and then the model-based error was added. For the greenhouse, a "control" level of 8.5 kJ $m^{-2}\ d^{-1}$ and a 20% depletion level of 12.5 kJ $m^{-2}\ d^{-1}$ were calculated from Green et al. (1980). Actual depletions given here are based on the Björn and Murphy (1985) model. $PAR/UV\text{-}B_{BE}$ was calculated only for the field simulation using $PAR_{MAX}$= 12.6 MJ $m^{-2}\ d^{-1}$. The actual normal ratio would be 1950, dropping to 1480 for a 20% ozone reduction.

| ($PAR/PAR_{MAX}$) 100 < | % of Total Days | Compound Simulated Ozone Depletion | | | |
|---|---|---|---|---|---|
| | | Field | $PAR/UV\text{-}B_{BE}$ | Greenhouse control | Greenhouse -20% |
| 100 % | - | 25 | 1313 | 20 | 36 |
| 90 % | 76 | 27 | 1258 | 24 | 40 |
| 75 % | 48 | 29 | 1153 | 31 | 45 |
| 69 % (Annual Mean) | 41 | 30 | 1106 | 33 | 47 |
| 50 % | 25 | 35 | 927 | 42 | 54 |
| 25 % | 10 | 44 | 584 | 56 | 65 |

same supplement represents a 35% depletion, and every 10 days the canopy can be expected to experience $UV\text{-}B_{BE}$ consistent with a 44% reduction in column ozone. Thus, in the field, treatment levels designed to approximate the worst case for ozone depletion over the next century will, on every fourth day, deliver radiation consistent with depletions of very nearly twice that level. In addition, PAR is reduced and the plants are presumably even more susceptible to UV-B damage.

In the greenhouse the choice of model and adjustments for weather become even more important since greenhouse glass does not normally transmit UV-B and the entire dose must be supplied by artificial means. Thus, if a clear-sky $UV\text{-}B_{BE}$ of 8.5 kJ $m^{-2}\ d^{-1}$ is calculated, as from the Green et al. [8] model, and the entire "ambient" level is supplied by lamp banks, then under clear-sky conditions this "control" level already represents a 20% column ozone reduction. Making additional corrections for average cloud cover

Table 3. Soybean yield responses to supplemental PAS300 UV-$B_{BE}$. Target simulations and supplements were taken from the papers. Clear sky simulations were calculated from Björn and Murphy with aerosol coefficient = 0 and the compound average simulations are discounted for aerosols and cloud cover as in Table 1. Yield changes are indicated as either increases (+), decreases (-) or no change (nc) if P>0.05. Notations in the yield change column identify putative sensitive (S) and resistant (R) cultivars. A ? indicates cultivars that have not been studied enough to classify.

| Source | Green et al. Simulation Target (%Depletion) | Supplement (kJ $m^{-2}$ $d^{-1}$) | Normal (kJ $m^{-2}$ $d^{-1}$) | Clear sky simulation (% Depletion) | Compound average simulation (% Depletion) | # CVYs | Yield Change |
|---|---|---|---|---|---|---|---|
| Teramura & Murali (1986) | 16 | 4.5 | 5.6 | 28 | 34 | 6 | 1 - (York)<br>5 nc (1S, 4?) |
| Teramura et al. (1990) | 16 | 3 | 5.6 | 22 | 30 | 12 | 5+ (2S, 3R)<br>2 - (1S, 1R)<br>5 nc (3S, 2R) |
| Sinclair et al. (1990) | 16 | 3 | 7.0 | 18 | 22 | 6 | 6 nc (1R, 5?) |
| Miller et al. (1993) | - | 2.35 | 5.6 | - | 22[1] | 7 | 7 nc (3S, 4?) |
| Teramura et al. (1990) | 25 | 5.1 | 5.6 | 30 | 36 | 10 | 1+ (1R)<br>5 - (4S, 1R)<br>4 nc (1S, 3R) |
| Sullivan & Teramura (1990) | 25 | 5.1 | 5.6 | 30 | 36 | 2 | 2 nc (2S) |
| Miller et al. (1993) | - | 4.81 | 5.6 | - | 36[1] | 7 | 7 nc (3S, 4?) |

1. Based on ground measurements throughout the season.

means on an average day this same 8.5 kJ $m^{-2}$ $d^{-1}$ represents a 33% ozone depletion. Taking this illustration one step farther, we can calculate the supplement necessary to simulate a 20% ozone depletion by the Green et al. model (+4 kJ $m^{-2}$ $d^{-1}$ ). By supplying the total daily dose from lamp banks, we calculate from Björn and Murphy [3] that the dose really simulates a 36% ozone depletion for clear-sky conditions and a 47% reduction for an average day . Strictly speaking, the values in Table 2 apply only to the summer solstice; thus the actual dosages may be even higher at different times of the year.

Having established the magnitude of possible dose reporting errors, we would like to examine previous yield studies within that context. In the experiments at Raleigh presented by Miller et al. [13], the irradiance levels were estimated by model prior to the experiments, but the reported dosages were calculated after conclusion of the experiment from actual ground measurements as shown in Figure 1. Adequate information, however, is not available for other studies on soybean [16,18,20,21).] On the basis of the Björn and Murphy model and assuming similar cloud cover and aerosol figures as our location at Raleigh, we can estimate some corrections to their reported dosages that can be used as a basis for comparison. In addition, we shall examine the available field data on the basis of cultivar-years (CVYs), wherein one CVY represents yield data from one cultivar collected during a single season. Thus 2 CVYs would represent the same cultivar tested during two years or for one year within different experiments or locations. On this basis, then, we can break down the available data as in Table 3.

There are a total of 50 CVYs included in Table 2 where "normal" values for PAS300 UV-$B_{BE}$ were calculated using Björn and Murphy P3] with an aerosol coefficient of 0. Where available in the literature, the target levels of ozone column depletion, generally calculated using the Green et al. [8] model, also are listed. Supplement levels were taken as given in the appropriate papers except for Teramura and Murali [20] where the supplement was calculated from the given total daily dose. The clear-sky simulations in Table 3 were calculated from Björn and Murphy [3] based on the sum of the normal daily flux and the supplement. The dosages then were discounted for likely cloud cover and aerosols to arrive at an estimate of the average dosages applied to the plants. In the case of Miller et al. [13] the values presented are based on actual ground measurements during the season. It also should be noted that Miller et al. [13] adjusted the treatment levels for seasonal changes. Since such adjustments were not performed in the other studies, the actual compound average simulations would be even somewhat higher than presented here.

Examination of the compound average simulation values reveals that actual treatment levels in two of the 16% target depletions were probably as high as twice the stated levels based on the Green et al. [8] model, while the third exceeded its target by an additional 6% depletion. The end result is that actual treatments were clustered into two groups, one representing a 22% ozone depletion and the other a 30 to 36% depletion level. These groups are approximately equal to and about twice the worst case scenario for ozone depletion for the next century. There are only 13 CVYs included in the 22% category, and none showed any statistically significant change in yield as a result of UV-B treatment. Of the remaining 37 CVYs, 6 (16%) showed yield increases, 8 (22%) yield decreases, and 23 (62%) no change at all. Looking at all 50 CVYs collectively, regardless of treatment level, we can see that 72% showed no change, 12% increased yield, and 16% decreased yield as a result of UV-B treatments. Perhaps even more revealing is to consider the case of the putative sensitive cultivar Essex, identified as the sensitive (S) in Table 2, which comprises 40% of the CVYs reported here. Surprisingly, this "sensitive" cultivar exhibited yield decreases in only 25% of the years tested, and none at the 22% ozone depletion level. The rest of the time Essex yield increased or remained unchanged in response to UV-B treatment.

A somewhat more liberal approach might be to examine the data on the basis of the clear-sky simulations. In this case the data may be segregated into treatments of <30% ozone depletion levels and 30% depletion, which puts 31 CVYs in the former category and 19 in the latter. For an ozone depletion level <30%, only 10% of the CVYs showed yield decreases while the remaining 90% either increased or did not change. The sensitive cultivar Essex showed a yield decrease in only 1 out of 10 CVYs at this level of treatment and a yield increase in 2 of those 10 CVYs. Of the 19 CVYs included in the 30% category of treatment, yield declined in 5 CVYs, remained the same in 13 and increased in 1.

The data in Table 2 can be summarized as follows: at treatment levels approximating the worst case scenario for the next century, there is very little evidence to suggest that UV-B threatens soybean yields and at treatment levels 1.5 to 2 times that level, the data indicate decreased yields only about 25% of the time in a sensitive cultivar.

## CONCLUSIONS

We have attempted to reconcile the results of our experiments with the popular notion that increased UV-B radiation may have a major effect on soybean production. Examination of the literature from this perspective shows that reported UV-B doses are frequently underestimated. In part, these differences may arise from the use of the Green et al. [8] predictive model that consistently overestimates ground-level UV-B. Calculated supplements therefore also are overestimated for a target ozone column depletion. In addition, cloud cover, atmospheric aerosols, and seasonal differences further contribute to the under-estimation of actual doses.

In the light of these findings, it appears that increases in ground-level UV-B, well in excess of current projections for the next century, will not constitute any direct hazard to soybean production.

**Acknowledgments**

For the TOMS data we would like to thank Drs. Richard D. McPeters and Arlin J. Krueger of NASA, GSFC, members of the TOMS Nimbus Experiment and Ozone Processing teams, and the National Space Science Data Center/World Data Center-A for Rockets and Satellites.

## REFERENCES

1. Booker, F.L., E.L. Fiscus, R.B. Philbeck, A.S. Heagle, J.E. Miller, and W.W. Heck. 1992a. A supplemental ultraviolet-B radiation system for open-top field chambers. J. Environ. Qual. 21:56-61.

2. Booker, F.L., J.E. Miller, and E.L. Fiscus. 1992b. Effects of ozone and Uv-B radiation on pigments, biomass and peroxidase activity in soybean. *In* R.L. Berglund (ed.), Tropospheric Ozone and the Environment: Effects, Modeling and Control II. Air and Waste Management Association, Pittsburgh, PA. p. 489-503.

3. Björn, L.O., and T.M. Murphy. 1985. Computer calculation of solar ultraviolet radiation at ground level. Physiol. vegetal 23:555-561.

4. Bruns, B., K. Hahlbrock, and E. Schafer. 1986. Fluence dependence of the ultraviolet-light-induced accumulation of chalcone synthase mRNA and effects of blue and far-red light in cultured parsley cells. Planta 169:393-398.

5. Caldwell, M.M. 1971. Solar ultraviolet radiation and the growth and development of higher plants. *In* A.C. Giese (ed.), Photophysiology. Academic Press, New York. pp.131-177.

6. Cen, Y-P., and J.F. Bornman. 1990. The response of bean plants to UV-B radiation under different irradiances of background visible light. J. Exp. Bot. 41:1489-1495.

7. Green, A.E.S., T. Sawada, and E.P. Shettle. 1974. The middle ultraviolet reaching the ground. Photochem. Photobiol. 19:251-259.

8. Green, A.E.S., K.R. Cross, and L.A. Smith. 1980. Improved analytical characterization of ultraviolet skylight. Photochem. Photobiol. 31:59-65.

9. Green, A.E.S. 1983. The penetration of ultraviolet radiation to the ground. Phys. Plant. 58:351-359.

10. Heck, W.W. 1989. Assessment of crop losses from air pollutants in the United States. *In* J.J. Mackenzie and M.T. El-Ashry (eds.). Air Pollution's Toll on Forests and Crops. Yale University Press, New Haven. pp.235-315.

11. Horton, P., and J.R. Bowyer. 1990. Chlorophyll fluorescence transients. *In* J.L. Harwood and J.R. Bowyer (eds.), Methods in Plant Biochemistry: Lipids, Membranes, and Aspects of Photobiology. Academic Press, London. pp.259-296.

12. Krizek, D.T. 1981. Plant physiology: Ultraviolet radiation. *In* S.B. Parker (ed.), 1981 Yearbook of Science and Technology. McGraw-Hill, New York. pp.306-309.

13. Miller, J.E., F.L. Booker, E.L. Fiscus, A.S. Heagle, W.A. Pursley, S.F. Vozzo, and W.W. Heck. 1993. Effects of ultraviolet-B radiation and ozone on growth, yield, and photosynthesis of soybean. J. Environ. Qual. in press.

14. Mirecki, R.M., and A.H. Teramura. 1984. Effects of ultraviolet-B irradiance on soybean V: The dependence of plant sensitivity on the photosynthetic photon flux density during and after leaf expansion. Plant Physiol. 74:475-780.

15. Ohl, S., K. Hahlbrock, and E. Schäfer. 1989. A stable blue-light-derived signal modulates ultraviolet-light-induced activation of the chalcone-synthase gene in cultured parsley cells. Planta 177:228-236.

16. Sinclair, T.R., O. N'Diaye, and R.H. Badges. 1990. Growth and yield of field-grown soybean in response to enhanced exposure to ultraviolet-B radiation, J. Environ. Qual. 19:478-481.

17. Stolarski, R., R. Bojkov, L. Bishop, C. Zerefos, J. Staehelin, and J. Zawodny. 1992. Measured trends in stratospheric ozone. Science 256:342-349.

18. Sullivan, J.H., and A.H. Teramura. 1990. Field study of the interaction between ultraviolet-B radiation and drought on photosynthesis and growth in soybean. Plant Physiol. 92:141-146.

19. Teramura, A.H. 1980. Effects of ultraviolet-B irradiances on soybean I: Importance of photosynthetically active radiation in evaluating ultraviolet-B irradiance effects on soybean and wheat growth. Physiol. Plant. 48:333-339.

20. Teramura, A.H., and N.S. Murali. 1986. Intraspe-

cific differences in growth and yield of soybean exposed to ultraviolet-B radiation under greenhouse and field conditions. Environ. and Exptl. Bot. 26:89-95

21. Teramura, A.H., J.H. Sullivan, and J. Lydon. 1990. Effects of UV-B radiation on soybean yield and seed quality: A 6-year field study. Physiol Plant 80:5-11.

22. Teramura, A.H., and J.H. Sullivan. 1991. Potential impacts of increased solar UV-B on global plant productivity. *In* E. Riklis (ed.), Photobiology. Plenum Press, New York. pp.625-633.

23. Teramura, A.H., R.H. Badges, and S.V. Kossuth. 1980. Effects of ultraviolet-B irradiances on soybean II: Interaction between ultraviolet-B and photosynthetically active radiation on net photosynthesis, dark respiration, and transpiration. Plant Physiol. 65:483-488.

24. Tevini, M., and A.H. Teramura. 1989. UV-B effects on terrestrial plants. Photochem. Photobiol. 50:479-487.

25. Warner, C.W., and M.M. Caldwell. 1983. Influence of photon flux density in the 400-700nm waveband on inhibition of photosynthesis by UV-B (280-320nm) irradiation in soybean leaves: Separation of indirect and immediate effects. Photochem. Photobiol. 38:341-346.

EFFECTS OF UV-B RADIATION
ON AQUATIC SYSTEMS

# AQUATIC SYSTEMS (FRESHWATER AND MARINE)

Robert C. Worrest

Consortium for International Earth Science Information Network (CIESIN)
1825 K Street NW, Suite 805
Washington, DC 20006

## SUMMARY

Stratospheric ozone depletion has the potential to exert very substantial effects on the environment. However, our ability to predict with confidence and precision the likelihood of occurrence of particular effects, and to quantify their anticipated ultimate scope, varies greatly due to differences in the current level of our state of knowledge about particular types of effects. A major dilemma is the fact that our state of knowledge, even now, is low with regard to certain effects that have the greatest potential for widespread global impacts. For example, the current state of knowledge concerning potential effects of increased UV-B radiation on aquatic systems is relatively low, but the resultant global impact on human health and the environment could be quite high (Table 1).

For aquatic systems, studies are needed to elucidate how UV-B effects are modified by additional biotic and abiotic factors, such as carbon dioxide, temperature, nutrient stress, heavy metals, diseases, and predators. The manner and magnitude of adaptation to increased UV-B radiation, such as by increased screening pigments or enhanced DNA-damage repair capacity, need to be investigated both as adaptation capacity of individual organisms and as genetically based changes in populations. In addition, there is a need for sufficient information on productivity and other responses using realistic UV-B enhancements under field conditions.

Research priorities continue to relate to UV-B penetration into the water column and through ice. This basic physical information should then be used both in laboratory and field research to determine effects on phytoplankton and zooplankton. In addition, adaptive strategies of organisms in freshwater and marine communities need to be studied over extended periods of time, as well as the effects of interactive stresses (temperature, salinity, etc.) and cumulative doses received over the lifetime of organisms. And finally, plankton effects should be modeled into broader ecosystem dynamic patterns, which will feed into models of global carbon and nitrogen cycling.

*Why Is a Potential Increase in UV-B Exposure of Aquatic Organisms Important for Society to Address?*

- More than 30% of the world's animal protein for human consumption comes from the sea. Human populations may be affected by the direct and indirect consequences of increased solar UV-B radiation on aquatic food webs.

- Aquatic ecosystems can account for about one-half the global production of organic carbon each year. A potential consequence of a decrease in freshwater and marine primary productivity would be a reduction in the capacity of the lakes, streams, and oceans to absorb carbon dioxide. This reduced sink

NATO ASI Series, Vol. I 18
Stratospheric Ozone Depletion/
UV-B Radiation in the Biosphere
Edited by R. H. Biggs and M. E. B. Joyner

Table 1. Potential aquatic effects of increased UV-B exposure resulting from decreased stratospheric ozone.

| |
|---|
| State of knowledge low: |
| field / community studies |
| interactive effects |
| manner and magnitude of adaptation |
| Potential for global impacts high: |
| ecosystem effects / species composition |
| biomass production / food supply |
| decreased nitrogen assimilation |
| reduced sink capacity for carbon dioxide |
| DMS / cloud condensation nuclei |

capacity for atmospheric carbon dioxide would augment the greenhouse effect.

- Decreased nitrogen assimilation by prokaryotic microorganisms could possibly lead to a drastic nitrogen deficiency for higher plant ecosystems, such as rice paddies. A large share of the nitrogen consumed by higher plants is made available by bacterial microorganisms, which have been found to be very sensitive to UV-B radiation. Losses in nitrogen fixation could be compensated by additional nitrogen fertilization; however, such actions could stress the capabilities of developing nations.

- The role of dimethyl sulfide (DMS), released from plankton and macroalgae as aerosol and cloud nuclei, is of major concern. Most importantly, a UV-B-induced decrease in phytoplankton populations may have an impact on cloud patterns and concomitant global climate changes.

- As a result of a differential sensitivity among species capable of occupying similar niches within aquatic ecosystems, enhanced levels of UV-B exposure will undoubtedly alter species and community composition, thereby altering biodiversity as well as ecosystem structure and function.

- The quantity and type of species within aquatic systems play a determinant role in water quality. Altering the community composition and abundance, as a result of exposure to enhanced levels of UV-B radiation, could significantly alter the water quality of the system.

*Reduction of Uncertainty in the Following Areas Is Required to Properly Address the Impact of Enhanced Exposure of Freshwater and Marine Ecosystems to UV-B Radiation:*

- UV-B penetration through the water column
- Temporal, spatial, and spectral variability of UV-B radiation within the water column
- Spectral irradiance profiles through the water column
- Vertical distribution of target species, including mixing effects

- Sensitivity of species, including both spectral and total exposure
- Adaptation and repair mechanisms of target species, and the time scales involved
- Trophic interactions
- Biomass production, carbon flux, and carbon deposition within and from the water column
- Abiotic interactive effects (e.g., temperature, carbon dioxide concentrations, pollutants)

## RECOMMENDED ACTIONS

- Initiate an intensive planning effort for projects that have a targeted suite of objectives with a focus on multidisciplinary projects that address multi-agency/organization issues that have the greatest potential for economic or ecological returns.
- Develop research approaches that will provide information that is meaningful at the societal level and is cost effective. These include:
    - develop an understanding of the critical processes and mechanisms that underlie species responses
    - use long-term studies
    - work *in situ* (includes mesocosms) and use cooperative, interdisciplinary efforts with common formats
    - interface with existing long-term "projects" (e.g., LTER sites, ARM sites, MAB biosphere reserves, other international activities)
    - develop quantitative, predictive capabilities for assessing not only UV-B effects at current levels of ozone but especially effects of enhanced levels of UV-B exposure resulting from predicted *changes* in ozone concentration
    - integrate UV-B projects with global change research programs
    - promote global monitoring of standing crop and productivity for baseline comparisons
- Develop a system of accountability for experimental design, including:
    - ability to intercalibrate results
    - detailed description of experimental design in publications
    - inclusion of spectral data in publications
    - access to data in a timely fashion
- Develop an education program both for scientists currently working in this area of research and for those just entering, including:
    - development of guidance/training brochures for instrumentation and experimental design
    - development of an electronic, on-line means of information exchange (e.g., bibliography, technological information, research results)

# UV-B EFFECTS ON AQUATIC SYSTEMS

Donat-P. Häder

Institut für Botanik und Pharmazeutische Biologie
Staudtstraße 5, D-91058 Erlangen, Germany

Key words: chromoproteins, gravitaxis, phototaxis, phytoplankton

## ABSTRACT

The marine phytoplankton represents the single most important ecosystem on the earth as it produces more biomass than all terrestrial ecosystems taken together. Thus, it also constitutes an important sink for atmospheric carbon dioxide. Earlier studies have shown that many phytoplankton organisms are under UV-B stress even at ambient levels. Phytoplankton is not equally distributed in the water column, but rather uses highly specific orientation strategies to move to and stay in zones of suitable light intensity for growth and survival.

The most important external cues for the orientation mechanisms are light (phototaxis) and gravity (gravitaxis). These orientation mechanisms, however, are impaired by UV-B radiation as are motility (percentage of motile cells) and the individual swimming velocities. As a consequence, the photosynthetic pigments as well as the structural proteins of the photosynthetic apparatus are affected by solar radiation as shown by FPLC, gel electrophoresis, and spectroscopic methods. Chromoproteins of the photoreceptor apparatus responsible for phototaxis have also been found to be affected. Any increase in solar UV-B radiation due to partial stratospheric ozone depletion caused by the anthropogenic emission of gaseous pollutants carries the risk of affecting the phytoplankton ecosystems.

## INTRODUCTION

The most drastic and best known example of stratospheric ozone depletion is the Antarctic ozone hole which was first documented in 1979 and has increased in size and depth since. In addition, satellite and other data have indicated a general decrease in the stratospheric ozone concentration on a global basis. Since ozone is an effective filter for solar shortwave ultraviolet radiation (UV-B range, 280–320 nm), ground-based devices have measured an increase in solar UV-B radition.

Solar ultraviolet radiation at current and enhanced levels has been found to directly or indirectly affect almost all forms of life. One concern about adverse effects of enhanced solar UV-B radiation is related to marine phytoplankton [8,33,36]. Because these organisms are not protected by an epidermal layer, they dwell in the top layers of the water column, and they have been found to be very sensitive to ultraviolet radiation [5,40], many aquatic ecosystems are under considerable UV-B stress even at current levels [38,45]. The question of whether this radiation significantly impairs phytoplankton in its natural habitat can only be solved when more data that allow a prediction on a global basis are available. To arrive at quantitative data three questions need to be solved:

1. What is the spectral distribution of solar radiation penetrating into the water column?
2. What is the vertical distribution of major biomass producers in the water column?
3. What are their biological sensitivities?

## SPECTRAL DISTRIBUTION OF UV-B RADIATION IN THE WATER COLUMN

The penetration of radiation into the water column strongly depends on the wavelength [35]: short and long wavelength radiation is attenuated more than

NATO ASI Series, Vol. I 18
Stratospheric Ozone Depletion/
UV-B Radiation in the Biosphere
Edited by R. H. Biggs and M. E. B. Joyner

green or blue radiation; UV-B is also strongly absorbed. The attenuation of ultraviolet radiation is largely controlled by the quality of water: In costal waters with high turbidity and large concentrations of particulate matter, UV-B penetrates only a few decimeters or meters into the column [44] while in clear oceanic waters, UV-B has been found to penetrate to depths of dozens of meters [1]. Gieskes and Kraay (1990) measured record penetration in Antarctic waters where 1% of the solar UV striking the surface has been detected at a depth of 65 m. Our group is currently using a double monochromator spectroradiometer (Optronic 752) equipped with a long quartz light guide to measure the spectral distribution of the penetrating radiation.

## PHYTOPLANKTON AND AQUATIC ECOSYSTEMS

Even though precise data are hard to obtain, the standing crop of phytoplankton is rather small compared to terrestrial systems. In contrast, the photosynthetic productivity of phytoplankton equals or exceeds that of terrestrial ecosystems. The incorporation of carbon into organic material on our planet has been estimated to be about $2 \times 10^{11}$ tons annually [34], half of which (104 gigatons of carbon annually) is due to the primary producers (phytoplankton) in the oceans. The biomass produced is then utilized by primary consumers (zooplankton), including unicellular and multicellular organisms, which in turn feed the secondary consumers. The end consumers are large fish, birds, and mammals, including humans.

Phytoplanktonic microorganisms are not equally distributed throughout the water column since they depend on the availability of solar energy for their growth and metabolism [16,29]. Depending on the water quality, phytoplankton is limited to the top layers between a few decimeters and dozens of meters, called the euphotic zone. In contrast to most terrestrial plants, most phytoplankton organisms are not adapted to unfiltered solar radiation, so the highest concentration of organisms is not found at or close to the surface. Using chlorophyll concentration, cell density, or carbon assimilation as an indicator shows characteristic vertical distribution patterns of phytoplankton with a maximum a certain distance below the surface [4] and only a few organisms in the subsequent attenuation zones, although the vertical distribution pattern of phytoplankton can be affected temporarily by the depth of the mixing layer.

Being limited to the euphotic zone close to the surface, phytoplankton is exposed to high levels of solar ultraviolet irradiation. The fact that phytoplankton is under considerable UV-B stress at ambient levels of solar radiation is indicated by two observations:

1. The large algal blooms in temperate zones occur in spring, when increasing temperatures allow growth. These blooms usually disappear when the level of solar ultraviolet irradiation increases during the summer. Often there is a second, smaller algal bloom in autumn, when the UV-B levels have decreased, if temperature and nutrients are favorable.

2. Phytoplankton is not uniformly distributed in the oceans of our planet, as documented by satellite pseudocolor images showing the surface chlorophyll concentration [37]. Rather, their density is highest in the circumpolar regions were lower UV-B levels are measured; in equatorial waters, where UV-B levels are much higher, the density is several orders of magnitude lower. Other factors, such as nutrient concentration, temperature, salinity, and so on, also affect growth and abundance of phytoplankton. Exceptions to the general latitudinal pattern include the large concentrations of phytoplankton in the upwelling areas along the continental shelves where high turbidity and particulate matter concentrations attenuate ultraviolet radiation.

## UV-B EFFECTS ON PHYTOPLANKTON

Solar UV-B radiation has been found to affect many physiological and biochemical reactions in microorganisms even though no systematic and long-term measurements are available. At lower doses, the radiation impairs nitrogen incorporation, growth, and the endogenous rhythms that occur in many microorganisms [56; Döhler, this volume]. Recent investigations in Antarctic waters have indicated that photosynthesis can be reduced by as much as 12–16% in the top 10 to 20 m due to increased UV-B radiation [46]. It is estimated that 10–20% of the global primary production of aquatic systems is located in the southern oceans [49], therefore, significant losses in productivity due to UV-B may have global consequences.

## EFFECTS ON PIGMENTATION

UV-B radiation bleaches the photosynthetic pigments [26], an effect that can be observed with the naked eye in some phytoplankton organisms in unfiltered solar radiation on a short time scale [29]. In order to quantify the bleaching effects, cells of the marine *Cryptomonas maculata* were distributed in a solid agar within a quartz cuvette and exposed to solar radiation. Different absorption spectra were measured at increasing exposure times using nonirradiated cells as a reference. These spectra show that the accessory biliprotein pigments are bleached first, next the carotenoids, and the chlorophylls were damaged last.

Fluorescence measurements have also been used to study detrimental effects of UV-B radiation. Under uninhibited conditions part of the excitation energy is lost by fluorescence, a radiative process. Any damage to the photosynthetic apparatus results in a higher fluorescence emission since this energy is not funnelled into the photochemical reactions. A strong increase in fluorescence can be detected after a short exposure to UV-B, indicating that the excitation energy cannot be utilized effectively by the photosynthetic apparatus.

After longer exposure times the fluorescence decreases again, indicating that the absorbing pigments are progressively destroyed by the radiation [9]. A biochemical comparison of the protein extracts from cells before and after UV-B exposure using SDS PAGE shows a drastic loss of several proteins. Further separation of the proteins on an FPLC anion exchange column indicates the loss of a chlorophyll *a*/*c* protein complex. As a result of the loss of proteins and pigments from the photosynthetic apparatus due to ultraviolet radiation, within a rather short time frame the oxygen production decreases drastically [12,51]. Also, since partial removal of the UV-B component still caused a decrease in oxygen production, the inhibitory role of other spectral bands of solar radiation cannot be excluded.

Laboratory investigations have shown short wavelength radiation to have multiple targets. UV-B damages the photosystem II reaction center and specifically affects the herbicide-binding protein associated with photosystem II, which decreases the noncyclic photosynthetic electron transport [41,48]. Another target is the water-splitting site of the photosynthetic apparatus. In addition, the integrity of the membranes is affected because of the decrease in the lipid content.

## UV-B EFFECTS ON MOTILITY, ORIENTATION, AND VERTICAL DISTRIBUTION

As mentioned above, phytoplankton organisms are not equally distributed within the water column; rather they populate zones where light conditions are optimal for growth and survival. Some organisms have flagella or cilia for active propulsion, others utilize buoyancy to adjust their position within the water column. The organisms orient themselves with respect to a number of external chemical and physical stimuli to control these vertical movements with light and gravity regarded as the main external factors [39]. Since light intensity changes with the time of day and is further modified by variations in cloud cover, the organisms move up and down in the water column as the cells constantly readjust to external conditions. Some dinoflagellates are known to migrate as much as 15 m up and down the water column [3]. The active movements are superimposed on the passive relocations within the mixing layer [43].

The mechanism for vertical orientation has been studied in several organisms. *Peridinium gatunense*, a freshwater dinoflagellate which dominates the phytoplankton in Lake Kinneret in Israel, moves to the surface using positive phototaxis at low fluence rates with an optimum at 1 klx. At fluence rates exceeding 16 klx, the cells move diaphototactically (perpendicular to the incident sun rays) which causes them to remain at a specific water depth. A similar behavior has been found in the marine *Peridinium faeroense*. Both in freshwater and in marine flagellates the precision of orientation is affected by even short exposure to solar radiation; histograms of phototaxis taken at increasing intervals during exposure show a decrease in the precision of orientation [22,25].

It is interesting to note that the cells are oriented far longer when UV-B radiation is partially excluded by inserting a UV-B cut-off filter or a layer of artificially-produced ozone. Similar effects using artificial ultraviolet radiation suggest that solar UV-B radiation exerts the strongest inhibition even though effects by UV-A and visible radiation cannot be excluded. Since the position in the water column is of considerable ecological significance, any inhibition of orientation with respect to external stimuli by increased UV-B radiation would negatively affect the chances for growth and survival. And indeed there is growing evidence that these important mechanisms are impaired by short wavelength solar radiation.

In the absence of a light stimulus many organisms orient with respect to the gravitational field of the earth [2]. Like phototaxis, gravitaxis also has been found to be impaired within a few minutes by solar and artificial ultraviolet radiation [23,24]. One interesting effect, however, was found in several dinoflagellates. They stopped moving after excessive UV irradiation and, since the cells are heavier than water, they sank in the water column to a lower depth. This behavior may be an effective escape mechanism to avoid stress situations such as excessive solar radiation [6,7,23]. One *Gymnodinium* spp. even showed positive gravitactic orientation after prolonged UV-B radiation.

To follow the vertical movement patterns of microorganisms, a transparent vertical Plexiglas column was designed with 18 outlets evenly spaced along its length. After thermal equilibration with the surrounding water, samples were taken at regular time intervals from outlets using a peristaltic pump which handled 18 samples in parallel [18]. The cell density in the samples was determined with a real-time image analysis system also used to track flagellates [27]. More recently a new device was used consisting of 20 submerged pumps evenly spaced throughout the top 6 m of the water column to obtain 1 l samples within 1 min from subsequent layers in the water column.

The vertical distribution of the marine dinoflagellate *Prorocentrum micans* during the day shows that the cells accumulate in the top layers in the afternoon, while they are almost randomly distributed during the rest of the day [6,7]. In the early afternoon the precision of negative phototaxis reached a maximum, while the precision of positive phototaxis was found to be highest in the morning and at night. The cells showed a clear negative gravitactic orientation, which was most precise in the early afternoon. Phototaxis and gravitaxis were also studied in the marine dinoflagellates *Amphidinium caterea* and *Peridinium faeroense* [7]. *A. caterea* showed a negative gravitaxis and *P. faeroense* a positive one. The precision of gravitaxis varied dependening on the time of the day and age of culture.

Actively moving phytoplankton organisms such as flagellates are capable of swimming with velocities of up to several body lengths per second. This motility has been shown to be impaired by both artificial and solar radiation [13-15,19,20]. In *Peridinium gatunese* the percentage of motile organisms decreased within a few hours when exposed to solar radiation [22]; blocking part of the UV-B radiation by inserting an artificially-produced layer of ozone prolonged the tolerated exposure times.

A similar behavior was found in marine and freshwater *Cryptomonas* spp., as well as in a number of other flagellates [17]. Also, the linear velocity of the fraction of still motile cells decreased drastically after an initial increase known as photokinesis. In these experiments, inhibition of motility and velocity cannot be attributed to thermal stress by the infrared component of solar radiation since exposure was carried out at constant temperature (20°) in growth chambers with double-layered Plexiglas tops which allowed solar radiation to penetrate, developed by Tevini and coworkers [47]. Similar inhibitory effects were found under artificial ultraviolet radiation that did not contain any visible radiation, indicating that UV-B is the major factor for inhibition induced by solar radiation.

The second argument for the inhibitory role of UV-B is the fact that the organisms tolerated solar radiation significantly longer when the short wavelength component of solar radiation was successively cut off by means of UV-B absorbing filters (Schott WG series) or an artificial layer of ozone. When motility or the swimming velocity are impaired, the organisms are limited in their capability to adapt to the constantly changing parameters in their environment. As a consequence, they find themselves in an environment where they are either exposed to either excessive or inadequate light intensities.

## TARGETS OF UV-B RADIATION AND PROTECTIVE STRATEGIES OF ORGANISMS

Action spectra have been measured for a number of UV-B effects in all kinds of organisms and for very different responses. Because of the close resemblence of some of the action spectra to the absorption spectrum of DNA, it was assumed that DNA is the main target for damaging UV-B radiation. However, some UV-B responses have been found not to be dependent on DNA damage [30]; in one instance, ultraviolet radiation impaired motility within 10 min, which is too short a time to account for protein resynthesis mediated by DNA [12]. Furthermore, photoreactivation could not be observed in some flagellates [31].

Similarly, in filamentous cyanobacteria, UV-B-induced inhibition of motility was very fast and could not be photorepaired [28].

Another potential mechanism by which ultraviolet radiation affects living cells is the involvement of photodynamic reactions. Photodynamic reactions have been found to play a role in some UV-B-induced types of damage in, for instance, the ciliate *Stentor coeruleus* [21], but these mechanisms could be excluded in others [28].

In those cases where DNA damage and photodynamic reactions can be excluded as mechanisms of the UV-B effects, it can be speculated that the target molecules are intrinsic components of the photoreceptor or the motor apparatus. Aromatic amino acids absorb strongly in the UV-B. Thus, proteins are specifically affected by short wavelength radiation. While at sublethal doses these may be regenerated, higher doses eventually cause irreversible damage. This notion was proven by biochemical analysis of the proteins involved in photoperception, motility, and the photosynthetic apparatus. The action spectrum for the inhibition of motility in the photosynthetic flagellate *Euglena* shows a major peak at about 270 nm, a smaller one at 305 nm, and a shoulder at 290 nm [24].

## CONSEQUENCES OF UV-B DAMAGE IN AQUATIC ECOSYSTEMS

Because of the size of the marine ecosystems, even a small loss has noticeable adverse effects [30,31]. Decreases related to increased solar UV-B radiation are difficult to assess since no reliable data are available for the era before the ozone layer started to decrease. Until recently, only sporadic measurements, such as those in Antarctic waters related to the occurrence of the ozone hole, were available. Thus, our present knowledge is far too limited to predict exact losses and damages on a global basis. However, any sizable reductions in primary productivity would lead to a significant reduction in commercial fishing yields [32].

Since some organisms are more sensitive to solar UV-B radiation than others, a change in species composition can be predicted. Any change in population at the base of the food chain is bound to have impacts on the higher trophic levels, especially a change in the size distribution of phytoplankton because the primary consumers select their food according to size rather than species. Thus, UV-B related changes in the phytoplankton populations will lead to altered patterns of predation, competition, diversity, and trophic dynamics.

Since marine phytoplankton is a major biological sink for atmospheric $CO_2$ [10], any decrease in the phytoplankton populations would result in an additional increase in the atmospheric $CO_2$ concentration. A hypothetical 10% decrease in phytoplankton productivity would result in additional atmospheric $CO_2$ equal to that resulting from fossil fuel burning (5 Gt), a factor not yet accounted for by the existing climate change models. This effect would have long-term consequences for the global climate and would enhance the predicted sea level rise [42].

**Acknowledgments**
This work was supported by financial aid from the Bundesminister für Forschung und Technologie (project KBF 57), the European Community (EV5V-CT91-0026) and the State of Bavaria (BayForKlim).

## REFERENCES

1. Baker, K.S., and R.C. Smith. 1982. Spectral irradiance penetration in natural waters. *In* The Role of Solar Ultraviolet Radiation in Marine Ecosystems, J. Calkins (ed.), New York: Plenum Press pp.233-246.

2. Bean, B. 1985. Microbial geotaxis. *In* G. Colombetti and F. Lenci (eds.), Membranes and Sensory Transduction, New York & London: Plenum Press, pp.163-198.

3. Burns, N.M., and F. Rosa. 1980. *In situ* measurements of the settling velocity of organic carbon particles and ten species of phytoplankton. Limnol. Oceanogr. 2:855-864.

4. Cabrera, S., and V. Montecino. 1987. Productividad primaria en ecosistemas limnicos. Arch. Biol. Med. Exp. 20:105-116.

5. Cullen, J.J., and M.P. Lesser. 1991. Inhibition of photosynthesis by ultraviolet radiation as a func-

tion of dose and dosage rate: Results for a marine diatom. Marine Biol. (in press).

6. Eggersdorfer, B., and D.-P. Häder. 1991. Phototaxis, gravitaxis and vertical migrations in the marine dinoflagellate, *Prorocentrum micans*. Eur. J. Biophys. 85:319-326.

7. Eggersdorfer, B., and D.-P. Häder. 1991. Phototaxis, gravitaxis and vertical migrations in the marine dinoflagellates, *Peridinium faeroense* and *Amphidinium caterii*. Acta Protozool. 30:63-71.

8. Ekelund, N.G.A. 1991. The effect of UV-B radiation on dinoflagellates. J. Plant Physiol. 138:274-278.

9. Fischer, M., and D.-P. Häder. 1992. UV effects on the pigmentation of the flagellate *Cynaophora paradoxa*—biochemical and spectroscopic analysis. Europ. J. Protist. 28:163-169.

10. Gaundry, A., P. Monfray, G. Polian, and G. Lanabert. 1987. The 1982-1983 El Nino: A 6-billion-ton $CO_2$ release. Tellus 39B:209-213.

11. Gieskes, W.C., and G.W. Kraay. 1990. Transmission of ultraviolet light in the Weddell Sea: Report on the first measurements made in Antarctic. Biomass Newsletter 12:12-14.

12. Häberlein, A., and D.-P. Häder. 1992. UV effects on photosynthetic oxygen production and chromoprotein composition in the freshwater flagellate *Cryptomonas* S2. Acta Protozool. 31:85-92.

13. Häder, D.-P. 1985. Effects of UV-B on motility and photobehavior in the green flagellate, *Euglena gracilis*. Arch. Microbiol. 141:159-163.

14. Häder, D.-P. 1986. Effects of solar and artificial UV irradiation on motility and phototaxis in the flagellate, *Euglena gracilis*. Photochem. Photobiol. 44:651-656.

15. Häder, D.-P. 1986. The effect of enhanced solar UV-B radiation on motile microorganisms. *In* R.C. Worrest and M.M. Caldwell (eds.), Stratospheric Ozone Reduction, Solar Ultraviolet Radiation and Plant Life, Berlin, Heidelberg, New York: Springer Verlag pp.223-233.

16. Häder, D.-P. 1991. Phototaxis and gravitaxis in *Euglena gracilis*. *In* F. Lenci, F. Ghetti, G. Colombetti, D.-P. Häder, and P.-S. Song (eds.), Biophysics of Photoreceptors and Photomovements in Microorganisms, New York & London: Plenum Press pp.203-221.

17. Häder, D.-P. 1993. Risks of enhanced solar ultraviolet radiation for aquatic ecosystems. Progr. Phycol. Res., in press.

18. Häder, D.-P., and K. Griebenow. 1988. Orientation of the green flagellate, *Euglena gracilis*, in a vertical column of water. FEMS Microbiol. Ecol. 53:159-167.

19. Häder, D.-P., and M. Häder. 1989. Effects of solar radiation on photoorientation, motility and pigmentation in a freshwater *Cryptomonas*. Botanica Acta 102:236-240.

20. Häder, D.-P., and M.A. Häder. 1989. Effects of solar and artificial radiation on motility and pigmentation in *Cyanophora paradoxa*. Arch. Microbiol. 152:453-457.

21. Häder, D.-P., and M.A. Häder. 1991. Effects of solar radiation on motility in *Stentor coeruleus*. Photochem. Photobiol. 54:423-428.

22. Häder, D.-P., M. Häder, S.-M. Liu, and W. Ullrich. 1990. Effects of solar radiation on photoorientation, motility and pigmentation in a freshwater *Peridinium*. BioSystems 23:335-343.

23. Häder, D.-P., and S.-L. Liu. 1990. Effects of artificial and solar UV-B radiation on the gravitactic orientation of the dinoflagellate, *Peridinium gatunense*. FEMS Microbiol. Ecol. 73:331-338.

24. Häder, D.-P., and S.-M. Liu. 1990. Motility and gravitactic orientation of the flagellate, *Euglena gracilis*, impaired by artificial and solar UV-B radiation. Curr. Microbiol. 21:161-168.

25. Häder, D.-P., S.-M. Liu, M. Häder, and W. Ullrich. 1990. Photoorientation, motility and pigmentation in a freshwater *Peridinium* affected by ultraviolet radiation. Gen. Physiol. Biophys. 9:361-371.

26. Häder, D.-P., E. Rhiel, and W. Wehrmeyer. 1988. Ecological consequences of photomovement and photobleaching in the marine flagellate *Cryptomonas maculata*. FEMS Microbiol. Ecol. 53:9-18.

27. Häder, D.-P., and K. Vogel. 1991. Simultaneous tracking of flagellates in real time by image analysis. J. Math. Biol. 30:63-72.

28. Häder, D.-P., M. Watanabe, and M. Furuya. 1986. Inhibition of motility in the cyanobacterium, *Phormidium uncinatum*, by solar and monochro-

matic UV irradiation. Plant Cell Physiol. 27:887-894.

29. Häder, D.-P., and R.C. Worrest. 1991. Effects of enhanced solar ultraviolet radiation on aquatic ecosystems. Photochem. Photobiol. 53:717-725.

30. Häder, D.-P., R.C. Worrest, and H.D. Kumar. 1989. Aquatic ecosystems. UNEP Environmental Effects Panel Report, pp.39-48.

31. Häder, D.-P., R.C. Worrest, and H.D. Kumar. 1991. Aquatic ecosystems. UNEP Environmental Effects Panel Report, pp.33-40.

32. Hardy, J., and H. Gucinski. 1989. Stratospheric ozone depletion: implications for marine ecosystems. Oceanogr. Mag. 2:18-21.

33. Helbing, E.W., V. Villafane, M. Ferrario, and O. Holm-Hansen. 1991. Impact of natural ultraviolet radiation on rates of photosynthesis and on specific marine phytoplankton species. Marine Ecology Progress Series.

34. Houghton, R.A., and G.M. Woodwell. 1989. Global climatic change. Scientific Amererican 261:18-26.

35. Jerlov, N.G. 1970. Light--general introduction. *In* O. Kinne (ed.), Marine Ecology, Vol. 1, pp.95-102.

36. Karentz, D. 1991. Ecological considerations of Antarctic ozone depletion. Antarctic Sci. 3:3-11.

37. Lohrenz, S.E., R.A. Arnone, D.A. Wiesenburg, and I.P. DePalma. 1988. Satellite detection of transient enhanced primary production in the western Mediterranean Sea. Nature 335:245-247.

38. Maske, H. 1984. Daylight ultraviolet radiation and the photoinhibition of phytoplankton carbon uptake. J. Plankton Res. 6:351-357.

39. Nultsch, W., and D.-P. Häder. 1988. Photomovement in motile microorganisms II. Photochem. Photobiol. 47:837-86.

40. Raven, J.A. 1991. Responses of aquatic photosynthetic organisms to increased solar UVB. J. Photochem. Photobiol. B: Biol. 9:239-244.

41. Renger, G., M. Völker, H.J. Eckert, R. Fromme, S. Hohm-Veit, and P. Gräber. 1989. On the mechanisms of photosystem II deterioration by UV-B irradiation. Photochem. Photobiol. 49:97-105.

42. Schneider, S.H. Sept. 1989. The changing climate. Sci. Am. 261:38-47.

43. Smith, R. 1989. Ozone, middle ultraviolet radiation and the aquatic environment. Photochem. Photobiol. 50:459-468.

44. Smith, R.C., and K.S. Baker. 1978. Penetration of UV-B and biologically effective dose-rates in natural waters. Photochem. Photobiol. 29:311-323.

45. Smith, R.C., K.S. Baker, O. Holm-Hansen, and R. Olson. 1980. Photoinhibition of photosynthesis in natural waters. Photochem. Photobiol. 31:585-592.

46. Smith, R.C., B.B. Prézelin, K.S. Baker, R.R. Bidigare, N.P. Boucher, T. Coley, D. Karentz, S. MacIntyre, H.A. Matlick, D. Menzies, M. Ondrusek, Z. Wan, K.J. Waters. 1992. Ozone depletion: UV radiation and phytoplankton biology in Antarctic waters. Science 255:952-959.

47. Tevini, M., P. Grusemann, and G. Fieser. 1988. Assessment of UV-B stress by chlorophyll fluorescence analysis. *In* H.K. Lichtenthaler (ed.), Applications of Chlorophyll Fluorescence, Kluwer Academic Publishers. Dordrecht, pp.229-238.

48. Tevini, M., A.H. Teramura, G. Kulandaivelu, M.M. Caldwell, and L.O. Björn. 1989. Terrestrial Plants. UNEP Environmental Effects Panel Report, pp.25-37.

49. Voytek M.A. 1990. Adressing the biological effects of decreasing ozone in the Antarctic environment. Ambio 19:52-61.

50. Worrest, R.C. 1982. Review of literature concerning the impact of UV-B radiation upon marine organisms. *In* J. Calkins (ed.), The Role of Solar Ultraviolet Radiation in Marine Ecosystems, New York: Plenum Press pp.429-457.

51. Zündorf, I., and D.-P. Häder. 1991. Biochemical and spectroscopic analysis of UV effects in the marine flagellate *Cryptomonas maculata*. Arch. Microbiol. 156:405-411.

# UV-EFFECTS ON THE NITROGEN METABOLISM OF MARINE PHYTOPLANKTON AND ADAPTATION TO UV RADIATION

G. Döhler

Botanisches Institut der J.W. Goethe-Universität, Siesmayerstraße 70
D-60323 Frankfurt a.M., Germany

Key words: Antarctic, nitrogen metabolism, phytoplankton, UV radiation

## ABSTRACT

The impact of UV-B (290-320 nm) radiation on uptake of $^{15}$N-ammonia and $^{15}$N-nitrate of natural phytoplankton and sea-ice algae of the Weddell Sea was studied during the Polarstern Cruise (EPOS III, Leg 3) 1989. Assimilation rate of $^{15}NH_4Cl$ was higher and more affected by UV-B radiation than the uptake of $K^{15}NO_3$ by natural phytoplankton populations and pure cultures. Pool sizes of glutamate and glutamine varied in dependence on the used inorganic source and after UV-B exposure. Utilization of $^{15}NO_3^-$ and $^{15}NH_4^+$ as well as the inhibitory effect of UV-B irradiance on sea-ice algae were dependent on the temperature: The damage by UV-B increased with higher temperatures. Batch cultures of the Antarctic diatom *Odontella weissflogii*, were less sensitive to UV-B irradiance than the temperate *Ditylum brightwellii*, *Lithodesmium variabile*, *Odontella sinensis*, and *Thalassiosira rotula* isolated from the North Sea, Germany. Fluorescence spectra of both tested diatoms indicate a damaging effect of the photosystem II reaction center by UV-B. All the tested diatom species can produce heat-shock proteins (HSPS) of the 70 kD family by *in vivo* labeling with [$^{35}$S]-methionine. *Odontella weissflogii*, a species relatively insensitive to UV-B irradiance, did not synthesize UV induced HSPS, whereas the UV sensitive diatom Odontella sinensis produced all the observed HSPS after UV-B exposure. A specific UV-stress protein of 43 kD was found. Results were discussed with reference to an UV-protecting system and the primary target of UV-B exposure.

## INTRODUCTION

Since 1979, a marked depletion of stratospheric ozone has been observed over Antarctica during austral summer from early October on. This caused an increase in incident UV radiation. Measurements of ambient solar spectra from the Palmer Station, Antarctica, during the "ozone hole" in 1988 showed an enhancement in radiation at wavelengths <310 nm [1, 2]. Solar UV radiation can penetrate seawater to depths that might affect natural phytoplankton communities [3, 4]. Recently, much work has been done to study the effect of ultraviolet radiation on various metabolic processes of marine phytoplankton mainly from the temperate region and partly from the Antarctic Ocean [4, 5, 6, 7, 8, 9]. Enhanced levels of UV-B resulted in a decrease of cellular pigment and protein contents as well as in primary production measured by the $^{14}$C-technique [6, 10, 11, 12, 13].

Similar results were obtained with Antarctic phy-

NATO ASI Series, Vol. I 18
Stratospheric Ozone Depletion/
UV-B Radiation in the Biosphere
Edited by R. H. Biggs and M. E. B. Joyner

toplankton at the Palmer Station [7,14]. These authors observed a significant reduction in photosynthetic pigmentation and rates of primary production of Antarctic phytoplankton exposed to ambient and enhanced UV conditions.

Variations in the light regime are known to cause a species-dependent response with regard to the pigmentation and biomass production [15], and this may also be true for UV radiation. A species-dependent reaction to UV-B exposure was, for instance, observed in microalgae [16, 17, 18], and this may cause variations in the species composition of the phytoplankton assemblage. Such UV-induced changes in the community composition of the aquatic ecosystem may result in a variation of the quantity and quality of the food for the primary consumers.

Less information is available on the impact of ambient or enhanced levels of UV radiation on the nitrogen metabolism of phytoplankton and ice-algae. Therefore, we have studied the inhibitory effect of UV-radiation at different wavebands on uptake and assimilation of $^{15}$N-ammonia and $^{15}$N-nitrate by natural phytoplankton and isolated species under laboratory and field conditions. Additionally, the impact of UV radiation on the Antarctic diatom *Odontella weissflogii* and temperate diatoms on synthesis of stress proteins was also investigated.

## MATERIALS AND METHODS

The Antarctic diatom *Odontella weissflogii* (Janisch) Grunow was collected at Station No. 265 of the Weddell Sea by Prof. Dr. G. Döhler during the Polarstern Cruise EPOS III, Leg 3 January/February 1989. The temperate species *Ditylum brightwellii* (West) Grunow, *Lithodesmium variabile* (Tanako), *Odontella sinensis* (Greville) Grunow, and *Thalassiosira rotula* (Meunier) were isolated from the North Sea near Heligoland, Germany. Algae were grown in an artifical seawater medium according to v. Stosch and Drebes [19] under controlled laboratory conditions: light/dark rhythm of 12:12 h (intensity 7.2 W $m^2$, Osram lamps L 40 W (25-1)), bubbling with normal air (0.035 vol.% $CO_2$) at 4°C or 18°C, respectively. Algae were harvested during exponential growth ($10^3$ cells $ml^{-1}$) for the experiments.

The samples were placed into special plexiglass cuvettes and bubbled with normal air (0.035 vol.% $CO_2$) at a temperature of 2° or 18°C. One cuvette was not transparent for UV-B (Plexiglass No. 233, Röhm GmbH, Darmstadt, Germany) and served as control whereas the other vessel allowed UV transmission (Plexiglass No. 2458, Röhm GmbH, Darmstadt, Germany). Slide projectors were used to provide white light (240 W $m^{-2}$). The light intensities were measured with an IL 1700 radiometer and detectors No. 5436 (UV-B) and No. 5470 (white light) of International Light, Inc., USA. UV irradiance has been performed with special Philips lamps (TL 20/12, TL 40/01) and cut-off filters (WG 295, 305, and 320). The different UV-B doses were obtained by changing the exposure time or distance from the light source to the algal sample.

After a preillumination period with white light the nitrogen compound ($^{15}NH_4Cl$, 96 atom% or $K^{15}NO_3$ 95.6 atom%) was added and UV-B exposure started. Samples were collected by a syringe at different light or dark periods, filtered on GF/C Whatman filters, and heated for 2h at 60°C. The samples were then combusted according to the Dumas method. The $^{15}$N content was determined with an atomic emission spectrometer (Jasco model N 150). All the procedures have been described in more details elsewhere [20, 21].

Chlorophyll a and chlorophyll $c_1 + c_2$ contents were measured in 80% acetone according to the method of Jeffrey and Humphrey [22] with a spectrophotometer Lambda 2 (Perkin Elmer). Estimation of protein was performed after sonification (2 x 30 sec, 20 kHz) and centrifugation (20 min, 30,000g) according to Bradford [23]. Fluorescence emission spectra of the algae were determined in liquid nitrogen (77°K) with special equipment (O-SMA) after exitation at 435 or 480 nm in cooperation with Dr. Marquardt. For more details see Rehm et al. [24]. Separation and analysis of free amino acids from algal extracts were carried out after precolumn derivatization with o-phtaldialdehyde by reversed-phase high-performance liquid chromatography (HPLC) with a Beckman (Model 342). For more details see Döhler [25].

The two-dimensional gel electrophoresis was performed according to the procedure described by O'Farrell [26] using the capillary system of Biometra (IEF Model G 43 and Minigel G 42, Biometra, Göttingen, Germany). Gels were processed for fluorography by the method described by Bonner and Laskey [27].

## RESULTS

### IMPACT OF UV-B ON PHYTOPLANKTON FROM THE WATER COLUMN

The influence of UV-B irradiance on uptake of inorganic nitrogen compounds has been studied under controlled laboratory conditions during the "Polarstern" Cruise EPOS III, Leg 3, 1989. Data presented in Figure 1 and Table 1 are representative of that for the other experiments. Generally, assimilation of $^{15}$N-ammonia was significantly higher than utilization of $^{15}$N-nitrate. This indicates an important role of $NH_4^+$ for the nitrogen metabolism of Antarctic phytoplankton. This can be also supported by our observation that enhanced concentrations of ammonia appeared after remineralization of krill feces and can affect the utilization of $^{15}$N-nitrate. Inhibitory effects by $NH_4^+$ are well known and species-dependent (see Dortch's review [28]). A predominant assimilation of ammonia has been described for phytoplankton from the North Sea (Döhler and Stolter [17]) and for Antarctic plankton of the Scotia Sea (Rönner et al. [29]).

The strong sensitivity to UV-B radiation of the nitrogen assimilation by phytoplankton species of large sizes (e.g. *Corethron*, *Chaetoceros criophilum*) is shown in Fig. 1. A marked depression of the uptake rates exists after half UV-B exposure. The $^{15}NO_3^-$

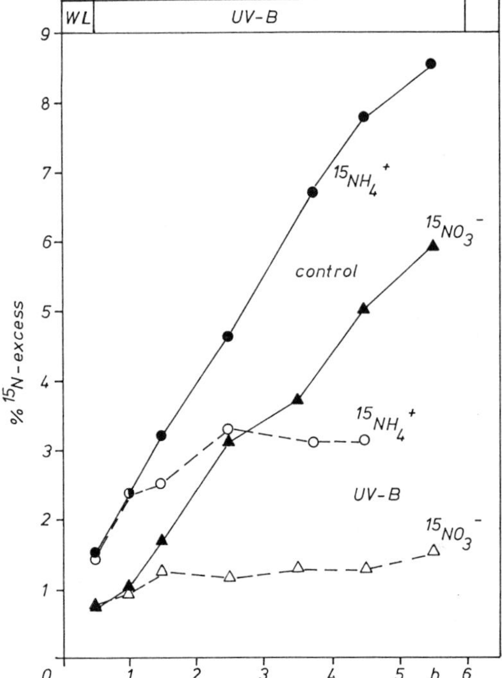


Fig. 1. Time course of $^{15}NH_4^+$ and $^{15}NO_3^-$ uptake of natural phytoplankton from Station 274 of the Weddell Sea (12°W, 71°S) at +2°C. White light intensity: 2.4 x 10-2 W cm$^{-2}$. Values are expressed as $^{15}$N enrichment (% $^{15}$N excess). Final concentrations of $^{15}$N labeled nitrogen were 60 mM. Salinity 34.4 . Main autotrophic organisms were *Corethron criophilum* (73%), *Chaetoceros criophilus* (21%), *Rhizosolenia alata* (3%), and *Dactyliosolen* (3%).

Table 1. Effect of UV-B radiation on pool sized of free amino acids and amides of extracts of sea ice microalgae from Station 219 (47°W, 61.5°S), Weddell Sea. Algae were collected after 2.5 h UV-B exposire at 2° and 10° C and extracted for HPLC analysis. For further details, see Materials and Methods.

| | +2°C | | | | +10° C | | | |
|---|---|---|---|---|---|---|---|---|
| | control | | control | | control | | control | |
| amino acid | μg/ml | % | μg/ml | % | μg/ml | % | μg/ml | % |
| aspartate | 2.0419 | 13.02 | 2.0105 | 12.74 | 1.7100 | 12.41 | 1.0713 | 13.87 |
| glutamate | 4.6058 | 29.38 | 4.4031 | 27.90 | 3.3769 | 24.50 | 4.4876 | 30.05 |
| asparagine | 0.1504 | 0.96 | 0.1075 | 0.68 | 0.0817 | 0.59 | 0.1647 | 1.10 |
| serine | 0.8414 | 5.37 | 1.4502 | 9.19 | 1.0966 | 7.96 | 0.1600 | 1.07 |
| glutamine | 1.6882 | 10.77 | 1.0263 | 6.50 | 1.1614 | 8.43 | 1.2621 | 8.45 |
| glycine | 0.6817 | 4.35 | 1.0322 | 6.54 | 0.7158 | 5.19 | 0.6308 | 4.22 |
| threonine | 0.3178 | 2.03 | 0.4855 | 3.08 | 0.4673 | 3.39 | 0.3982 | 2.67 |
| arginine | 2.1579 | 13.76 | 1.6898 | 10.71 | 1.9437 | 14.10 | 2.0868 | 13.97 |
| alanine | 1.2652 | 8.13 | 1.2785 | 8.10 | 0.9863 | 7.16 | 1.2243 | 8.20 |
| tyrosine | 0.2618 | 1.67 | 0.3521 | 2.23 | 0.2939 | 2.13 | 0.3271 | 2.19 |
| methionine | 0.0806 | 0.51 | 0.0626 | 0.40 | 0.0441 | 0.32 | 0.0341 | 0.23 |
| valine | 0.3501 | 2.23 | 0.4078 | 2.58 | 0.3714 | 2.69 | 0.4216 | 2.82 |
| phenylalanine | 0.1897 | 2.23 | 0.1862 | 1.18 | 0.1482 | 1.08 | 0.1438 | 0.96 |
| isoleucine | 0.1952 | 1.24 | 0.2587 | 1.64 | 0.2468 | 1.79 | 0.2675 | 1.79 |
| leucine | 0.2795 | 1.78 | 0.3453 | 2.19 | 0.3496 | 2.54 | 0.3933 | 2.63 |
| lysine | 0.5617 | 3.58 | 0.6832 | 4.33 | 0.7901 | 5.73 | 0.8593 | 5.75 |
| Total | 15.6789 | | 15.7795 | | 13.7838 | | 14.9325 | |

utilization was more affected than that of $^{15}NH_4^+$. In contradiction to findings of other experiments, a high nitrate assimilation could be measured. Pools of glutamate and glutamine are reduced after UV-B. The different response to UV-B exposure can be attributed to the different composition of the phytoplankton assemblages. Similar data were observed with pure cultures of *Phaeocystis pouchetii* and of phytoplankton consisting mainly of this haptophycean alga [30].

## IMPACT OF UV-B RADIATION ON ICE ALGAE

The dominant sea-ice algae were *Nitzschia* sp. (7-18% cells), *Dactyliosolen* sp. (48-53% cells), *Eucampia antarctica* (3-5% cells), and *Thalassiosira antarctica* (2% cells). Main photoautotrophic organisms were small species of *Rhizosolenia*, *Nitzschia*, *Dactyliosolen*, *Phaeocystis pouchetii*, *Eucampia antarctica*, and *Coscinodiscus* as well as nanoplankton collected at Station 265.

Significant damaging effects by UV-B irradiance on uptake of $^{15}$N-ammonia and $^{15}$N-nitrate in ice-algae from Station 265 has been observed. The strongest inhibition was detected under low white light conditions. Results indicate a predominant uptake of $^{15}NH_4^+$ which is in agreement with findings of other microalgae [17 and others]. However, higher uptake rates of $^{15}NO_3^-$ by ice-algae were measured compared to those of the phytoplankton collected at the same station in the water column.

Figure 2 presents data of UV-B radiation on $^{15}$N-ammonia uptake by ice algae assemblages (Station 219) at different temperatures. Results indicate an inhibitory effect of UV-B with increasing temperature: no reduction was detectable at 2°C whereas a decrease of $^{15}NH_4^+$ uptake was found at 5°C and 8°C. Maximal uptake rates were observed at 5°C. This is in agreement with studies of pure cultures of the Antarctic *Rhizosolenia truncata* [31].

Pool sizes of free amino acids were estimated after 3h of UV-B exposure from ice-microalgae at temperatures of 2°C and 10°C (Table 1). Total concentration of free amino acids of the algae exposed to UV-B at 2°C did not change compared to values of non-UV-B treated cells, whereas a slight increase was found at 10°C. After UV-B radiation at 2°C, an increase in the pools of glycine and serine, and a decrease in glutamine were measured. No effect of UV-B radiation was found on the glutamine pool at 10°C, but slightly enhanced values of glutamate, asparagine, and serine were measured. The observed variations in the pool sizes indicate damage to the enzymes of the nitrogen metabolism by UV-B irradiance.

## EFFECT OF UV-B RADIATION OF DIFFERENT UV WAVEBANDS ON UPTAKE OF $^{15}$N-AMMONIUM

In another series of experiments the effect of UV radiation of different wavebands on the uptake of $^{15}$N-ammonium of several marine diatoms has been tested using cut-off filters. The findings illustrated in Figure 3 are representative of those obtained in other studies. The greatest damage was found after irradiation with UV at the shorter wavelenghths (WG 295). The reduction of $^{15}$N-ammonium uptake by *Lithodesmium variabile* was less pronounced after radiation with special UV-B lamps (Philips TL 40/01) or using cut-off filters WG 320 in connection with TL 20/12 Philips lamps. Summarizing, the strongest inhibition could be observed after irradiation with UV-B combined with part UV-A. Practically no damage was found after UV-A exposure and only moderate damage after UV-B exposure.

## EFFECT OF UV-B RADIATION ON FLUORESCENCE SPECTRA

In another series of experiments, the impact of UV-B radiation (5 h) on fluorescence emission spectra of *O. sinensis* and *O. weissflogii* was studied. Results presented in Figure 4 show a shift of the maximum to the longer wavelengths of *O. sinensis* cells treated with UV-B; fluorescence maximum of non-UV-B-exposed algae (=control) was at 689 nm and that of UV-B irradiated cells (=UV) at 693 nm. However, no change of the fluorescence maximum of *O. weissflogii* was found after UV-B irradiation under the same conditions. These findings and the results on the nitrogen metabolism indicate a different response to UV-B radiation by the tested *Odontella* species.

Additionally, the absorption spectra of both species were estimated (Figure5). Absorption spectra of the acetone extracts show a peak at 337 nm; this maximum is significantly pronounced in the Antarctic *O. weissflogii* and only slightly pronounced after UV-B irradiation of *O. sinensis*. This can be attributed to a mycosporine-like amino acid (probably a UV protecting substance) [32, 33].

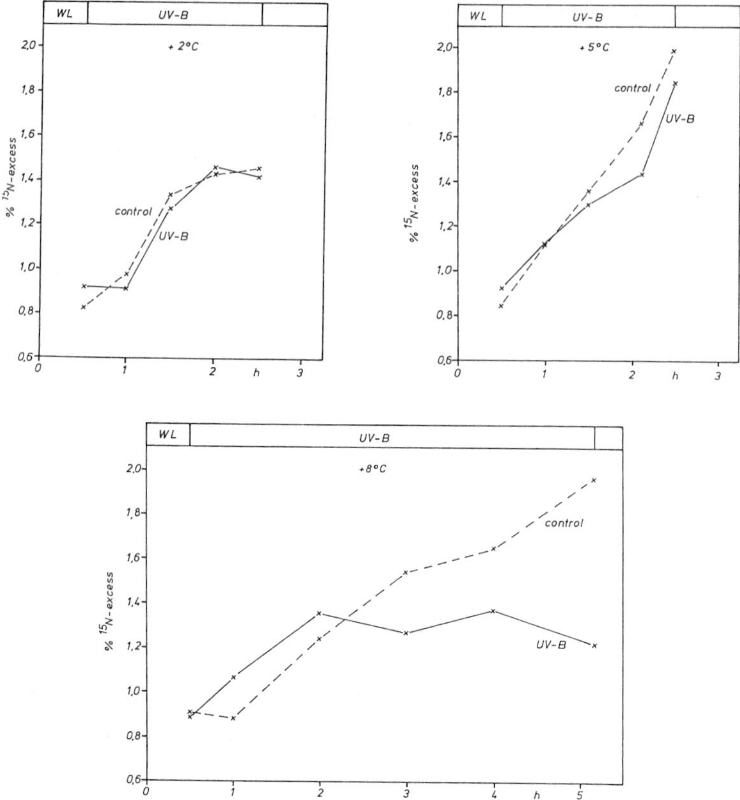


Fig. 2. Effect of UV-B radiation on $^{15}NH_4^+$ uptake of sea-ice algae from Station 219 (47°W, 61.5°S) of the Weddell Sea at +2, +5°, and +8°C. Control not UV-B-irradiated or UV-B-exposed cells. Final concentration of $^{15}NH_4Cl$ was 50 mM. Main autotrophic organisms were *Nitzschia* sp. (15%), *Dactyliosolen* (35%), *Coscinodiscus* sp. (30%), *Rhizosolenia* sp. (9%), and *Eucampia antarctica* (3%). Salinity 29 , intensities of white light 2.4 x $10^{-2}$ W $cm^{-2}$ and UV-B 1.4 x $10^{-2}$ W $cm^{-4}$.

## EFFECT OF UV-B RADIATION ON SYNTHESIS OF SPECIAL PROTEINS

After inoculation with L-[$^{35}$S]-methionine, SDS-PAGE gel electrophoresis showed an increase of labelling in 71 kD, 50 kD, and 40 kD proteins of *O. sinensis* after heat shock or UV-B irradiation. A 30 min treatment at 30°C or 3 h UV-B irradiation at 18°C led to synthesis of all the known heat shock proteins in *O. sinensis*. The 50 kD protein might be the LSU of Rubisco. Labeling was enhanced after 3 h UV-B exposure, whereas a reduction was found after 3 h heat treatment. Similar results were obtained with synthesis of the 40 kD protein which could not be detected after 3 h heat shock. The pattern of proteins, separated by two-dimensional electrophoresis, supported our preliminary findings. The response to heat and UV stress of the temperate *Ditylum brightwellii* was similar to that of *O. sinensis*. In another series of experiments, the Antarctic diatom *O. weissflogii* was exposed to heat shock (16°C) and UV-B irradiation under conditions similar to those of *O. sinensis*. Labeling of a 75 kD protein was detected only after heat treatment. No further radioactivity was observed in other proteins; UV-B irradiance had no effect on additional protein biosynthesis (see Table 2). The same response to the stress conditions existed in *Thalassiosira rotula*, a less sensitive temperate diatom isolated from samples from the North Sea.

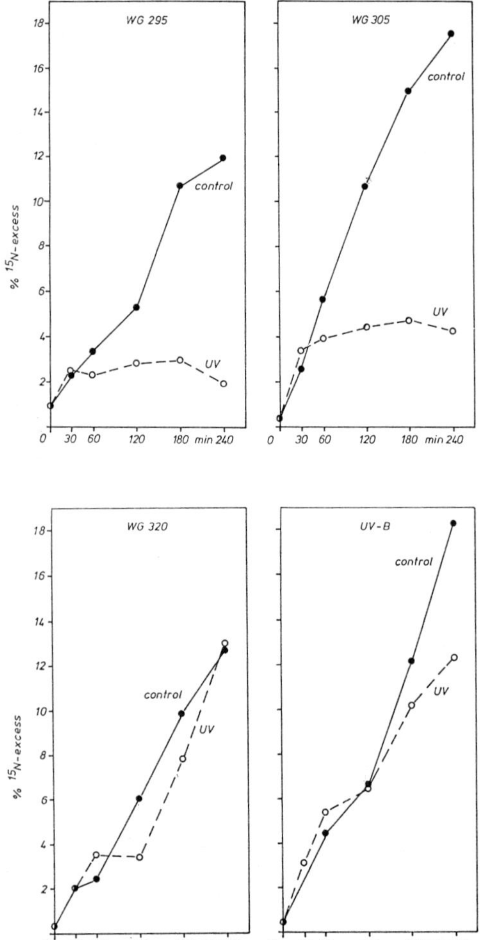


Fig. 3. Impact of UV radiation of different wavebands on uptake of $^{15}$N-ammonium of *Lithodesmium variabile* under controlled laboratory conditions. Special UV-B lamps (Philips TL 20/01) or Philips 40/12 with cut-off filters WG 295, WG 305, and WG 320 were used. Further details in Material and Methods.

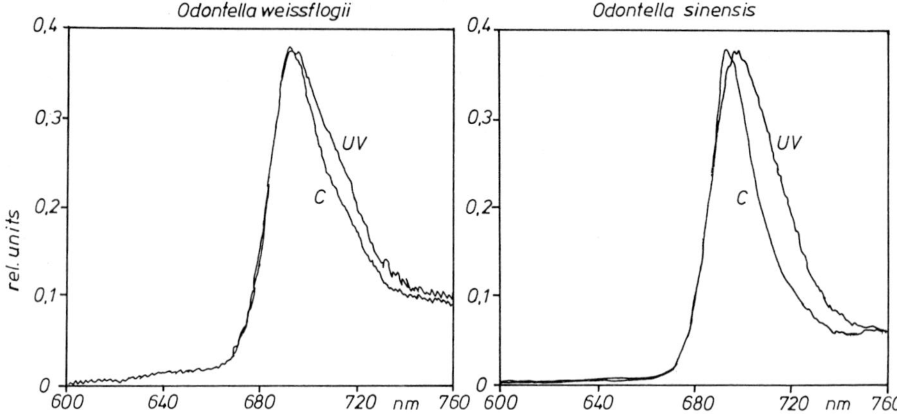


Fig. 4. Impact of UV-B radiation (h) on fluorescence spectra of *Odontella weissflogii* and *O. sinensis*. C: not UV-B exposed cells. UV: cells harvested after 5 h UV-B irradiance (1.2 W $m^{-2}$). Exitation at 480 nm and measurement of emission spectra under liquid nitrogen (77°K). Fluorescence spectrum of UV-B treated *O. sinensis* was shifted to longer wavelengths (maximum of control: 689 nm; exposed cells: 693 nm).

## DISCUSSION

It is well known that uptake rates of $^{15}$N-ammonium by natural phytoplankton from different habitats and unialgal cultures of several microalgae are usually much higher than uptake of $^{15}$N-nitrate [11, 17]. Contradictory results were observed with Antarctic phytoplankton from the water column (Fig. 1). Utilization of ammonia by pure cultures was more sensitive to solar UV and UV-B radiation than that of nitrate. The different response to UV-B of *O. weissflogii* compared to *O. sinensis* can also be

Table 2. Pattern of [$^{35}$S]-labeled proteins of several marine diatoms induced by heat shock (30 or 16oC) or UV-B radiation (3 hr duration). O.s.= *Odontella sinensis*; O.w. - *O. weissfolgii*; T.r. = *Thalassiosira rotula*; D.b. = *Ditylum brightwellii*. For further details see Materials and Methods.

| TRT | O.s. | O.w. | T.r. | D.b. |
|---|---|---|---|---|
| hsp | 71 kD | 75 kD | 72 kD | not tested |
| | 25 kD | — | — | not tested |
| UV-B | 71 kD | — | — | 80 kD |
| | 40 kD | — | — | 40 kD |
| | 25 kD | — | — | 25 kD |

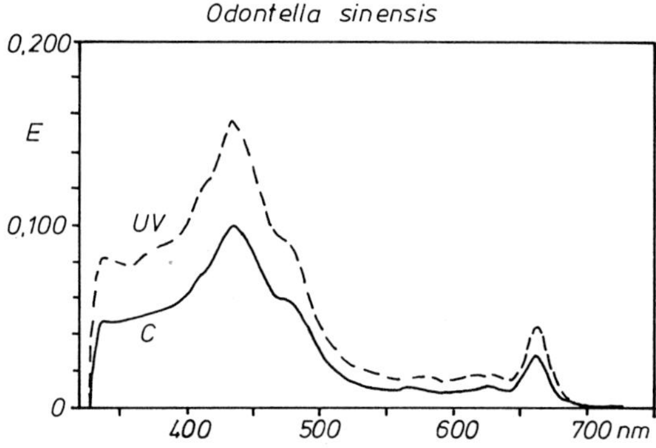


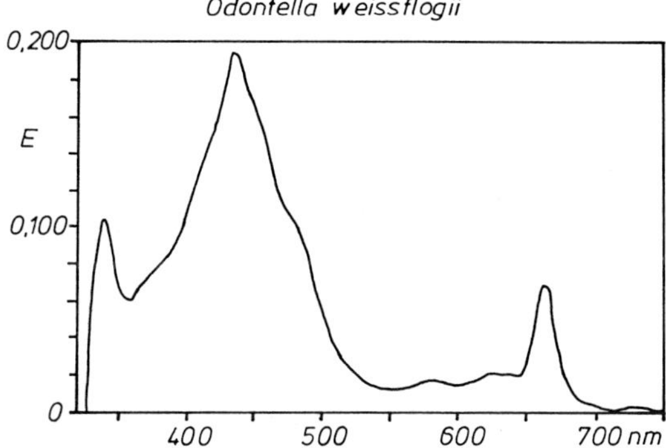


Fig. 5. Absorption spectra of acetone extracts (80%) of *Odontella sinensis* after 5 h UV-B irradiance (UV), not UV radiated cells (C) and *O. weissflogii* not UV-B exposed.

documented by the differences of the Lineweaver-Burk plot: non-competitive inhibition for $^{15}NH_4^+$ and competitive inhibition for $^{15}NO_3^-$. The pattern of $^{15}N$ labeled free amino acids and the pool sizes were different after additional irradiation with blue and red light. These findings can be interpreted as distinct uptake systems for these inorganic nitrogens. The variation of uptake of $^{15}NH_4^+$ and $^{15}NO_3^-$ as an effect of UV-B irradiance, depending on the wavelengths (red or blue light), can probably be explained by that way, too.

The reason for the reduced assimilation of inorganic nitrogen compounds by phytoplankton after UV exposure might include:

1. reduction in the supply of energy (ATP) and carbon skeletons for amino acid biosynthesis,
2. damage to protein synthesis and activities of key enzymes of the carbon and nitrogen metabolism
3. variation in the pattern of amino acids, lipids, and fatty acids as well as inhibition of biosynthesis
4. inhibition of the regulatory mechanisms.

The ATP content of Antarctic phytoplankton of

the Weddell Sea was markedly reduced with increasing exposure time to UV-B irradiance [34]. A parallel reduction in the uptake of inorganic nitrogen was observed using the same phytoplankton samples (Figure 1).

Results of the fluorescence studies (Figure 4) also showed the different UV-B sensitivity of the tested *Odontella* species. There was practically no effect on *O. weissflogii* and a shift of the fluorescence (77°K) maximum from *O. sinensis*. These findings can be interpreted as a damaging effect of UV-B radiation on the donor/acceptor system of the photosystem II reaction center. This is in agreement with the results of Renger et al. [35].

Generally, the marine diatoms isolated from the North Sea were more affected by UV-B irradiance under laboratory conditions than the Antarctic species *O. weissflogii* from the Weddell Sea. A possible explanation for the species-dependent response to UV-B radiation might be a special adaptation to the different environmental conditions, for example, by way of a UV-protecting system. Figure 5 indicates that the less-UV-sensitive diatom *O. weissflogii* exhibits the mycosporine-like amino acid palythenic acid. Karentz et al. [36] found mycosporine-like amino acids in Antarctic phytoplankton populations. The UV-sensitive temperate species *O. sinensis* can synthesize specific UV stress proteins (43 kD) and by that way presumably survive under enhanced UV levels. The function of UV stress proteins are not yet known.

The impact of UV-B on phytoplankton and microalgae is complex and the question arises, what might be the primary target. The photosystem II reaction center as well as the biosynthesis of DNA/RNA may be of importance. Further studies should elucidate the unanswered questions.

The *in vivo* labeling experiments (Table 1) clearly show variations in the pattern of proteins; no HSPS or 40 kD protein was produced by the less-UV-sensitive Antarctic *O. weissflogii*. UV-B radiation induced synthesis of HSPS of the 70 kD family in those species sensitive to UV-B irradiance. Nicholson and Howe [37] found cytoplasmatic proteins of 71 kD and 65 kD after UV irradiation of *Chlamydomonas reinhardtii*. Gromoff et al. [38] observed light-induced 68 kD, 70 kD, and 80 kD proteins, which could be attributed to heat shock proteins, in dark-grown *Chlamydomonas* cells.

Summarizing our results, the 40/43 kD protein could be detected after UV-B exposure only and might be a specific UV stress protein. This protein is probably identical to the 38 kD protein of prokaryotes induced by UV irradiance. Because of this, *Odontella sinensis* is a suitable organism for identifying the damaging effects of UV-B.

The species-dependent response to UV-B irradiance might be due to a specific adaptation to the environmental conditions. The Antarctic *Odontella weissflogii* is probably adapted to enhanced UV level found during the Austral spring via synthesis of LMW stress proteins found in regular UV radiation. On the other hand, mycosporine-like amino acids detected in several organisms can probably act as a protective system against the effects of UV-B. The *de novo* synthesis of the specific UV stress proteins might be a protecting mechanism for the PS II center. Further studies should give more information on the role of the stress proteins in diatoms.

## CONCLUSIONS

The change in assimilation of inorganic nitrogen after exposure to UV-B radiation was dependent on the species composition of the plankton assemblages. Shade-adapted microalgae were more sensitive to UV-B than those under strong white light. The different inhibitory effects on ammonia and nitrate indicate different uptake systems for each compound [39, 40].

The possible reasons for damage by UV-B could be a reduction in the supply of energy and carbon skeletons. Vosjan et al. [34] found a marked depression of the ATP content under conditions presented in Fig. 1. On the other hand, an inhibition of key enzymes of the carbon and nitrogen metabolism was also observed. Fluorescence studies showed that the photosystem II reaction center was affected by UV-B irradiance. Summarizing, the target of UV might be more complex than simply the photoinhibition by strong white light.

The size and developmental stage of algae cells are of importance when considering possible damage from UV radiation because light is more scattered by small species. A variation in the response to UV-B irradiance exists during the division cycle of marine diatoms. The different sensitivity of microalgae might lead to a change in the species composition in the aquatic ecosystem and in the quantity and quality of the nutrients in the food web.

**Acknowledgments.**

Data presented here were obtained during the European "Polarstern" Study (EPOS) sponsored by the European Science Foundation and the Alfred Wegener Institute for Polar and Marine Research. This work was supported by a grant from the Bundesministerium für Forschung und Technologie, Bonn (07 UV-B 04). Thanks are expressed to Mrs. Iris Schall, Rosemarie Reuter, Liselotte Tramp, and Dipl.-Biol. P. Hansen for their technical help in the laboratory, Frankfurt a. M.

## REFERENCES

1. Lubin, D., and J.E. Frederick. 1989. Measurements of enhanced springtime ultraviolet at Palmer Station, Antarctica, Geophys. Res. Letters 16:783-785.

2. Lubin, D., and J.E. Frederick. 1991. The ultraviolet radiation environment of the Antarctic peninsula: The role of ozone and cloud cover, J. Appl. Meteorol. 30:478-493.

3. Gieskes, W.C., and G.W. Kraay. 1990. Transmission of ultraviolet light in the Weddell Sea: Report of the first measurements made in Antarctic, Biomass Newsletter 12:12-14.

4. Smith, R. 1989. Ozone, middle ultraviolet radiation and the aquatic environment, Photochem. Photobiol. 50:459-468.

5. Bidigare, R.R. 1989. Potential effects of UV-B radiation on marine organisms of the Southern Ocean: Distributions of phytoplankton and krill during austral spring. Photochem. Photobiol. 50:469-477.

6. Döhler, G. 1989. Influence of UV-B (290-320 nm) radiation on photosynthetic $^{14}CO_2$ fixation of *Thalassiosira rotula* Meunier. Biochem. Physiol. Pflanzen 185:221-226.

7. Helbling, E.W., V. Villafane, M. Ferrario, and O. Holm-Hansen. 1992. Impact of natural ultraviolet radiation on rates of photosynthesis and on specific marine phytoplankton species, Mar. Ecol. Prog. Ser. 80:89-100.

8. Worrest, R.C. 1982. Review of literature concerning the impact of UV-B radiation upon marine organisms. *In* The Role of Solar Ultraviolet Radiation in Marine Ecosystems, edited by J. Calkins, pp.429-457, Plenum Publishing Corp., New York.

9. Worrest, R.C. 1986. The effect of solar UV-B radiation on aquatic systems: an overview. *In* Effects of Changes in Stratospheric Ozone and Global Climate. Overview, edited by J.G. Titus, pp.175-191, US Environmental Protection Agency and United Nations Environmental Progr. 1.

10. Bühlmann, B., P. Bossard, and U. Uehlinger. 1976. The influence of longwave ultraviolet (UV-A) on the photosynthetic activity ($^{14}C$-assimilation) of phytoplankton. J. Plankton Res. 9:935-943.

11. Jitts, H.R., A. Morel, and Y. Saijo. 1976. The relation of oceanic primary production to available photosynthetic irradiance, Aust. J. Mar. Freshwater Res. 27:441-454.

12. Lorenzen, C.J. 1979. Ultraviolet radiation and phytoplankton photosynthesis (3), Limnol. Oceanogr. 24:1117-1120.

13. Maske, H. 1984. Daylight ultraviolet radiation and the photoinhibition of phytoplankton carbon uptake, J. Plankton Res. 6:351-357.

14. El Sayed, S.Z., F.C. Stephens, R.R. Bidigare, and M.E. Ondrusek. 1990. Effect of ultraviolet radiation on Antarctic marine phytoplankton, in Antarctic Ecosystems, edited by K.R.Kerry, G. Hempel, pp.379-385, Springer-Verlag Berlin, Heidelberg, New York.

15. Sakshaug, E., and O. Holm-Hansen. 1986. Photoadaptation in Antarctic phytoplankton: Variations in growth rate, chemical composition and $P_1$ versus I curves, J. Plankton Res. 8:459-473.

16. Calkins, J., and T. Thodardottir. 1980. The ecological significance of solar UV radiation on aquatic organisms. Nature 283:563-566.

17. Döhler, G. and H. Stolter. 1986. Impact of UV-B radiation on photosynthesis-mediated uptake of $^{15}N$-ammonia and $^{15}N$-nitrate of several marine diatoms, Biochem. Physiol. Pflanzen 181:533-539.

18. Worrest, R.C., H. van Dyke, and B.E. Thomson. 1978. Impact of enhanced simulated solar ultra-

violet radiation upon a marine community, Photochem. Photobiol. 17:471-478.

19. von Stosch, H.A., und G. Drebes. 1964. Entwick lungsgeschichtliche Untersuchungen an zentrischen Diatomeen. IV. Die Planktondiatomee Stephanopyxis turrisihre Behandlung und Entwicklungsgeschichte. Helgol. wiss. Meeresunters. 11:209-257.

20. Döhler, G., and H.-J. Roßlenbroich. 1981. Photosynthetic assimilation of $^{15}$N-nitrate uptake in the marine diatoms *Bellerochea yucatanensis* and *Skeletonema costatum*, Z. Naturforsch. 36c:834-839.

21. Döhler, G., and I. Biermann. 1985. Effect of salinity on $^{15}$N-ammonia and $^{15}$N-nitrate assimilation of *Bellerochea yucatanensis* and *Thalassiosira rotula*. Biochem. Physiol. Pflanzen 180:589-598.

22. Jeffrey, S.W., and G.F. Humphrey. 1975. New spectrometric equations for determining chlorophyll a, b, $c_1$ in higher plants, algae and natural phytoplankton, Biochem. Physiol. Pflanzen 167:191-194.

23. Bradford, M.M. 1976. A rapid and sensitive method for the quantitation of microgram quantities of protein utilizing the principle of proteindye binding. Analyt. Biochem. 72:248-254.

24. Rehm, A.M., M. Gülzow, J. Marquardt, and A. Ried. 1990. Changes in the photosynthetic apparatus of red algae induced by spectral alteration of the light field. II. Further characterization of the light-dependent regulation of the apparent quantum yield of PS I, Biochimica et Biophisica Acta 1016:127-135.

25. Döhler, G. 1985. Effect of UV-B radiation (290-320 nm) on the nitrogen metabolism of several marine diatoms. J. Plant Physiol. 118:391-400.

26. O'Farrel, P.H. 1975. High resolution two-dimensional electrophoresis of proteins, J. Biol. Chem. 250:4007-4021.

27. Bonner, W.M., and R.A. Laskey. 1974. A film detection method for tritium-labelled proteins and nucleic acids in polyacrylamide gels, Eur. J. Biochem. 46:83-88.

28. Dortch, Q. 1990. The interaction between ammonium and nitrate uptake in phytoplankton, Mar. Ecol. Prog. Ser. 61:183-201.

29. Rönner, U., F. Sörensson, and O. Holm-Hansen. 1983. Nitrogen assimilation by phytoplankton in the Scotia Sea, Polar. Biol. 2:137-147.

30. Döhler, G. 1992. Impact of UV-B radiation on uptake of $^{15}$N-ammonia and $^{15}$N-nitrate by phytoplankton of the Wadden Sea, Mar. Biology 112:485-489.

31. Döhler, G., E. Hagmeier, E. Grigoleit, and K.-D. Krause. 1991. Impact of solar UV radiation on uptake of $^{15}$N-ammonia and $^{15}$N-nitrate by marine diatoms and natural phytoplankton, Biochem. Physiol. Pflanzen 187:293-303.

32. Carreto, J.I., M.O. Carignan, G. Daleo, and S.G. DeMarco. 1990. Occurrence of mycosporine-like amino acids in the redtide dinoflagellate *Alexandrinum excavatum*: UV-photoprotective compounds? J. Plankton Research 12:909-921.

33. Nakamura, H., J. Kobayashi, and Y. Hirata. 1982. Separation of mycosporine-like amino acids in marine organisms using reversed-phase high-performance liquid chromatography, J. Chromatography 250:113-118.

34. Vosjan, J.H., G. Döhler, and G. Nieuwland. 1990. Effect of UV-B irradiance on the ATP content of microorganisms of the Weddell Sea (Antarctic), Netherlands J. Sea Research 25:391-393.

35. Renger, G., M. Voelker, H.J. Eckert, R. Fromme, S. Hohm-Veit, and P. Graeber. 1989. On the mechanism of photosystem II deterioration by UV-B irradiation, Photochem. Photobiol. 49:97-106.

36. Karentz, D., F.S. Mc Euen, M.C. Land, and W.C. Dunlap. 1991. Survey of mycosporine-like amino acid compounds in Antarctic marine organisms: Potential protection from ultraviolet exposure, Mar. Biology 108:157-166.

37. Nicholsen, P., and C.J. Howe. 1986. Ultraviolet-inductible proteins in *Clamydomonasreinherdtii*, Biochem. Soc. Trans. 14:1098.

38. Gromoff, E.D., U. Treiber, and C.F. Beck. 1989. Three light inductible heat shock genes of *Chlamydomonas reinhardtii*, Mol. Cell. Biol. 9:3911-3918.

39. Ullrich, W.R. 1991. Transport of nitrate and ammonium through plant membranes. *In* Nitrogen metabolism in plants, edited by K. Mengel, D. Pilbeam, pp.121-137, Oxford Univ. Press, Oxford.

40. Wheeler, P.A. 1983. Phytoplankton nitrogen metabolism. *In* Nitrogen in the marine environment, edited by E.J. Carpenter, D.G. Copone, pp.309-346, Academic Press, New York.

# PREVENTION OF ULTRAVIOLET RADIATION DAMAGE IN ANTARCTIC MARINE INVERTEBRATES

Deneb Karentz

Department of Biology, University of San Francisco
San Francisco, CA 94117-1080

Key words: Antarctic, ozone depletion, marine invertebrates, plankton, UV radiation

## ABSTRACT

One aspect of investigating the ecological effects of ozone depletion on marine communities is to identify species-specific characteristics relative to strategies for dealing with prevention and repair of ultraviolet B (UV-B, 280-320 nm) damage. In adult marine invertebrates protection from UV-B exposure can be provided by either external coverings (e.g., teste, shell, body wall) or by the presence of UV-absorbing molecules (e.g., mycosporine-like amino acid compounds) or by both in cells and tissues. External coverings are not opaque to UV-B. Using biological dosimetry, penetration of UV-B wavelengths through the exteriors of adult invertebrate has been observed. In addition, selective partitioning of UV-absorbing compounds occurs in adult invertebrate tissues. The highest concentrations of these substances are found in ovaries and eggs. Planktonic larval stages receive higher UV exposures than benthic adults and have considerably less optical shielding. Since egg cells tend to have higher concentrations of UV-absorbing compounds than other body tissues, this may provide for increased protection of embryos and larvae during planktonic development.

## INTRODUCTION

In the two decades since scientists first predicted CFC-mediated destruction of ozone in the stratosphere [5,29], the most drastic instance of ozone depletion has been occurring over the Antarctic continent and adjacent regions of the Southern Ocean [10,14,15,24,35]. Since the late 1970s, predictable losses of over 50% of column ozone have occurred during spring [10]. These changes in atmospheric chemistry have caused shifts in the spectral distribution and increases in intensity of incident UV-B radiation (280-320 nm) [11,25-27]. Antarctic organisms are now exposed to larger amounts of biologically harmful radiation. In order to assess the impact of this environmental perturbation on the local ecosystem, species responses to UV-B exposure are being evaluated relative to protection and recovery [17,18,22].

One of the most distinctive aspects of Antarctic ozone depletion is that this phenomenon is directly linked to seasonal changes in the physical structure of the atmosphere [35]. It is the springtime formation and subsequent disintegration of the polar vortex that provides the unique physical and chemical environment in which stratospheric ozone is destroyed and not regenerated. This process relies on a series of photochemical reactions that are initiated as the sun appears above the horizon after the dark winter period. In terms of biology, the ozone depletion cycle begins at the worse possible time, a time when organisms are expected to have the highest vulnerability to UV exposure. They are emerging from dark winter days into a springtime environment where intensities of UV-B under an ozone-depleted column can be as high as UV-B levels that occur during mid-summer [11]. If organisms have the ability to adapt to changes in UV exposure, it is not clear if the time frame presented by the Antarctic depletion cycle is sufficient

NATO ASI Series, Vol. I 18
Stratospheric Ozone Depletion/
UV-B Radiation in the Biosphere
Edited by R. H. Biggs and M. E. B. Joyner

to allow for effective development of adaptive mechanisms.

An additional aspect confounding the ecological evaluation of ozone depletion over the Antarctic is the lack of prior data. There is no information available on the UV-photobiology of endemic species before the ozone depletion cycle began. Nor is there a sufficiently consistent data set from which to examine historical trends that might be related to the changes in UV-B that have recently been occurring. If ecosystem alterations can be caused by ozone depletion, those relevant to the Antarctic were initiated during the first occurrence of the springtime ozone losses and subsequent annual occurrences have continuously modified Antarctic biology. The diversity of life forms and the complexity of trophic interactions also contribute to the difficulties associated with evaluating the impact of ozone depletion in this (or any) region.

## CONSIDERATIONS IN EXPERIMENTAL DESIGN

For the reasons stated above, it has been difficult to design appropriate field experiments to determine what the long-term effects of springtime ozone depletion have been in the Antarctic. However, several investigations have examined the response of Antarctic organisms to recent levels of ambient UV-B, as modified by seasonal changes in ozone concentrations [13,17,19,34]. Laboratory studies have also been conducted to examine specific physiological processes related to UV-B exposure [18]. The majority of these research efforts in the Antarctic have been directed towards understanding phytoplankton responses because of their importance as primary producers and their role in biogeochemical cycling. Research on the effects of UV-B on Antarctic marine invertebrates has only recently been initiated.

### IDENTIFYING SPECIES-SPECIFIC RESPONSES

One aspect of investigating the ecological effects of ozone depletion on marine communities has been to identify species-specific responses relative to strategies for dealing with protection from UV-B exposure [28]. In terms of UV exposure in marine habitats, intertidal organisms would be considered to have the highest exposures as they are periodically exposed to the full solar spectrum during low tide periods. Neustonic communities, made up of organisms that float on the surface of the water, also experience relatively high UV-B exposure levels. These assemblages can often include floating eggs and small larval stages.

It is well established that permanently submerged and water column habitats in the Antarctic are exposed to UV-B [21,34]. However, there are at least two factors that can reduce the amount of UV-B entering the water column in the Antarctic during spring: clouds and ice cover [1,25,26,36]. Springtime weather is usually cloudy in coastal areas and ice cover can be extensive, but does vary greatly on both short- and long-term temporal scales.

The penetration of UV-B into Antarctic waters during springtime ozone depletion has been measured spectroradiometrically to depths of greater than 50 m [34], although observable biological effects have only been documented in the upper 20 m of the water column [13,21,34]. Organisms living in the interstitial waters of the sea ice, the water column, and benthic habitats are potentially exposed to harmful UV-B radiation.

The primary means of avoiding UV-induced damage is to avoid exposure to sunlight. Many organisms accomplish this by various behaviors that may or may not be in response to the ambient light field. For example, behaviors such as living under a rock for physical protection or undergoing nocturnal migrations to the surface to optimize feeding efficiency, also serve to effectively reduce the UV exposure experienced by an organism.

The first line of defense that an organism has to UV exposure is its outer body covering. Several Antarctic invertebrate species have been evaluated to determine the amount of UV protection provided by this outer layer [20]. Animals were collected in Arthur Harbor (Antarctic Peninsula) and immediately dissected. Four species were examined: the sea urchin (*Sterechinus neumayeri*), the limpet (*Nacella concinna*), the seastar (*Odontaster validus*), and the tunicate (*Cnemidocarpa verrucosa*).

UV transmission was monitored using a biological dosimeter based on a DNA-repair deficient strain of the common bacterium *Escherichia coli* (strain CSR06) [21]. The dosimeter consists of a liquid culture of CSRO6 cells packaged in UV transparent polyethylene bags (Whirl-Pak). Half of the dosimeters were wrapped in one layer of polyester film (Mylar D) that filters out wavelengths in the UV-B

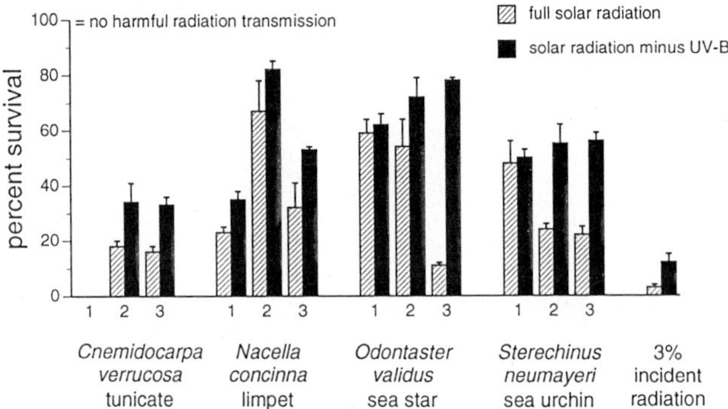


Figure 1. UV transmission through exteriors of four Antarctic invertebrate species as determined by biological dosimetry. Each paired set represents one animal (three animals tested for each species). Error bars calculated from triplicate assays of the dosimeter. Absence of bars indicates 100% killing of dosimeter cells (e.g., *Cnemidocarpa verrucosa* 1). Data at far right represent lethality of 3% incident radiation on the dosimeter cells during the course of the experiment. (From [20]).

region. Dosimeters, with and without UV-B filters, were incubated under sections of outer walls for four hours during midday. The dosimeters and animal exteriors were held in the same plane and submerged 4 cm below the surface of the water in a flowing seawater tank. Water was pumped directly from Arthur Harbor to maintain ambient seawater temperature of -1.0°C. For incident radiation comparison, one pair of dosimeters (with and without UV-B filters) was shielded by neutral density screening and otherwise exposed directly to sunlight. The screening used reduced exposure intensity to 3% of the incident radiation. This was necessary because of the high sensitivity of the CSR06 cells to UV-B exposure. Without neutral density shielding, survival time of cells is less than 30 min.

Survival was calculated relative to a dark control. Because CSR06 cells are DNA repair deficient, they produce a typically linear exposure-response curve [21]. When the UV-B portion of the solar spectrum is removed, survival is greatly enhanced. By comparing the full solar radiation exposure to the solar radiation minus UV-B (with Mylar) exposure, it is possible to determine the following: 1) if UV-B is penetrating and 2) what the contribution of UV-B is to the killing of cells, thus allowing for a relative measure of the biologically damaging potential of the transmitted radiation.

Analysis of results of dosimetry under the various animal exteriors indicated distinct differences between individuals and between species (Figure 1). With no penetration of harmful solar radiation, cells would have 100% survival. In all specimens, enough solar radiation penetrated to cause killing of cells. Data on right side of the plot represents the lethality of ambient light at 3% intensity for comparison. The removal of UV-B wavelengths with Mylar filters increased survival, clearly indicating that UV-B does penetrate the outer defenses of these organisms and causes biological damage in excess of that caused by 3% of the ambient radiation field. This does not equate to the same lethal effect measured for the dosimeter cells because the test organisms are capable of repairing some or all of the damage induced. Future research will examine the rate of damage relative to repair and assess the effects relative to short-term (hours) and long-term (days and months) exposures.

## MYCOSPORINE-LIKE AMINO ACIDS (MAAs)

The blockage of UV is partially due to presence of UV-absorbing compounds in body walls and shells

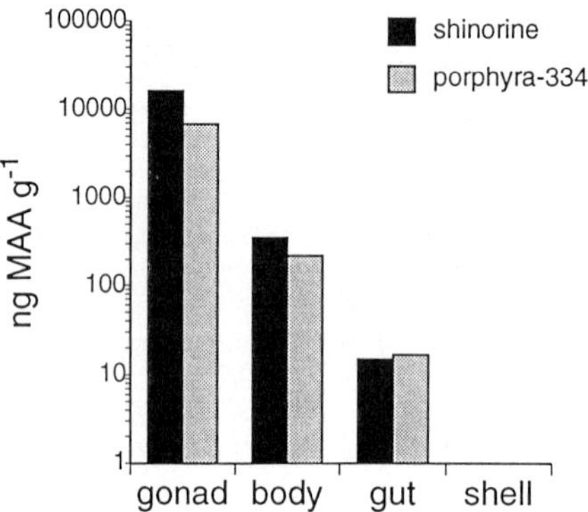


Figure 2. Distribution of shinorine and porphyra-334 in various tissues of *Nacella concinna*. Data presented are from one female animal collected in the intertidal. Concentrations are standardized as ng MAA per g dry weight.

of these species. Of particular interest are a group of UV-absorbing compounds known as mycosporine-like amino acids (MAAs). These compounds are very common in marine organisms from all latitudes and have been found in nearly all species examined. MAAs have been reported from corals [6,7], starfish [30], ascidians [23], zoanthids [16], brine shrimp [12], mussels [3], fish [4,9], holothuroids [31], dinoflagellates [2] and red algae [33,37].

Over 20 MAAs have been isolated and identified in marine organisms. These compounds are composed of a cyclohexenone ring with an amino acid side group. Maximum absorption wavelengths range between 310 and 365 nm, spanning both UV-B and UV-A portions of the solar spectrum. Most organisms contain two or more MAAs, so the combination increases the screening band width [2]. Much of the published work on MAAs pertains to chemical structures and there has been some interest in using these compounds as human sunscreens [7]. Although MAAs are widespread in marine organisms, relatively little is known about the physiological or ecological function of these compounds. There is no definitive evidence that these compounds serve primarily as UV protectants. Their role as UV protectants is inferred from observations that organisms living in shallow areas tend to have higher concentrations than those living in deeper water [8]. There is also experimental evidence that suggests that UV exposure may regulate MAA concentration in tissues [32], but these compounds may have another primary physiological function.

It is generally accepted that photosynthetic organisms synthesize MAAs and that their occurrence in animals is through ingestion of plants (or prey) containing MAAs or through translocation from symbionts [22,32]. It is also recognized that MAAs can be physiologically altered in animals (perhaps by microbial gut flora) and can accumulate as a different MAA than that ingested [22,32]. Research on these physiological aspects has only recently been initiated, and specific data are not yet available.

Eight MAAs have been identified from Antarctic species (Table 1) [22]. Seven of these compounds have been previously documented from temperate and tropical marine species, one (mycosporine-glycine:valine) has not yet been reported from other regions.

While the primary function of MAAs has not yet been conclusively demonstrated, an unequal partitioning of MAAs between different body parts of invertebrates has been documented [19,22,31]. MAA content of gonad, gut, body wall, and shell can differ by orders of magnitude (Figure 2) [19]. For example, in the Antarctic limpet *Nacella concinna*, female reproductive tissues (ovaries) and cells (eggs) tend to have the highest concentrations of MAAs. Testes and sperm have very low levels. This partitioning of MAAs occurs also in the Antarctic sea urchin *Sterechinus neumayeri*. Recent observations of tropical urchin species indicates a similar pattern of MAA distribution within individuals (I. Bosch pers. comm.).

*N. concinna*, along with several other dominant benthic invertebrates (e.g., *S. neumayeri* and *Odontaster validus*) have planktonic larvae whose occurrence in the water column coincides with the springtime interval of ozone depletion. Although adults are benthic and protected by an outer layer, the juvenile forms are transparent and must survive in a higher intensity UV-B environment. Therefore, the selective packaging of MAAs into eggs could be a significant strategy for UV protection relative to life history and reproductive success. It is not yet known how adequate or successful this strategy is under the potential increase in UV-B stress caused by ozone depletion.

Table 1. Mycosporine-like amino acid compounds (MAAs) found in Antarctic invertebrate species. Wavelengths of maximum absorbance ($l_{max}$) are indicated (From [22]).

| MAA | $l_{max}$ (nm) |
|---|---|
| Mycosporine-glycine | 310 |
| Palythine | 320 |
| Asterina-330 | 330 |
| Palythinol | 332 |
| Shinorine | 334 |
| Porphyra-334 | 334 |
| Mycosporine-glycine:valine | 335 |
| Palythene | 360 |

**Acknowledgments**

I. Bosch, M. Slattery, T. Gast and W.C. Dunlap were involved with various aspects of this work which was supported by the Office of Polar Programs/U.S. National Science Foundation.

## REFERENCES

1. Buckley, R.G., and H.J. Trodahl. 1987. Scattering and absorption of visible light by sea ice. Nature 326:867-869.

2. Carreto, J.I., M.O. Carignan, G. Daleo, and S.G.d. Marco. 1990. Occurrence of mycosporine-like amino acids in the red tide dinoflagellate *Alexandrium excavatum*: UV-photoprotective compounds? J. Plank. Res. 12:909-921.

3. Chioccara, F., A.D. Gala, M.d. Rosa, E. Novelline, and G. Prota. 1979. Occurrence of two new mycosporine-like amino acids, mytilins A and B in the edible mussel, *Mytilis galloprovincialis*. Tetrahedron. Lett. 34:3181-3182.

4. Chioccara, F., A.D. Gala, M.d. Rosa, E. Novelline, and G. Prota. 1980. Mycosporine amino acids and related compounds from the eggs of fishes. Bull. Soc. Chim. Belg. 89:101-1106.

5. Cutchis, P. 1974. Stratospheric ozone depletion and solar ultraviolet radiation on earth. Science 184:13-19.

6. Dunlap, W.C., and B.E. Chalker. 1986. Identification and quantitation of near-UV absorbing compounds (S-320) in a hermatypic scleractinian. Coral Reefs 5:155-159.

7. Dunlap, W.C., B.E. Chalker, and W.M. Banaranayake. 1988. New sunscreening agents derived from tropical marine organisms of the Great Barrier Reef, Australia. Int. Patent Application Pub. #WO 86/02350.

8. Dunlap, W.C., B.E. Chalker, and J.K. Oliver. 1986. Bathymetric adaptations of reef-building corals at Davies Reef, Great Barrier Reef, Australia. III. UV-B absorbing compounds. J. Exp. Mar. Biol. Ecol. 104:1-10.

9. Dunlap, W.C., D.M. Williams, B. Chalker, and A. Banaszak. 1989. Biochemical photo-adaptation in vision: UV-absorbing pigments in fish eye tissues. Comp. Biochem. Physiol. 93B:601-607.

10. Farman, J.C., B.G. Gardiner, and J.D. Shanklin. 1985. Large losses of total ozone in Antarctica reveal seasonal $ClO_x/NO_x$ interaction. Nature 315:207-210.

11. Frederick, J.E., and H.E. Snell. 1988. Ultraviolet radiation levels during the Antarctic spring. Science 241:438-440.

12. Grant, P.T., C. Middleton, P.A. Plack, and R.H. Thomson. 1985. The isolation of four aminocyclohexenimines (mycosporines) and a structurally related derivative of cyclohexane-1:3-dione (gadusol) from the brine shrimp, *Artemia*. Comp. Biochem. Physiol. 80B:755-759.

13. Helbling, E.W., V. Villafañe, M. Ferrario, and O. Holm-Hansen. 1992. Impact of natural ultraviolet radiation on rates of photosynthesis and on specific marine phytoplankton species. Mar. Ecol. Prog. Ser. 80:89-100.

14. Hofmann, D.J. 1989. Direct ozone depletion in springtime Antarctic lower stratospheric clouds. Nature 337:447-449.

15. Hofmann, D.J., J.W. Harder, S.R. Rolf, and J.M. Rosen. 1987. Balloon-borne observations of the development and vertical structure of the Antarctic ozone hole in 1986. Nature 326:59-62.

16. Ito, S., and S. Hirata. 1977. Isolation and structure of a mycosporine from the zoanthid *Palythoa tuberculosa*. Tetrahedron Lett. 28:2429-2430.

17. Karentz, D. in press. Ultraviolet tolerance mechanisms in Antarctic marine organisms. *In* C.S.

Weiler and P.A. Penhale (eds.), Ultraviolet Radiation and Biological Research in Antarctica. American Geophysical Union. Washington, DC.

18. Karentz, D., J.E. Cleaver, and D.M. Mitchell. 1991. Cell survival characteristics and molecular responses of Antarctic phytoplankton to ultraviolet-B radiation exposure. J. Phycol. 27:326-341.

19. Karentz, D., W.C. Dunlap, and I. Bosch. 1992. Distribution of UV-absorbing compounds in the Antarctic limpet, *Nacella concinna*. Antarctic J. U.S.

20. Karentz, D., and T. Gast. in press. Transmission of solar ultraviolet radiation through invertebrate exteriors. Antarctic J. U.S.

21. Karentz, D., and L.H. Lutze. 1990. Evaluation of biologically harmful ultraviolet radiation in Antarctica with a biological dosimeter designed for aquatic environments. Limn. Ocean. 35:549-561.

22. Karentz, D., F.S. McEuen, K.M. Land, and W.C. Dunlap. 1991. Survey of mycosporine-like amino acid compounds in Antarctic marine organisms: potential protection from ultraviolet exposure. Mar. Biol. 108:157-166.

23. Kobayashi, J., H. Nakamura, and Y. Hirata. 1981. Isolation and structure of a UV-absorbing substance 337 from the ascidian *Halocynthia roretzi*. Tetrahedron Lett. 22:3001-3002.

24. Krueger, A.J., M.R. Schoeberl, R.S. Stolarski, and F.S. Sechrist. 1988. The 1987 Antarctic ozone hole: a new record low. Geophys. Res. Lett. 15:1365-1368.

25. Lubin, D., and J.E. Frederick. 1991. The ultraviolet radiation environment of the Antarctic Peninsula: the roles of ozone and cloud cover. J. Appl. Meterol. 30:478-493.

26. Lubin, D., J.E. Frederick, C.R. Booth, T. Lucas, and D. Neuschuler. 1989. Measurements of enhanced springtime ultraviolet radiation at Palmer Station, Antarctica. Geophys. Res. Lett. 16:783-785.

27. Lubin, D., J.E. Frederick, and A.J. Krueger. 1989. The ultraviolet radiation environment of Antarctica: McMurdo Station during September-October 1987. J. Geophys. Res. 94:8491-8496.

28. Mitchell, D.L., and D. Karentz. 1993. The induction and repair of DNA photodamage in the environment. *In* A.R. Young, L.O. Björn, J. Moan, and W. Nultsch (eds.), Environmental UV Photobiology. Plenum Press. New York. pp. 345-377.

29. Molina, M.J., and F.S. Rowland. 1974. Stratospheric sink for chlorofluoromethanes: chlorine atom-catalysed destruction of ozone. Nature 249:810-812.

30. Nakamura, H., J. Kobayashi, and Y. Hirata. 1981. Isolation and structure of a 330 nm UV-absorbing substance, asterina-330 from the starfish *Asterina pectinifera*. Chem. Lett. 1413-1414.

31. Shick, J.M., W.C. Dunlap, B.E. Chalker, A.T. Banaszak, and T.K. Rosenzweig. 1992. Survey of ultraviolet radiation-absorbing mycosporine-like amino acids in organs of coral reef holothuroids. Marine Ecol. Prog. Ser. 90:139-148.

32. Shick, J.M., M.P. Lesser, and W.R. Stochaj. 1991. Ultraviolet radiation and photooxidative stress in zooxanthellate Anthozoa: The sea anemone *Phyllodiscus semoni* and the octocoral *Clavularia* sp. Symbiosis 10:145-173.

33. Sivalingham, P.M., T. Ikawa, Y. Yokohama, and K. Nisizawa. 1974. Distribution of a 334 UV-absorbing substance in algae, with special regard of its possible physiological roles. Botan. Mar. 17:23-29.

34. Smith, R.C., B.B. Prézelin, K.S. Baker, R.R. Bidigare, N.P. Boucher, T. Coley, D. Karentz, S. MacIntyre, H.A. Matlick, D. Menzies, M. Ondrusek, Z. Wan, and K.J. Waters. 1992. Ozone depletion: Ultraviolet radiation and phytoplankton biology in Antarctic waters. Science 255:952-959.

35. Solomon, S. 1990. Progress towards a quantitative understanding of Antarctic ozone depletion. Nature 347:347-354.

36. Trodahl, H.J., and R.G. Buckley. 1990. Enhanced ultraviolet levels under sea ice during the Antarctic spring. Geophys. Res. Lett. 17:2177-2179.

37. Tsujino, I., K. Yabe, and I. Sekikawa. 1980. Isolation and structure of a new amino acid, shinorine, from the red alga *Chondrus yendoi* Yamada et Mikami. Bot. Mar. 23:65-68.

# EVALUATION OF FIELD STUDIES OF UVB RADIATION EFFECTS ON ANTARCTIC MARINE PRIMARY PRODUCTIVITY

Barbara B. Prézelin, Nicolas P. Boucher, and Oscar Schofield

Department of Biological Sciences and The Center for Remote Sensing and Environmental Optics, University of California, Santa Barbara, California, USA 93106

Key words: Antarctica, carbon fixation, ozone, phytoplankton, Southern Ocean, UV-B radiation

**ABSTRACT**

Phytoplankton productivity can account for half the global production of organic carbon each year. Serious concerns exist that global thinning of atmospheric ozone ($O_3$) and associated enhancement of UV-B radiation, may impair phytoplankton primary productivity. Thus it is essential to quantify $O_3$-related effects on primary production and elucidate regulating mechanisms underlying any impairment of phytoplankton biology by UV-B under natural conditions. Field studies conducted under the influence of the severe Antarctic $O_3$ hole can provide the sensitivity required to quantify $O_3$-dependent impacts on marine primary production. Given the further reduction in global stratospheric $O_3$ expected to occur at all latitudes over the next century, the findings of these field studies may also be relevant outside polar regions.

The objectives of the paper are:

1) to provide an ecological framework in which to judge the findings of Antarctic field studies conducted to date;
2) to evaluate experimental designs, controls, and instrumentation suitable to address the $O_3$ question of "whether $O_3$-depletion, and associated enhancement of UV-B radiation, over Antarctica is reducing marine primary production in natural phytoplankton communities?";
3) present relevant findings that may improve substantially our ability to predict changes in rates of carbon fixation within phytoplankton communities experiencing enhanced UV-B radiation.

Findings are reported which indicate:

1) that controversial results between studies can be resolved once spectral intercalibration of data is achieved;
2) that public use of both laboratory and field-based studies is fraught with misstatements and a clear lack of understanding as to what field questions can be answered with which data bases;
3) that significant $O_3$-dependent UV-B inhibition of rates of primary production do occur in the Southern Ocean;
4) that successful modeling of the biological consequences requires a full spectral approach based on precise knowledge of the fluxes in UV-B, UV-A, and photosynthetically-available radiation.

## INTRODUCTION

Regarding the Antarctic ozone ($O_3$) "hole" and possible impairment of marine primary productivity due to enhanced UV radiation, we find that recent public comment on laboratory and field studies are often fraught with scientific misstatements. We believe that some misunderstandings are due, in part, to an inability to distinguish environmentally relevant studies of $O_3$-dependent UV effects from those where either or both 1) supplemental UV radiation was artificially provided and unable to mimic the

NATO ASI Series, Vol. I 18
Stratospheric Ozone Depletion/
UV-B Radiation in the Biosphere
Edited by R. H. Biggs and M. E. B. Joyner

spectral effects of $O_3$ depletion or 2) experimental samples were either not representative of phytoplankton communities or environmental conditions, or both, that dominate the Antarctic marine ecosystems during the austral spring when loss of stratospheric $O_3$ over Antarctica is accelerated by anthropogenic chemicals. Some progress has also been lost while the dispute over the use of nontoxic polyethylene bags as incubation containers was settled [4,6,7].

Here, our tutorial attempts to clarify the environmental relevancy of recent Antarctic studies of UV effects on primary productivity by providing an ecological framework by which to judge their experimental design. From these considerations arise recommendations for basic experimental criteria and control measurements to consider when attempting to quantify the extent to which $O_3$-depletion over Antarctica is reducing marine primary production in the Southern Ocean. In this discussion, we present findings of the *Icecolors '90* expedition (Figure 1) which improved abilities to predict changes in rates of primary production within phytoplankton communities experiencing $O_3$-related enhanced UV radiation [1,6,10]. We hope these general comments also prove useful to others pursuing environmentally-relevant studies of UV effects on at locations outside the Southern Ocean.

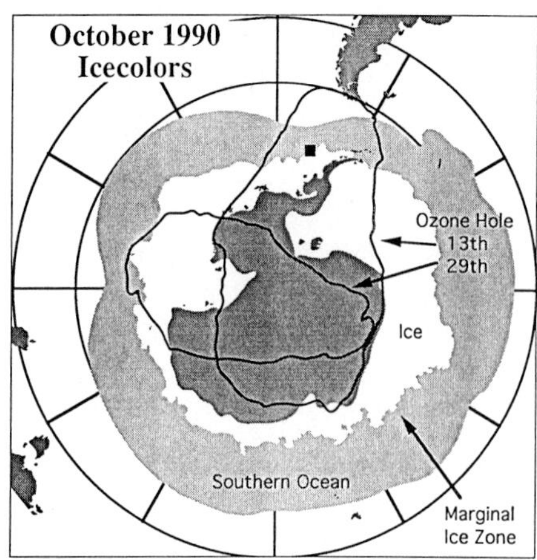


Figure 1. Ozone hole positions (depicted by 200 DU contours) with respect to the Antarctic Continent (dark area) for 13 and 29 October 1990. The position of the pack ice (light area), the edge of the MIZ (outer edge of the light area) during the *Icecolors '90* cruise, and the mean position of the Antarctic Convergence within the Southern Ocean (outer edge of the gray area) are also shown. Cruise location, centered at about 64°S, 72°W, is indicated by the black box. (Modified after Smith, Prézelin et al., 1992).

## THE ECOLOGICAL FRAMEWORK

The Antarctic Ocean, also known as the Southern Ocean, is unique in that it completely surrounds a continent, Antarctica (Figure 1). The Antarctic Convergence (or polar front) at about 60°S is the major northern boundary current of the Southern Ocean. Its surface currents flow clockwise around the continent and are considered the great "mix masters" of the Atlantic, Pacific, and Indian Oceans. Antarctic phytoplankton ecosystems lie between this polar front and the Antarctic continent and are always very much colder and richer in inorganic nutrients than those just north of the Antarctic Convergence. The Antarctic marine ecosystems are characteristically ice-dominated, with enormous quantities of freezing and thawing sea ice having major influences on the physical, chemical, and biological properties of phytoplankton communities. In addition, these communities receive sufficient sunlight for plant growth only during about 4 to 5 months of the year and much of that time the sky is overcast. Together, these environmental conditions have selected for distinctly polar phytoplankton communities where many unique species have adaptations specialized for survival in this extreme environment [2]. We contend that the uniqueness of the Southern Ocean, as well as the distinct nature of phytoplankton communities within its environment, preclude field studies conducted outside the region (or in laboratories with nonpolar species of phytoplankton) from being directly relevant to immediate efforts to quantify the ecological consequences of atmospheric $O_3$ depletion on Antarctic phytoplankton productivity.

It is only during the austral spring months (September-December) that atmospheric conditions exist that accelerate rates of $O_3$ photolysis in the stratospheric polar vortex ($O_3$ hole) over Antarctica (Figure 1). Accelerated in large part by the accumu-

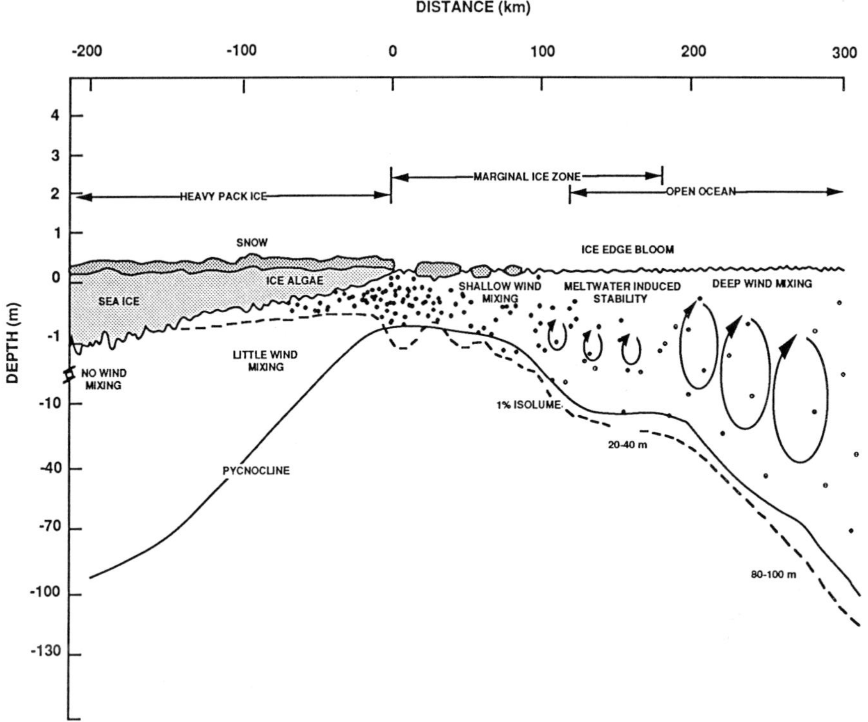


Figure 2. Conceptual model of the development of springtime phytoplankton blooms from sea ice and under ice (dots) microalgal communities along the MIZ. (Modified from Sullivan et al., 1988).

lation of anthropogenic polluting chemicals (primarily chlorofluorocarbons [CFCs]) within the vortex, the thinning of the $O_3$ layer has become much more acute in recent years. The $O_3$ hole is forming earlier and lasting longer in the spring while covering a greater area of the Southern hemisphere than at anytime since its discovery 20 years ago (*v.* [5,13]). For instance, the $O_3$ concentration over Antarctica has declined since before 1975 from a monthly mean average for October that was well above 300 DU (Dobson Units=matm-cm) to less than 150 DU every year since 1989. And, as we write up this paper while in the Antarctic at Palmer Station in August of 1993, there is every indication that the trend toward earlier thinning of the $O_3$ layer continues even into this late winter period.

At the same time that the $O_3$ hole now develops to cover much of the Southern Ocean, unique hydrographic conditions are being established which historically have given rise to spring phytoplankton blooms. As the sun comes up in the spring and days get longer, the sea ice melts along the marginal ice zone (MIZ, Figure 1) and forms a highly stable, relatively fresh, lens of water at the sea surface (Figure 2). "Seed" phytoplankton for the spring bloom may be released from thawing over winter ice algal communities, populations surviving under the ice and perhaps concentrated in surface waters by brash and frazil ice formation during late winter/ early spring, or transported as resting stages from shallow sediments to the surface by wind mixing events. The spring phytoplankton accumulate along

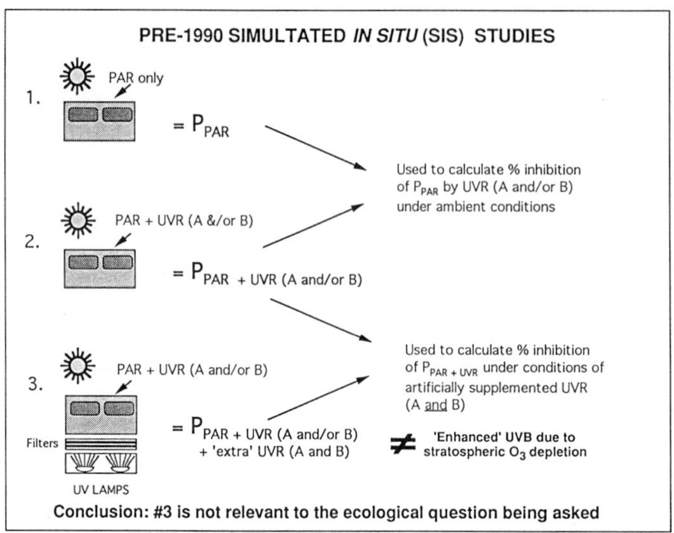


Figure 3. Experimental designs for field studies to determine rates of primary production in natural phytoplankton communities in 1) the absence of UV radiation, 2) in the presence of ambient levels of radiation and 3) in the presence of both ambient levels of UV radiation and artificially supplemented UV radiation. Small black pairs of rectangles within each incubation box represents a phytoplankton sample incubated in either quartz or polyethylene containers with radiolabeled $^{14}C$—sodium bicarbonate for determination of carbon fixation rates during the incubation time and as a function of light dose.

the MIZ and appear to become physically trapped in more buoyant surface waters and grow rapidly within this nutrient-rich, warmer, and sunlit domain. This springtime band of increased marine productivity presumably surrounds Antarctica and appears to moves south as the ice edge recedes with the season. The seaward edge of the bloom is diluted by deeper wind-mixing processes and perhaps provides the "seeds" for increased open water productivity during late spring and early summer (*v.* [2,6,11]. It is at the receding edge of the MIZ that many zooplankton, most notably krill, develop and feed on the phytoplankton in the spring and provide attractive feeding ground for a variety of seabirds, including penguins. It now appears, however, that the same physical conditions which have promoted the development of MIZ communities may become increasingly detrimental as phytoplankton are restricted to its shallow depths where $O_3$-dependent UV-B radiation (280-320 nm) increases most significantly when atmospheric $O_3$ is diminished [6,10].

The day-to-day size, shape, and position of the $O_3$ hole (polar vortex) changes continuously. Figure 1 shows changes in the spatial orientation of the $O_3$ hole with respect to the Antarctic continent for two dates, 16 days apart, in October 1990. The edge of the $O_3$ hole is largely an oscillating atmospheric front of UV-B radiation which drives major day-to-day fluctuations in UV-B irradiance (Quvb) at the earth's surface and in underwater light fields [10]. At the same time that Quvb fluctuates several fold in response to movement of the $O_3$ hole, the relative flux of UV-A radiation (Quva, 320-400 nm) and Qpar (PAR=photosynthetic available radiation, 400-700 nm) changes only slightly [9,10]. Thus the thinning of atmospheric $O_3$ not only increases the UV-B challenge to plants but also alters the spectral balance between Quvb:Quva:Qpar. We have hypothesized that upsetting the natural spectral balance of the light field would alter the evolved balance between different light-dependent processes in phytoplankton and have published field results which show that the ratio of Quvb: Qtot (Quvb+Quva+Qpar) is a critical determinant of the $O_3$-dependent susceptibility of marine primary production [6,11].

## EXPERIMENTAL CONSIDERATIONS

Experimental reality is that springtime marine phytoplankton under the influence of the $O_3$ hole are now subject to artificially large and sudden fluctuations in UV-B radiation and the spectral ratio of Quvb:Quva:Qpar [10] that are presently impossible to mimic in any other experimental design than what is provided by nature itself. Field studies have thus far attempted to assess the potential effects of $O_3$ depletion by 1) quantifying the decrease in primary production when phytoplankton are exposed to supplemental UV-B radiation (Figure 3) [2,12]; 2) quantifying the increase in primary production when ambient UV-B radiation is removed from the light field of discrete samples (Figure 3) which, until most recently, had been collected entirely outside the $O_3$ hole [2,4,6,10,12]; and 3) by comparing the UV sensitivity of primary production rates inside and outside the $O_3$ hole (Figure 1) [6,10]. We believe the first approach to be ecologically irrelevant, the second approach to provide information only on the background levels of UV inhibition of primary production in the absence of an $O_3$ hole, and the third approach to be the only one which quantifies the *additional* UV inhibition of primary production that occurs as a result of $O_3$ depletion inside the Antarctic polar vortex.

### FIELD INCUBATION CONTAINERS

Before detailing the experimental design and logic behind each kind of field approach mentioned above, we must first digress into the relative merits of using quartz versus polyethylene bags as field incubation containers for phytoplankton (depicted as the smallest pairs of dark rectangles in Figure 3). Holm-Hansen and Helbling (1993) have recently criticized the use by us as well as El-Sayed and Stephens' (1992) of polyethylene bags as incubation containers, contending that these bags become toxic when exposed to UV-B radiation and thus give artificially low productivity values that exaggerate the potential effect of $O_3$ depletion on biological processes. We have tested polyethylene bags for

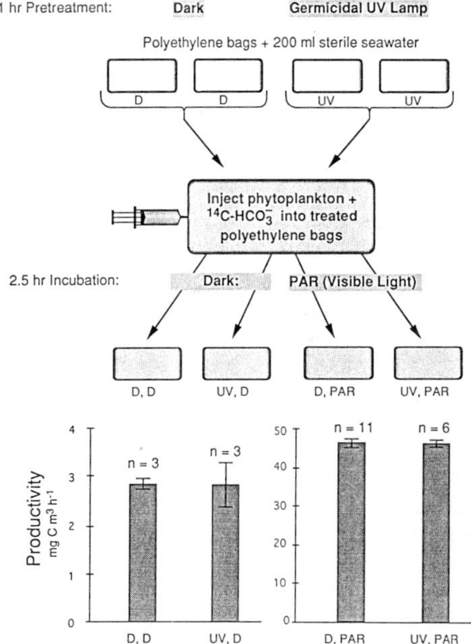


Figure 4. Schematic of a laboratory test of the effect of UV radiation on polyethylene bags used for the measurement of phytoplankton photosynthetic rates during *Icecolors '90*. 200 ml of sterile seawater medium was placed in untreated bags (400 ml capacity). Pretreatment consisted of incubating medium-containing bags for either 1 hr in darkness or 1 hr under a germicidal lamp (GE G30T8). Following pretreatment, rates of radiolabelled carbon fixation was measured by adding a 50 ml mixed sample of the diatom *Thalassiosira weisflogii* and the dinoflagellate *Hetercapsa pygmae* recently inoculated with $^{14}C$-sodium bicarbonate (final 250 ml sample conc = 0.05 µCi/ml). Replicate pretreated samples were incubated in both PAR-only light and in the dark. In sample designations (D,D; UV,D; D,PAR; UV,PAR) the first part refers to the pretreatment conditions (D = dark, UV = germicidal lamp) and the second part to the incubation conditions (D = dark, PAR = white light only). Five replicate Rts were determined for each bag. At the end of the incubations, samples were fixed with formalin and 30 ml samples were filtered onto 0.4 mm Nuclepore filters. (From Prézelin et al., 1993, with permission).

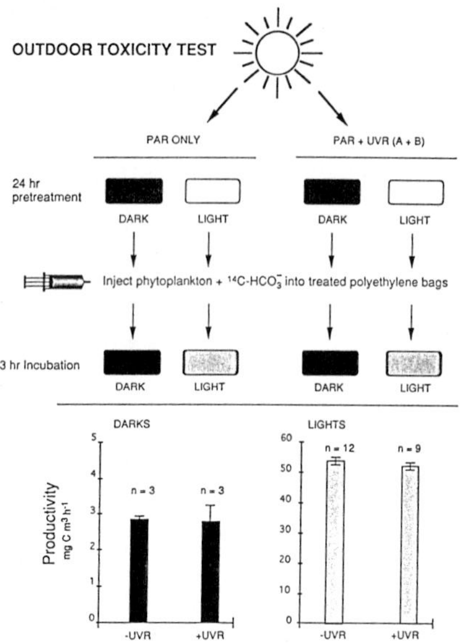


Figure 5. Schematic of a outdoor test of the effect of UV radiation on polyethylene bags used for the measurement of phytoplankton photosynthetic rates during *Icecolors '90*. 200 ml of sterile seawater medium was placed in untreated bags (400 ml capacity). Pretreatment consisted of incubating medium-containing bags outdoors for 24 hrs in temperature-controlled incubators in the light and the dark, in presence and absence of natural UVR. Following pretreatment, rates of radiolabelled carbon fixation were measured by adding a 50 ml mixed sample of the diatom *Thalassiosira weisflogii* and the dinoflagellate *Hetercapsa pygmae* recently inoculated with $^{14}$C-sodium bicarbonate (final 250 ml sample conc.=0.05 μCi/ml). Replicate pretreated samples were incubated pretreatment conditions for 3 hrs. Five replicate Rts were determined for each bag. At the end of the incubations, samples were fixed with formalin and 30 ml samples were filtered onto 0.4 mm Nuclepore filters. (From Prézelin et al., 1993, with permission.)

UV toxicity, both in the laboratory and in the Southern Ocean, and have reported the findings of several tests and field observations that showed they are nontoxic for phytoplankton production measurements [6,7]. Figures 4 and 5 summarize the part of our control work that indicated that pretreatment of the bags with anomalously high amounts of UV-B radiation or exposure to natural light fields for periods much longer than we employed in our *Icecolors '90* studies did not induce a toxic effect.

To see if we could reproduce and understand the reported results of Holm-Hansen and Helbling (1993), we recently made a side-by-side comparison of UV sensitivity of primary production in similarly sized, quartz, flat bottom, 500 ml, sealed beakers (Pyromatic, Cleveland, OH) and in polyethylene bags (Figure 6). One difference in our exercise was that the round-bottomed flasks of Holm-Hansen and Helbling (1993), hand-blown several years ago (Holm-Hansen, pers. comm.), were not commercially available. Comparisons were

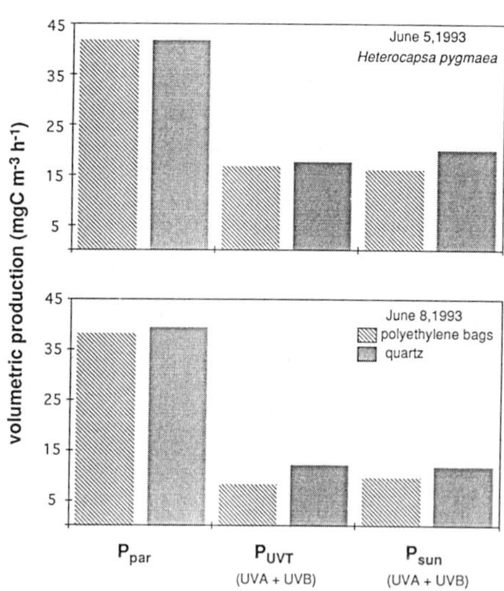


Figure 6. Side-by-side comparison of the rates of primary production measured in similarly-sized quartz and polyethylene containers. Replicate samples of the dinoflagellate *Heterocapsa pygmaea* were incubated outdoors under direct sunlight (Psun), under direct sunlight filtered through a UV-transparent Plexiglas (Puvt), and UNDER indirect sunlight filtered through a UV-blocking filter that only allowed photosynthetically available radiation (PAR = 400-700 nm) to reach the samples (Ppar). Measurements were done in Santa Barbara, California, on June 5th (a cloudy day) and June 8th (a mostly clear day).

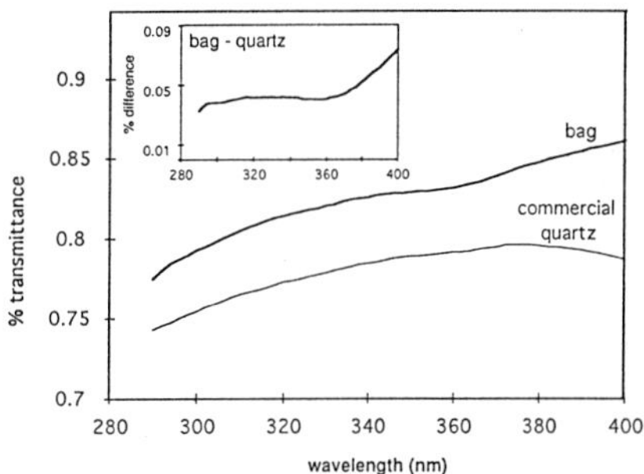


Figure 7. Comparison of the spectral transmission properties of commercially available quartz containers and polyethylene bags. See text for details.

made in Santa Barbara on samples in polyethylene and quartz containers incubated in temperature-controlled outdoor incubator systems and exposed to 24 h of daylight that was either unfiltered (direct sun), filtered through UV-transparent Plexiglas, or filtered through UV-blocking Plexiglas commonly used in field incubators [Prezelin et al., in press]. Measurements were made on two days in early summer 1993 using low-light adapted, thus physiologically very light sensitive, cultures of the dinoflagellate *Heterocapsa pygmaea*. On one partially overcast day (June 5th, Figure 6), the UV inhibition results were dramatic but not statistically different for samples in UV-transparent quartz or polyethylene containers. Rates of primary production (P) measured in the presence of UV radiation (Puv = Puvt, Psun) were only ~ 40% of those measured in the absence of UV radiation (Ppar), giving a value of ~ 0.6 for the fractional inhibition of photosynthesis by UV radiation (FIuv =1-Puv/Par). On a subsequent mostly sunny day (June 8th, Figure 6), we measured in triplicate a FIuv = 0.8 for photosynthesis in polyethylene bags and a Fiuv = 0.7 for photosynthesis in the quartz beaker, a difference that bordered on being statistically significant. Since we knew the transmittance spectra of the incubation containers, we could provide evidence for the increased UV transmission of polyethylene bags relative to the quartz container (Figure 7, discussion below) as well as the nontoxicity of the bags used in the experiment. Thus, unlike Holm-Hansen and Helbling (1993), we concluded that increased UV dosage led to lower UV production rates inside the bags than in quartz containers. This reinterpretation does not disallow the Holm-Hansen and Helbling (1993) observation that pretreating the bags with HCl did not change the behavior of the bags as incubation containers. If the bags are nontoxic, there is nothing to leach out with HCl; if the problem is lack of optical controls in their experiment, treating the containers with HCl will make no difference to the outcome of the measurements.

Given that some details of Holm-Hansen and Helbling's test of quartz vs. polyethylene containers are not available (i.e., the underwater spectral irradiance, the geometry of their containers, the degree to which containers were submerged, orientation of the neck on the quartz container, the spectral transmissional properties of their containers), we cannot offer quantitative comments on possible UV dosages within their incubation chambers. We do know, however, that the transmission spectra of commercially available quartz containers are variable within and between containers (Schofield, unpubl.) and are less transparent to UV radiation than commercial Whirlpac polyethylene bags (Figure 7). The percent transmission of the bags was 4 to 8% greater than the commercially available quartz beaker at all UV-B and UV-A wavelengths (Prézelin et al., in press). If this or any other aspect of the optical geometry of two or more kinds of UV transparent containers differ, then the exposure per unit volume (and area) would also differ significantly. And since the biological effects we are discussing are a function of exposure (whether Quvb, Quva, and/or Qpar), a study with uncontrolled optical geometry makes intercomparison of data problematic. In conclusion, we believe both quartz and polyethylene bags to be nontoxic and suitable for UV field work [6.7] although, given their different optical properties, researchers should be warned not to use them as equivalent containers in any instance. Reasons to choose one or the other might include experimental design concerns for quantifying internal spectral irradiances, optical uniformity, cost, availability, silicate contamination from quartz, durability in the field, required sample volume, and so on.

Table 1. Austral spring data bases of UV-B inhibition of marine primary production in the Southern Ocean.

| Published Work | Time | Location | Incubator | Dobson Units | # of treatments |
|---|---|---|---|---|---|
| Stephens 1989 El-Sayed & Stephens 1992 | 1987 (Nov-Dec ) | coastal bay | SIS[1] | 320-360 | 9 treatments |
| Helbing et al.[2] 1992 | 1988,1989, & 1991 (Oct-March) | coastal bay | SIS | NA[3] | uncertain 21 treatments≤ |
| Smith et al. 1992 | 1990 (Sep-Nov) | MIZ | IS[4] | 175-375 | 17 treatments |
| Prézelin et al. 1993 | 1990 (Sep-Nov) | MIZ | SIS | 175-375 | 35 treatments |
| Moline, Prézelin unpubl. | 1992 (Nov-Dec) | coastal bay | SIS | 290-337 | 3 treatments |

1 Simulated in situ measurement
2 spring and summer data are mixed; no $O_3$ data available within publication
3 Not available
4 In situ measurement

CONSIDERATIONS IN ASSESSING DATA

Concerning the topic of Antarctic UV field experiments in relationship to the $O_3$ hole, we considered it relevant to examine only data that were collected during the austral spring and within the Southern Ocean (Table 1). This was a straightforward exercise for Stephens' experiments [12], although we had to extract information on incident $O_3$ concentrations from the NASA TOMS database. The Helbling et al. (1992) publication is a largely undefined blend of spring and summer data from different locations in different years. Little $O_3$ data was provided and, since we were unsure of time and location of various data points in composite plots, we were unable to piece together an $O_3$ history for much of the database. While the band pass characteristics were available for UV filters used to screen solar radiation in both of these simulated *in situ* (SIS) studies, neither study provided quantitative information on the underwater spectral irradiances (280-700 nm) from which their samples were actually collected or under which their samples were incubated. We were fortunate during *Icecolors '90* to have precise unbroken knowledge of overhead $O_3$ concentrations and the full spectral irradiances (290-700 nm) for underwater light fields, largely due to the development of a Light and UltraViolet Submersible Spectroradiometer (LUVSS) [10].

Considering all data that met the criteria of being collected in springtime within the Southern Ocean (Table 1), we chose to eliminate any that had used supplemental sources of UV radiation so we might question whether $O_3$ depletion had an inhibitory effects on Antarctic marine primary production. Supplemental UV radiation is generally used to impose an increasing UV challenge to organisms already being incubated under ambient light conditions in hopes of simulating the potential increases in UV-B radiation that come about by thinning of the atmospheric $O_3$ layer. A major defect in this approach is that no spectral filter combinations have yet been found which closely mimic the $O_3$ absorption of sunlight through the atmosphere (Figure 8). For instance, as part of careful spectral studies conducted by Stephens and coworkers [10] in the austral spring of 1987, a supplemental source of UV radiation was provided to natural samples incubating under sunlight in outdoor incubators at Palmer Station, Antarctica [12]. The supplemental light contained minor but detectable levels of UV-C radiation below 280 nm, which is not $O_3$-dependent but highly mutagenic. In fact, wavelengths below 295 nm do not reach the surface of the earth. Also, the spectral output of the supplemental UV light had disproportionately smaller Quva:Quvb ratio than that absorbed by $O_3$ molecules in the atmosphere and included some Quva at wavelengths not absorbed by $O_3$ molecules (Figure 8). While an admirable first attempt to demonstrate the photoinhibitory effect of UV radiation on natural

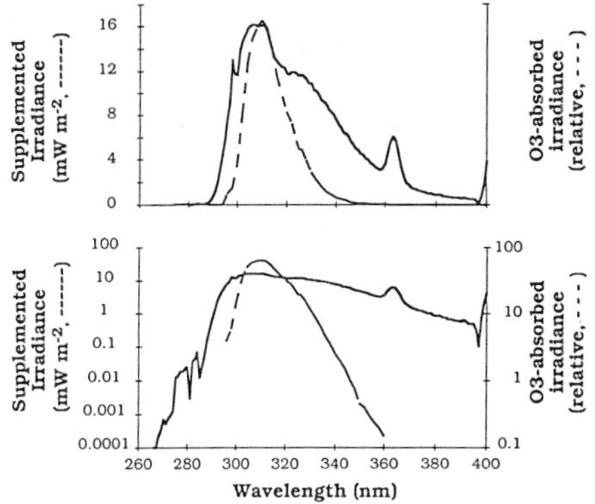


Figure 8. Comparison of the spectral output of a typical filtered light system (Stephens, 1989) employed to provide artificially supplemented UV radiation (solid heavy line) with a relative absorbed irradiance spectrum for enhanced UV radiation reaching the surface of the earth due to the depletion of atmospheric ozone, where irradiance is plotted arithmetically (inset) and semi-exponentially.

Antarctic marine phytoplankton, especially given the paucity of environmental UV spectral data that was then available for real time experimentation, this part of Stephens' studies is inadequate to quantify the biological effect of an $O_3$-dependent increases in UV radiation in nature.

There are other drawbacks to using only supplemented UV radiation as an experimental tool for ecologically relevant studies of plant physiology. For instance, there are several-fold changes in Quvb, Quva, and Qpar over the course of one day, modulated by changes in sea state, cloud cover, sun angle, and other factors. Also, there are predictable changes in the Quvb/Quva ratio that occur naturally with changes in sun angle over the day and with changes in sun angle with changes in seasons and which appear amplified and less predictable when under the influence of the $O_3$ hole [6].

ABSENCE OF UV-B RADIATION

The second kind of field measurement to quantify UV-B sensitivities in natural communities, but not always $O_3$-dependent UV-B effects, are those in which UV-B radiation is removed from the natural light field (Figure 3). Productivity in the absence of UV-B radiation is generally greater than in the presence of UV-B radiation and the difference is attributed to UV-B photoinhibition. With the exception of those of *Icecolors '90*, such field measurements have only been made outside the $O_3$ hole (Table 1) and, as such, can only provide information on the ambient levels of UV-B photoinhibition in the absence of the $O_3$ hole. However, this is important data to intercompare as it provides an estimate of background UV-B photoinhibition in the Southern Ocean prior to exposure or re-exposure to the elevated UV-B radiation inside the $O_3$ hole. Equally important, these limited data do allows us to assess to what extent the findings of different investigators agree.

If we do not take into account the spectral differences between the different incubator systems employed by different investigators, then field results suggest that UV-B inhibition outside the $O_3$ hole ranged from 8 to 15% for surface waters and 20-23% for subsurface waters (Table 2). However, the spectral properties of various field incubators to measure UV effects differ significantly (Figure 9A-C) and, as a result, researchers conducting identical experiments may thus come up with different quantitative results. Figure 9D illustrates the spectral composition of light which might be deemed "UV-B" exposure, E, for each incubation system. These spectra result from multiplying the difference spectra respectively for Figure 9 A, B, and C by the spectral irradiance of sunlight outside the $O_3$ hole (DU=350, data not shown). It is evident that the spectral irradiances for $E_{inc,uvb}$, are not identical between different studies. They all contain significantly more Quva than is attributable to increased spectral irradiance due to loss of $O_3$ (Figure 8). The inclusion of some "stray" Quva in $E_{inc,uvb}$ would result in an overestimation of FIuvb (Boucher & Prezelin, 1994). It should be noted, however, that the measurement of a strictly defined Fiuvb, one that allowed for no irradiances outside the 280-320 nm waveband, would be an underestimation of the $O_3$-dependent changes in UV radiation ($FI_{03}$) (Figure 8).

NORMALIZATION OF DATA

So how does one normalize for these differences in spectral irradiances between "UV-B" exposures in different incubation systems to derive a

Table 2. A comparison between the fractional inhibition of $P_{par}$ by UV-B radiation ($FI_{uvb}$) and the fractional inhibition of $P_{par}$ measured within the various UV-B incubators ($FI_{inc,UVB}$) for surface and subsurface concentration. All experiments were conducted under background concentrations (> 320 DU).

| | Surface | | Subsurface | |
|---|---|---|---|---|
| Original data source | $FI_{UVB}$ (%) | $FI_{inc, UVB}$ (%) | $FI_{UVB}$ (%) | $FI_{inc, UVB}$ (%) |
| El-Sayed and Stephens, 1991 | 4.0 | 8.0 | 11.6 | 22.6 |
| Helbling, Villafañe, Ferrario, and Holm-Hansen, 1992 | 7.1 | 15.3 | | |
| Smith, Prézelin, et al., 1992 | 4.8 | 8.1 | 11.6 | 20.0 |
| Moline and Prézelin, 1992 (unpubl.) | 6.5 | 14.1 | | |

better comparison of all relevant field results? In our case, we tried the following. If the spectra in Figure 9D are multiplied by the action spectrum for UV (280-400 nm) inhibition of photosynthesis for either the diatom *Phaeodactylum* sp. or the dinoflagellate *Prorocentrum micans* (Figure 9E), then a normalized biological weighted irradiance spectrum for UV photoinhibition of carbon rates is estimated (Figure 9F). [Note: This procedure is used only to normalize results from different studies to a single standard and is not, at this time, an endorsement of the relevancy of these particular action spectra to natural communities of Antarctic phytoplankton. Such biologically-weighted irradiance spectra are very dependent upon the spectral composition of the incident light field and the shape, as well as the accuracy of the shape, of the action spectrum derived from a limited number of discrete points often covering rather broad bandwidths (see Coohill, this volume).] With regards to UV action spectra, less attention has been paid to assessing the effect of UV-A radiation as on the effect of the highly damaging UV-B radiation. Thus there is some uncertainty associated with our attempts to portion the biological-weighted irradiance spectra between UV-A and UV-B radiation.

From the resulting spectra (Figure 9F), however, the fractional inhibition (FI) by any spectral bandwidth can be determined by portioning its area under the curve to the total area. For instance, the Fluvb is calculated by subtracting the contribution due to UV-A inhibition. Results for the different incubator systems are shown in Table 2. Available data is actually in considerable agreement. In the absence of an $O_3$ hole, ambient concentrations of springtime UV-B radiation routinely suppress daily rates of Antarctic primary production by at least 4% to 7% in surface waters (2 m) and by about 12% in near surface waters (5-10 m). [Further documentation of the subsurface increase in UV-B inhibition of photosynthesis and the possible underlying mechanisms are found in Prézelin et al., 1993. In general, it appears that UV-B inhibition of primary production for any given time period is a reflection of the combined kinetics of UV-B photodamage and protective mechanisms occurring during the same time period.]

The final experimental design question of this paper is how to quantify the $O_3$-dependent UV effects on marine primary production under the Antarctic $O_3$. Obviously any measure of production under the $O_3$ hole, where all incoming UV-B radiation is removed from the incubation chambers, will provide an estimate of UV-B photoinhibition that combines both the effects of background levels of UV-B normally present plus the additional UV-B radiation due to the thinning of the stratospheric $O_3$ layer. By subtracting the springtime value for the background level of UV-B inhibition, which can be measured quite consistently outside the hole (Table

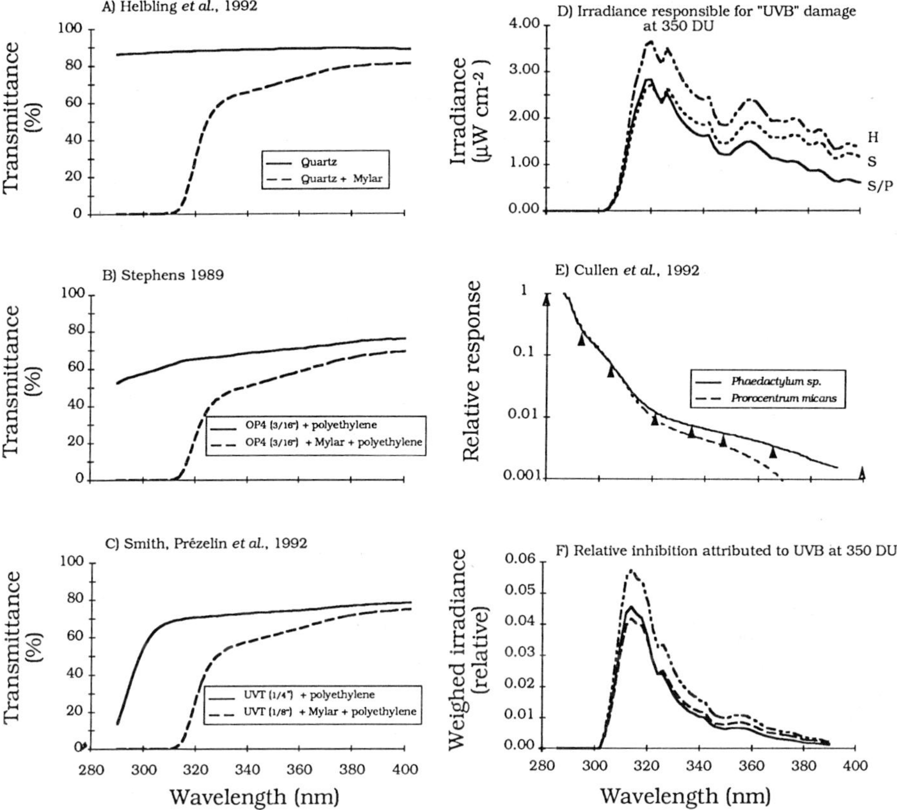


Figure 9. Comparison of the spectral transmission of UV-transparent incubators (solid line and UV-B selective incubators (dashed line) used to detect UV-B effects on Antarctic marine phytoplankton in independent studies by A) Helbling et al, 1992; B) Stephens, 1989; and C) Smith et al. 1992. D) Spectral irradiances deemed "UV-B" exposure (Einc,uvb) for each of the UV-B incubation systems presented in A-C. E) The biological weighting functions for UV-inhibition of rates of carbon fixation (primary production) in laboratory cultures of the diatom *Phaeodactylum* sp. (solid line) and the dinoflagellate *Prorocentrum micans* (dashed line) (Cullen et al. 1992). Closed triangles represent cutoff wavelengths for 50% transmission by filters used to determine the wavelength dependency of photosynthesis in these phytoplankers. F) Calculated as the product of individual spectrum in D) and the biological weighting function for *Phaeodactylum* sp. shown in E), a comparison is made of the biologically-weighted irradiance spectra for UV-inhibition of photosynthesis within different UV-B incubation systems used for field studies of the sensitivity of Antarctic phytoplankter to $O_3$-dependent UV-B inhibition of primary production.

2), an estimate of the level $O_3$-dependent UV-B inhibition can be derived. To date only the *Icecolors '90* expedition has made such an attempt [10]. Results for our 16 *in situ* moorings deployed in and out of the $O_3$ hole are shown in Figure 10, while supporting results for shorter term simulated *in situ* incubations to determine whether there was a diurnal periodicity or reciprocity or both in UV-B inhibitory effects on marine primary production are presented elsewhere [6].

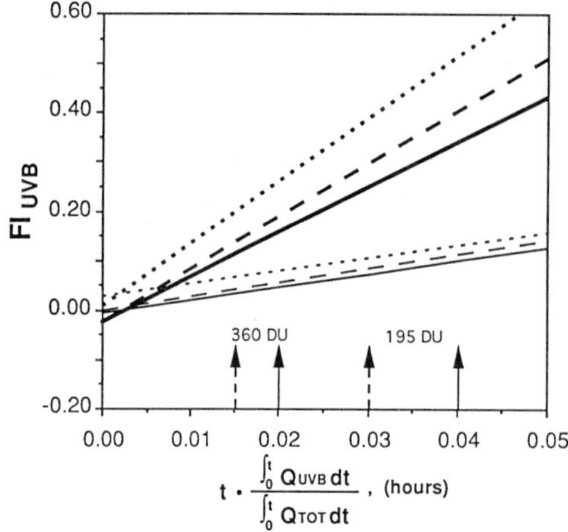


| | Surface $FI_{UVB}$ (%) | | | Subsurface $FI_{UVB}$ (%) | | |
|---|---|---|---|---|---|---|
| spectral calculation | none | (Ps) | (Pm) | none | (Ps) | (Pm) |
| $r^2$ | 0.35 | 0.49 | 0.47 | 0.74 | 0.75 | 0.75 |
| (p) | (0.165) | (0.081) | (0.089) | (0.001) | (0.001) | (0.001) |
| Low $O_3$ (195 DU) | 13.4 | 10.0 | 11.2 | 39.3 | 25.2 | 29.8 |
| High $O_3$ (360 DU) | 8.1 | 4.8 | 5.5 | 20.0 | 11.6 | 13.9 |
| change | 5.3 | 5.2 | 5.7 | 19.3 | 13.6 | 15.8 |

Figure 10. TOP: *In situ* fractional inhibition of PAR rates of photosynthesis due to UV-B radiation ($FI_{UVB}$) versus the length of exposure multiplied by the *in situ* UV-B radiation ($Q_{UVB}$) integrated over the incubation time and normalized to the integrated *in situ* total radiation ($Q_{TOT}$) for the same time period for surface (thin lines) and subsurface (bold lines) samples. The $FI_{UVB}$ were calculated without (dotted lines) or with a correction for "stray" UV-A light, the later being estimated using the biological weighting function for phytoplankton photosynthesis for *Phaeodactylum* sp. (solid lines). Arrows represent typical exposure inside (195 DU) the $O_3$ hole at the surface (solid arrows) and at 5 m (dashed arrows). Lower curves are for surface waters; upper curves for subsurface waters. (From Boucher and Prézelin, 1994, with permission). BOTTOM: Estimates of the fractional inhibition of UV-B radiation on primary production inside and outside the Antarctic ozone hole. Spectral corrections for possible UV-A "leakage" into the UV-B calculations were either ignored (none) or based on biological weighting functions (Cullen et al., 1992) determined for laboratory cultures of the diatom *Phaeodactylum* sp. (Ps) or the dinoflagellate *Prorocentrum micans* (Pm).

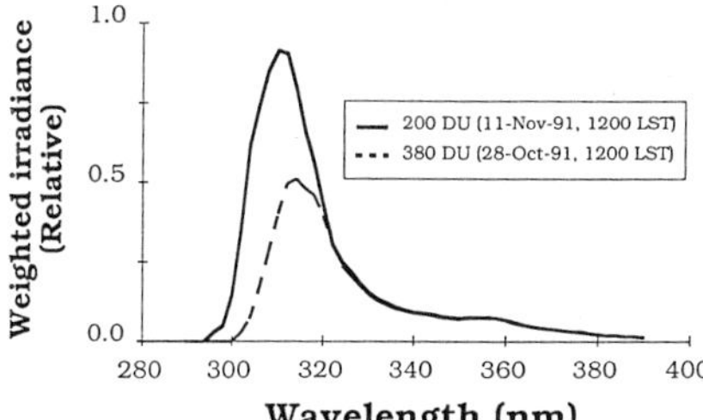


Figure 11. Comparison of estimates of the biologically-weighted spectral irradiances for UV inhibition of surface primary production for midday clear skies inside (200 DU) and outside (380 DU) during the 1990 $O_3$ hole. The biological weighting function for this calculation was determined for *Phaeodactylum* sp. (Cullen et al., 1992) and shown in Figure 9E.

We were able to detect UV-B-dependent inhibition of direct field measurements of *in situ* primary production down to 25 m (4% $Q_{PAR}$) (Figure 10). UV-B inhibition increased with increasing UV-B dose and with $O_3$-dependent increases in the ratio of $Quvb/Q_{tot}$. On occasion, UV-B radiation could lead to a 60% reduction in the potential primary productivity of near surface waters of the MIZ (Prézelin et al. 1993). UV-B inhibition of photosynthesis differed for surface and subsurface phytoplankton communities deployed at a fixed depth *in situ* for 9 to 12 hours of daylight [10]. Full-day exposures to a higher UV-B dosage rate in surface waters led to a *lower* photoinhibitory effect on rates of photosynthesis than did full day exposures to a lower UV-B dosage rate in subsurface waters. Furthermore, transfer experiments [6], where samples taken from depth were incubated at the surface and vice versa, showed UV-B-inhibition effects were independent of the sampled depth and depended only on dose at the depth incubated.

Most recently we discussed how and when to consider a UV-A spectral correction of the *Icecolors '90* database [1]. The only difference between these calculations and those describe above (Figure 9) for data collected outside the $O_3$ hole (DU=350) was that we recognized that the percentage of UV-B radiation within $E_{inc,uvb}$ will co-vary with $Q_{uvb}$ to $Q_{total}$ ratio and will therefore be inversely proportional to atmospheric $O_3$ concentration (Figure 11), zenith angle, and depth in the water column. As a consequence, the overestimation of UV-B inhibition (and the subsequent underestimation of UV-A inhibition) will be more pronounced inside the $O_3$ hole than outside, at the surface than at depth and at the beginning of the day than at midday. Again, any algorithm tempting to correct for this UV-A light "leakage" into UV-B incubators must take into account the spectral shape of the irradiance at the depth of incubation and over the incubation period, as well whether or not the biological weighting functions used in these analyses are applicable to field communities under investigation. Using the spectral algorithm for both UV action spectra in Figure 9E, we calculated the $FI_{UVB}$ for the *Icecolors '90 in situ* productivity measurements. Figure 10 shows the comparison between the dose response curves so obtained and the non-spectrally corrected dose response curves [1]. The results show that there is an additional 5%-16% decrease in rates of primary production in the upper part of the water column due $O_3$ depletion from 360 to 190 DU. Furthermore, application of these possible corrections to our *Icecolors '90* database does not change the overall conclusion that, during the austral spring of 1990, atmospheric $O_3$ depletion and associated increases in $O_3$-dependent UV-B radiation over the Southern Ocean resulted in at least a 6-12% loss of primary productivity in the marginal ice zone [1,6,10]. Again, this spectral correction is valuable only if one wishes to quantify UV-B effects alone. Since $O_3$ molecules absorb strongly in the UV-B and weakly in the UV-A, it may be valid not to correct for minor amounts of stray UV-A.

We believe that our careful consideration of the data available should be taken as encouragement in the sense that it shows that independent investigations are converging toward similar values for background levels of UV inhibition in the Southern Ocean and that one study [6.10] has provided the first conclusive evidence that reduction in atmospheric $O_3$ does increases the penetration of damaging UV radiation into the Southern Ocean and impairs the vitality of natural phytoplankton communities. The good news is that these studies provide a common base on which to pursue many pressing questions regarding the environmental consequences of atmospheric $O_3$ depletion to aquatic phytoplankton communities. The better news is that, during the present 1993 austral spring, there will be three additional NSF-funded studies (including *Icecolors '93)* in Antarctic waters to assess the level and underlying mechanisms of $O_3$-dependent UV photoinhibition of marine primary productivity, each with a strong emphasis on bio-optical models that can accurately predict observed changes.

We look forward to the time when we can once again synthesize our data and promote progress in scientific understanding within this area of environmental concern.

**Acknowledgments**

This work was supported by NSF grant DPP 8917076 and represents *Icecolors '90* contribution #7. We thank Sasha Madronich for helpful discussions and for providing data on the absorption spectrum for ozone; Pat Neale and John Cullen provided data on the biological weighting functions for photosynthesis in marine phytoplankton; and Ray Smith and Kirk Waters provided three fully spectral (280-700 nm) datasets which typify clear midday incident irradiances inside and outside the $O_3$ hole at the sea surface of the Southern Ocean during the *Icecolors '90*. We dedicate this paper to the women and men of Palmer Station, Antarctica, and the R/V Polar Duke who have contributed so much to our scientific achievements.

**REFERENCES**

1. Boucher, N.P., and B.B. Prézelin. 1994. Ozone-dependent UV effects versus UV-B specific effects on primary productivity in the Southern Ocean. How and when to Consider an UV-A spectral correction of direct field measurement. Antarctic Journal of the United States, in press.

2. El-Sayed, S.Z., and G.A. Fryxell. 1993. Phytoplankton. *In* Antarctic Microbiology, Wiley-Liss, Inc., pp.65-122.

3. Helbling, W.E., V. Villafane, M. Ferrario, and O. Holm-Hansen. 1992. Impact of natural ultraviolet radiation on rates of photosynthesis and on specific marine phytoplankton species. Mar. Ecol. Prog. Ser. 80:89-100.

4. Holm-Hansen, O., and E.W. Helbling. 1993. Polyethylene bags and solar ultraviolet radiation. Science 259:533.

5. Madronich, S. 1993. UV radiation in the natural and perturbed atmosphere. *In* M. Tevini (ed.), UV-B Radiation and Ozone Depletion: Effects on Humans, Animals, Plants, Microorganism, and Materials, CRC Press, Boca Raton, Florida, pp.17-69.

6. Prézelin, B.B., N.P. Boucher, and R.C. Smith. 1993. Marine primary production under the influence of the Antarctic ozone hole. *In* S. Weiler and P. Penhale (eds.), Ultraviolet Radiation and Biological Research in Antarctica, in press.

7. Prézelin, B.B., and R.C. Smith. 1993. Polyethylene bags and solar ultraviolet radiation: Response. Science 259:534-535.

8. Smith, K.C. (ed.). 1989. The Science of Photobiology. Plenum Press, New York, 246 pp.

9. Smith, R.C., and K.S. Baker. 1989. Stratospheric ozone, middle ultraviolet radiation and phytoplankton productivity. Oceanography 2:4-10.

10. Smith, R.C., B.B. Prézelin, K.S. Baker, R.R. Bidigare, N.P. Boucher, T. Coley, D. Karentz, S. MacIntyre, H.A. Matlick, D. Menzies, M. Ondrusek, Z. Wan, and K.J. Waters. 1992. Ozone depletion: Ultraviolet radiation and phytoplankton biology in Antarctic waters. Science 255:952-959.

11. Smith, W.O., Jr., and E. Sakshaug. 1990. Polar phytoplankton. *In* W.O. Smith, Jr. (ed.), Polar Oceanography, Part B: Chemistry, Biology and Geology, Academic Press, NY, pp.477-526.

12. Stephens, F.C. 1989. Effects of ultraviolet light on photosynthesis and pigments of antarctic marine phytoplankton. PhD thesis, Texas A & M University, 121p.

13. Stolarski, R.S., R. Bojkov, L. Bishop, C. Zerefos, J. Staehelin, and J. Zawondny. 1992. Measured trends in stratospheric ozone. Sci. 256:342-349.

14. Sullivan, C.W., C.R. McCain, J.C. Comiso, and W.O. Smith, Jr. 1988. Phytoplankton standing crops within an Antarctic ice edge assessed by satellite remote sensing. J. Geophys. Res. 93:487-498.

# UV-B AND HUMAN HEALTH

# EFFECTS OF UV-B RADIATION ON HUMAN AND ANIMAL HEALTH

Edward C. DeFabo

Department of Dermatology
The George Washington University Medical Center
Washington, DC 20037

## SUMMARY

In most effects of sunlight on the human body, UV-B radiation is the predominant component. Increased UV-B irradiance is, therefore, expected to have several consequences for human health. Experimental studies on animals and humans have shown that UV-B radiation may specifically suppress functions of the immune system. Increased UV-B irradiance might, therefore, have consequences on the incidence and severity of certain infectious diseases, for the effectiveness of vaccination programs and for the formation of cancers of the skin.

It is now clear that UV-B radiation can suppress immune functions not only in animal models but also in humans. Indeed, according to recently published reports, the laboratory mouse appears to be very adequate as a model for human immune suppression. Thus extrapolations may be possible in terms of risk assessment since UV activation of the human immune system is established and the proposed UV-B-activated initiator of immune suppression exists in both species, is found in the same tissue, and shows identical properties of photoisomerization by UV-B radiation.

## TYPES OF STUDIES

Two of the most commonly studies immune responses in the laboratory are suppression of the contact hypersensitivity (CHS), or delayed-type hypersensitivity response (DTH), and the suppression of the rejection of transplanted tumors. The photoreceptor for the suppression of the CHS response appears to be a natural component of skin, a deanimated form of histidine. Because of similar photoimmunologic responses between the suppression of the CHS and suppression of tumor rejection, this same photoreceptor may be involved in skin cancer development. Given such a wide diversity in suppression by UV-B radiation, the question now is raised whether other types of immune suppression are possible and if this might not be related to other types of human health problems. To answer this question, we suggest the following studies:

1. *Infectious diseases*. Experimental studies are required to determine which infectious diseases might be exacerbated by UV-B through ozone depletion. Based on what is known, certain criteria can be suggested. As a first approximation, these are 1) the infectious agent must come into contact with skin for some period of time and 2) UV-B exposure must be such that activation of immune suppression occurs while the infectious agent or its antigenic component is skin resident. The interaction of these two events must be such that suppressor T cells can be formed and their concentration accumulated in such a way that rejection of the infectious agent by immune cells is down modulated or switched off. Some experimental evidence is available indicating that infectious agents such as Leishmania parasite and Herpes virus are affected by UV-B radiation and, in the case of Herpes virus, this effect is known to involve immune

NATO ASI Series, Vol. I 18
Stratospheric Ozone Depletion/
UV-B Radiation in the Biosphere
Edited by R. H. Biggs and M. E. B. Joyner

suppression with the formation on antigen-specific suppressor T cells. To determine which other diseases might be involved, epidemiological studies are needed and could play an important role by determining if a strong latitudinal (or altitudinal) gradient exists for various diseases. For example, although diabetes and multiple sclerosis are not known to be infectious diseases, a latitudinal gradient has been shown to exist in the case of diabetes and, in the case of multiple sclerosis in the United States, a correlation exists showing that the northern half of the country shows a much higher incidence than the southern half. Although, in both cases, other parameters can be postulated to explain the correlation, it is also possible that UV-B may play a role. For these two diseases, this is particularly interesting because both diabetes and multiple sclerosis are now believed to have a strong autoimmunologic component. Autoimmune regulation is such an important component of maintaining immunologic balance or homeostasis that any potential modulation of this component by UV-B needs to be considered carefully in terms of human health perturbations. Thus, it is proposed that, where possible, comparable populations with different exposures to sunlight be used to ascertain correlation between UV-B and disease, autoimmune or otherwise. Such populations, for example, can be found in the UK and Australia. Additionally, existing disease registries could be used to screen for latitudinal or altitudinal correlations.

Another related component of UV-B and infectious disease is vaccination. Again, according to experimental data, the interrelationship between the sensitizing antigen, whether it be viral, tumor, or chemical, and UV-B appears to be such that prior exposure to UV-B, followed by sensitization with antigen, leads to suppressor T cell formation specific to that antigen. Thus, if such a mechanism is working in humans, the possibility exists that the efficacy of a particular vaccination with live or attenuated organisms following exposure to UV-B may be affected. The probability that this will happen, that is, some quantitative value is needed to evaluate this risk, can only be determined through additional research in this area. Initially this should be carried out in experimental animal systems then later through human epidemiologic analysis.

2. *Cataract*. Another potentially important effect of increased UV-B irradiance is an increase in the incidence of cataracts. Cataracts are opacities of the eye lens that leads to impaired vision; it can be surgically treated, but is a leading cause of blindness in the world. There are indications from epidemiology that sunlight is involved, but there are also contradictory observations. Animal models are available, but until now only scarcely used. Further studies, especially studies including diet, are urgently needed. An impact on important farm animals can lead to an impact on food production, especially in countries that depend heavily on the raising of such species as cattle and sheep.

3. *Skin cancer*. The best documented effect of increasing UV-B irradiance is an increase in the incidence of skin cancer. Skin cancer is the most investigated ramification of increased UV-B irradiance and has led to the best-developed theoretical work on the predictions of effects. This is the area where concepts such as the radiation amplification and biological amplification factors were defined, and so leads the way for study of other biological effects. The main uncertainty in this area is about malignant melanomas, cancers of the pigment cells of the skin. This is the type with the highest mortality. There are indications from epidemiology for an involvement with UV-B radiation. After long searching, two experimental models have become available in recent years: a South American *Monodelphus domestica*, the other a platyfish-swordtail hybrid fish.

There is little doubt in the scientific community that increased UV-B irradiance, other things being equal, will lead to an increased incidence of nonmelanoma skin cancers. These skin cancers have a high incidence but a comparatively low mortality of about 1%.

4. *Other issues*. Virtually nothing is known about the relationship between UV-B and metastasis. What is well know from animal studies is that UV-B immune suppression plays a major role in suppressing the destruction of initiated skin turmors. Thus a potential role of UV-B immune suppression and cancer metastasis needs investigation since it is the metastasis of melanoma and other cancers that represents the fatal aspect of this disease. There is some limited experimental data indicating that UV-B may indeed exacerbate this phenomenon although much more research in this area is needed. Using animal models to study the mechanism of metastasis and the effect of diet and nutrition on it are examples of at least one new direction to take in assessing the overall impact of UV-B on immune suppression and cancer.

# EPIDEMIOLOGY

C.R. Gillis and D.J. Hole

West of Scotland Cancer Surveillance Unit
Ruchill Hospital, Glasgow G209NB, Scotland

Key words: cataract, cancer, epidemiology

## ABSTRACT

Epidemiology is the study of the distribution and determinants of human disease and is the basic science of public health medicine.

The principles of investigating the effects of global climate change are no different from any other environmental hazard. The key issue is how to conclude that, though real, the hazard causes, rather than is, associated with the diseases suspected. Answers depend on the quality of the data used and the types of study carried out.

Those with responsibility for health require clear goals which the public and policy makers can relate to whether from "top-down" or "bottom-up." These are:

policy on health effects determined from data
monitoring the outcome of health interventions
conduct of local epidemiological studies
telling the public

These goals can can only be achieved and will only be achieved when there is confidence in the data. The route to such confidence lies in the sequential collection and examination of data on morbidity and mortality; the formation of multidisciplinary groups to review and interpret these; the conduct of case control and, ultimately, population cohort studied (including those on occupation; and, importantly, the use of this information in education for prevention and care.

The above is based on the authors' experience of carrying out this work in the West of Scotland over the past twenty years in the field of cancer studies.

## INTRODUCTION

The most useful definition of epidemiology is that it is "the study of the distribution and determinants of disease frequency in man" [1]. In other words, the where and why of disease.

Chronic disease epidemiology has been practiced since biblical times. Although it has only been written about systematically in the latter half of this century, in 1872 the first Registrar General for England and Wales, William Farr, wrote, "Without embarassing ourselves with the difficulties the vast theories of life present, there is a definite task before us—to determine from observation the sources of health and the direct causes of death in the two sexes at different ages and under different conditions. The exact determination of evils is the first step toward their remedies." So how exactly are "evils" determined?

Figure 1, a scene of Glasgow in 1952, shows atmospheric pollution, a dense population living in unattractive substandard housing—but as yet no problem with UV-B radiation. These were the famous urban slums of Glasgow. The housing has been removed and their population how moved to new

NATO ASI Series, Vol. I 18
Stratospheric Ozone Depletion/
UV-B Radiation in the Biosphere
Edited by R. H. Biggs and M. E. B. Joyner

Figure 1. The now-demolished Glasgow slums, shown in 1952.

towns on green field sites. The major causes of disease in these populations turned out not to be external atmospheric pollution, but cigarette smoking, malnutrition, and infection. Atmospheric pollution had a much lesser role.

When the first anti-tuberculosis drugs were used after the Second World War, mortality from infection, in particular tuberculosis, was at its lowest point in the last hundred years. Today, nearly fifty years later, we are again experiencing a rise in tuberculosis. Is it oversimplistic to relate this increase in the rate of tuberculosis to the consequences of drug abuse or AIDS or even, as some might claim, to global climate change altering the way in which diseases may occur. The late Alex Haddow from Glasgow discussed the similarity of the distribution of the map of Burkitt's lymphoma in Africa with the map of yellow fever and other viruses. He realized that if he deleted from the maps all areas in which either the temperature fell below 60°F or the rainfall below 20" per year [2], a perfect fit of the lymphoma and infection map occurred. This observation led to the discovery of the Epstein-Barr (EB) virus and the first identification of a cancer caused by a virus in man. The EB virus has been associated with glandular fever in the US and Europe, nasopharyngeal cancer in China, and recently with lymphoma in the UK. The question, could global climate change influence any of these, suggests itself.

In Figure 2, the "evils" we are talking about are shown as they relate to one other. The questions implied by Figure 2 include:

a. do they actually cause changes in the im mune response?
b. do they cause changes in the occurrence of any cancer?
c. is infection a marker of any of this?
d. are any other diseases involved?

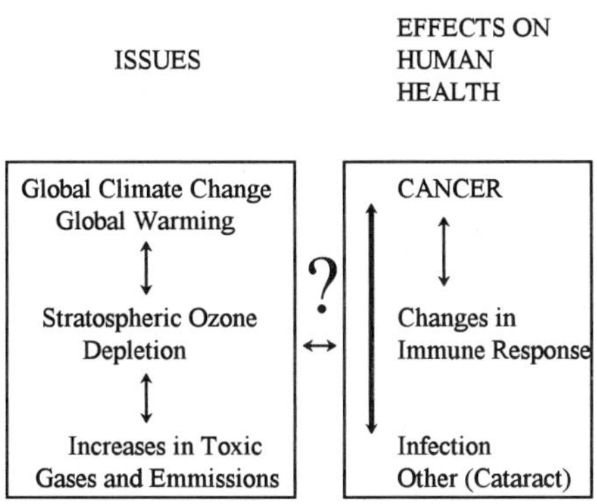


Figure 2.

## HOW EPIDEMIOLOGICAL STUDIES ARE CARRIED OUT

Other participants may tell you of the state of knowledge so far revealed by research. We now consider what is involved in the determination of causality from an epidemiological point of view. The evidence used comes from the examination of data on human populations.

The first source of data in human populations is the census which occurs in many countries usually every ten years and tells us the populations at risk, who they are, and where they live.

The second source is mortality data which record at least the fact and, though not always with accuracy, the cause of death.

The third source is disease registers which, if population-based, give some idea of the burden of disease people are suffering. Cancer and communicable disease registers are good examples but of course there are others for specialist purposes. As yet few of these are related to the general population. In some countries the data on populations is sufficient to give some idea of social class or economic deprivation by using types of housing, car ownership, or other data as surrogates.

It would be satisfying to believe that by examination of these data, hypotheses could be generated that could then be examined in the population using different types of epidemiological studies. In fact, all the best epidemiological investigations we have been involved with and know about have come from clinical colleagues with a hunch about a disease or, where a hazard has occurred in the population, the public have come and asked "does this hazard cause that disease?" It is equally true that many other investigations of this nature have come to nothing.

The first step is to see whether those exposed to the hazard have more disease than those who are not and that it is not a chance finding (association). The problem is that simple entry into a four-fold table (Figure 3) can lead to the risk of quite spurious associations between the hazard and the disease being taken seriously. One example of a spurious association occurred when a distinguished scientist mapped the distribution of television sets in a well-known city and, because owners of TV sets at that time lived in a certain part of that city where the occurrence of coronary heart disease was high, seriously considered that television was a major cause of coronary heart disease.

ESTABLISHING POSSIBLE LINKS

In 1965 Sir Austin Bradford Hill, a well-known statistician, suggested nine criteria to establish whether a possible link between a hazard and a disease was causal [3]. These criteria were:

1. consistency and lack of bias of the findings
2. strength of the association
3. temporal sequence

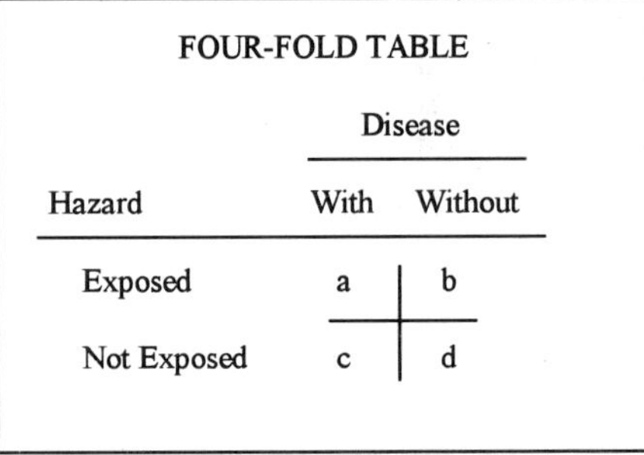

FOUR-FOLD TABLE

| | Disease | |
|---|---|---|
| Hazard | With | Without |
| Exposed | a | b |
| Not Exposed | c | d |

Figure 3.

4. biological gradient (dose-response relation ship)
5. specificity
6. coherence of the biological background and previous knowledge
7. biological plausibility
8. reasoning by analogy
9. experimental evidence

In practice, when related to disease and UV-B, this means:

a. Is there evidence of consistent increases in cancer or other diseases in areas where UV-B has increase compared to those where it was not?

b. Does the magnitude of UV-B dose to the population relate to the size of observed increases in disease?

c. Does any increase in disease in particular areas relate to the time an increase in exposure to UV-B radiation was first detected?

The answers to these questions are based on statistical evidence. If supported by biological evidence, such as on how ozone depletion or UV-B radiation in the biosphere could cause disease in animals or man, the statistical evidence becomes biologically plausible. In the case of cancer no single hazard causes an increase in all forms of cancer; all cancers have many causes. The most information is known about the causes of lung cancer yet some non-smokers develop it. So too other cancers are not expected to have any single cause.

Thus if a statistical association is shown to exist between increased UV-B and skin cancer, it should not be assumed that increased UV-B is the cause—only that it is associated and a major candidate in the study of causality.

## STRATEGIES FOR IDENTIFYING AND COPING WITH HEALTH EFFECTS

To take this further, clear goals are required. These are:

1. Local epidemiological studies of health effects;
2. Monitoring the outcome of specific interventions;
3. Policy on health effects managed from data;
4. Telling the public.

The extent to which they are related to each other depends on progress in the types of studies illustrated in Figure 4.

We have found through the study of cancer that local organizations are helpful in achieving these goals by bringing individuals of relevant disciplines together and coordinating their activities. Figure 4 illustrates the order in which such organizations should review what data they can. Each of the steps illustrated is necessary to ultimately link information on disease with data on environmental changes and produce confidence in the results.

## MORTALITY

Mortality data are often criticized usually because, in many countries, they are of variable or questionable quality. Indeed, in some countries it is difficult to link morbidity with mortality data for reasons of confidentiality. The fact of death is one of the few total population measures in existence and documented increases in deaths that can be related to specific events in time have been used as a starting point for investigations about cause.

## INCIDENCE

International incidence data on cancer have well-known problems in relation to completeness, accuracy, diagnostic technique, and classification. These data are published, mainly through the efforts of the International Agency for Research on Cancer in Lyon, in "Cancer Incidence in Five Continents" [3]. Over time, the quality of these data has been improving. Each successive volume produces further improvements and incorporates more registries. In addition, the ability to study time trends for a number of cancers now exists although non-melanoma skin cancer, the one where UV-B radiation is believed to be the primary cause, is of concern because of variations in completeness and accuracy. In some areas it may be necessary to set up special population-based registers.

In the field of infection, surveillance can be more difficult to relate to defined populations in the same way as cancer because of the lack of use of population denominator information. Again, in many countries, the situation is improving. Indeed, we have linked data on infection by hepatitis B with the occurance of primary liver cancer using routinely collected data from the West of Scotland Cancer Registry and the Communicable Diseases (Scotland) Unit [4].

In many countries, record linkage routinely takes place between cancer registries and the holders of

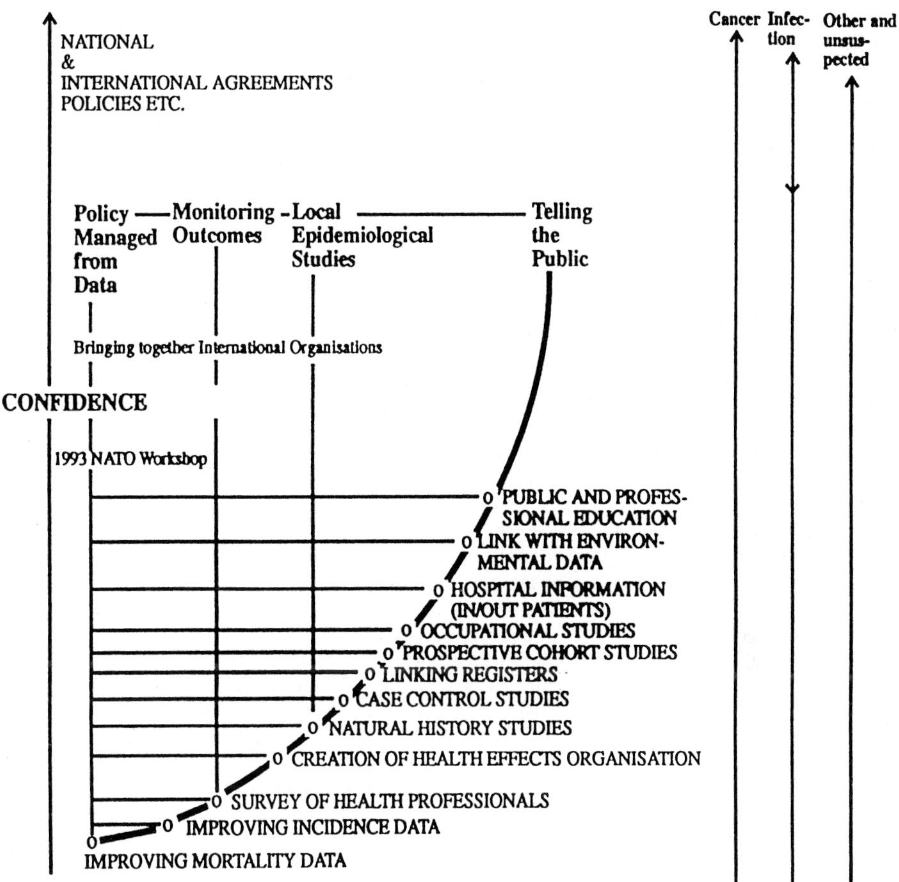


Figure 4. Strategy for coping with health effects.

mortality data. The step of linking the cancer registry with the communicable diseases registry was taken because a baseline could be drawn in a human population to assess the future consequences of exposure to hepatitis B and on the progress of future immunization strategies against this virus.

## HEALTH PROFESSIONALS

Discovery of the virus etiology of Burkitt's lymphoma was accomplished by strongly motivated health professionals coming together. If we are to understand health effects, we need organizations of motivated professionals from a wide variety of backgrounds who can be brought together to provide a basis for local studies and to contribute to an international program of research—in other words, a health effects organization. That has been out experience in cancer studies; it has also been our experience that such professionals will not stay together unless they are offered a service to feed back results. We found that the offer to professionals in both basic and applied medicine of statistical and computing services to be of considerable value and consequence.

Such basic studies need not be epidemiologically based, but in response to the ideas of local health professionals. In medicine, such studies are often called natural history studies.

### CASE-CONTROL STUDIES

A case-control study is a comparison of the level of exposure to the hazard among those who have the disease compared to the level of exposure to the hazard among those who do not have the disease. This approach seems to be the way in which most future studies of health effects will be carried out in different parts of the globe. The problem is that as the concentration of UV-B changes, so may the lifestyles of the population exposed. It is important in this instance that a clear understanding of potential confounders is achieved. Confounders are unrelated factors that may mask or exaggerate health effects thought to be due to changes, in this case changes in UV-B radiation. One problem with case-control studies lies in the choice of the control group; today's control may be tomorrow's case.

### COHORT STUDIES

One method of avoiding many of the problems of case-control studies is to carry out cohort studies. A cohort study is one where a group of individuals is identified at a particular time and carefully followed for their experience of disease prospectively. Advantage could be taken of the various prospective cohort studies in existence around the world so as not to duplicate the considerable cost and difficulty of setting up these studies and following members of the cohort for long periods of time.

### OCCUPATIONAL STUDIES

Another type of study is that of occupational groups with particular exposures such as those who work primarily outdoors including farmers, fishermen, and sportsmen.

### HOSPITAL INFORMATION

The ability to link information from case-control, cohort, and occupational studies with hospital in- and out-patient data often can be carried out by cancer registries and is a useful tool in the follow-up of these studies.

### LINK WITH ENVIRONMENTAL DATA

Figure 4 shows the point where linkage with environmental data is of most value in terms of being confident about the nature of the scientific findings.

## CONCLUSIONS

The use of these types of information in professional and public education goes without saying. The sequence of types of study is no more that what is in most textbooks of epidemiology. The point this paper makes is that, as appropriate data are collected, analyzed, and studied in the order shown, progress toward coping with health effects could be exponential. We can say from our experience over the past twenty years in cancer that it has been a strategy worth adopting.

## REFERENCES

1. MacMahon, B., and T.F. Pugh. 1970. Epidemiology: Principles and methods. Little Brown and Company, Boston.

2. Haddow, A.J. 1964. Age incidence in Brukitts Lymphoma Syndrome. E. Afr. Med. J. 41:1.

3. Hill, A.B. 1065. The environment and disease: Association or causation? Proc. R. Soc. Med. 58:295–300.

4. Lamont, D.W., K.A. Buchan, C.R. Gillis, and D. Reid. 1991. Primary hepatocellular carcinoma in an area of low incidence: Evidence for a viral aetiology from routinely collected data. In. J. Epidemiol. 20:60–67.

# ENVIRONMENTAL EFFECTS ON HUMAN HEALTH AND IMMUNE RESPONSES

Amminikutty Jeevan and Margaret L. Kripke

Department of Immunology, The University of Texas
M.D. Anderson Cancer Center, Houston, Texas.

**ABSTRACT**

Depletion of stratospheric ozone is expected to lead to an increase in the amount of UV-B radiation present in sunlight. In addition to its well known ability to cause skin cancer, UV-B radiation has been shown to alter the immune system. The immune system plays an essential role in defending the body against invading organisms and provides resistance against the development of certain types of cancer. All of the known effects of UV-B radiation on the immune system are mediated through the skin. Because skin is our primary organ that comes in contact with the environment, changes that occur within the environment can alter the immune response by means of their interaction with the skin. Such an impairment of immune function may jeopardize health by increasing susceptibility to infectious diseases, increasing the severity of infections, or delaying recovery from infections. In addition, impaired immune function can increase the incidence of certain cancers, particularly cancers of the skin.

It is well known that exposure of the skin to UV-B radiation can suppress certain types of immune responses in laboratory animals. These include rejection of UV-induced skin cancers and melanomas, contact allergy reactions to chemicals, delayed-type hypersensitivity responses to microbial and other antigens, and phagocytosis and elimination of certain bacteria from lymphoid tissues. Recent studies with mycobacterial infection of mice demonstrated that exposure to UV-B radiation decreased the delayed hypersensitivity response to mycobacterial antigens, increased the severity of infection and reduced the survival time of mice from lethal infection. In humans, UV-B radiation has also been shown to impair the contact allergy response. These studies demonstrate that UV radiation can decrease immune responses in humans and laboratory animals and raise the possibility that increased exposure to UV-B radiation could adversely affect human health by increasing the incidence or severity of certain infectious diseases.

NATO ASI Series, Vol. I 18
Stratospheric Ozone Depletion/
UV-B Radiation in the Biosphere
Edited by R. H. Biggs and M. E. B. Joyner

# UV-B RADIATION AND PHOTOIMMUNOLOGY

Frances P. Noonan and Edward C. DeFabo

Department of Dermatology
The George Washington University Medical Center
Washington, DC 20037

**ABSTRACT**

Irradiation with ultraviolet-B (UV-B, 280-320 nm) initiates a selective immunosuppression which results in a down regulation of cell-mediated immunity. This form of immune suppression may be a fundamental regulatory mechanism, controlling the interaction between mammals and potentially deleterious environmental radiation. Immunosuppression by UV-B is mediated via a functional alteration to antigen-presenting cells, specialized cells that initiate immune responses by processing foreign proteins (antigens) and "presenting" them to lymphocytes which carry out the immune response. Most antigen-presenting cells cannot be reached by UV-B photons *in vivo* and UV-B therefore affects their function indirectly.

We have suggested that the mechanism for UV-B suppression is initiated by the photoisomerization of a photoreceptor molecule, urocanic acid (UCA), in the skin from the trans to the cis isomer. Cis UCA initiates a series of events *in vivo* that result in altered antigen-presenting cell function and immunosuppression. The sequence of events by which cis UCA acts is incompletely understood. Cis UCA prevents the induction of cAMP in skin fibroblasts, a critical signaling step in activation of these cells, and may thus prevent the elaboration of critical regulatory molecules. Furthermore, cis UCA at higher concentrations stimulates the formation from blood monocytes of prostaglandin $E_2$, an immunosuppressant. A number of factors regulate UV suppression, including a genetically-determined susceptibility in which an autosomal dominant gene determining high susceptibility may interact with sex-linked factors.

The predicted increase in immunosuppressing radiation resulting from ozone depletion can be calculated; therefore, quatification of the risk to human populations of this enhanced immunosuppression may be possible as more data become available.

See also Drs. DeFabo and Noonan's contribution in the Appendix.

NATO ASI Series, Vol. I 18
Stratospheric Ozone Depletion/
UV-B Radiation in the Biosphere
Edited by R. H. Biggs and M. E. B. Joyner

# INSTRUMENTATION AND METHODS TO MEASURE RESPONSE

# COMPARISON OF THE RESPONSE OF SOYBEAN TO SUPPLEMENTAL UV-B RADIATION SUPPLIED BY EITHER SQUARE-WAVE OR MODULATED IRRADIATION SYSTEMS

Joseph H. Sullivan,[1*] Alan H. Teramura,[1] Paulien Adamse,[2]
George F. Kramer,[2] Abha Upadhyaya,[2] Steven J. Britz,[2]
Donald T. Krizek,[2] and Roman M. Mirecki[2]

[1]Department of Botany, University of Maryland, College Park, MD 20742
[2]Climate Stress Laboratory, USDA, Agricultural Research Service, Beltsville, MD 20705

Key words: *Glycine max*, UV-B radiation, UV-B exposure systems, photosynthesis

**ABSTRACT**

Soybean [*Glycine max* (L.) Merr.] cv CNS was grown in a field study at the Beltsville Agricultural Research Center, USDA, Beltsville, MD. Supplemental UV-B radiation was provided by two contrasting UV-B delivery systems, both of which were intended to simulate a 25% depletion of stratospheric ozone as estimated by an empirical model. In the first system, a seasonally-based supplemental UV-B irradiance was provided in a square-wave fashion (SQ) and in the second, a state-of-the-art modulated (MOD) UV-B supplementation system was employed.

Since the MOD system but not the SQ system compensates for local deviation from clear sky conditions, the daily average total UV-B irradiance received by the plants was up to 30% greater in the SQ than in the MOD system. In addition to peak irradiance, the spectral ratios of UV-B, UV-A and visible light varied diurnally between the two systems. The difference in total UV-B exposure was reflected in several plant responses. These included an increase in the accumulation of UV-absorbing compounds, increased leaf thickness, reduction in photosynthetic capacity, and reductions in leaf area and seed yield under SQ conditions. However, increases in photosynthetic pigments, reduced dark respiration and light compensation points, and changes in carbohydrate metabolism were observed only in the MOD system. It is unclear whether the differences in the responses were due solely to differences in total UV-B exposure or were the result of subtle differences in the timing of exposure or spectral differences between the two delivery systems.

## INTRODUCTION

The fluxes of solar irradiance reaching the earth's surface are dependent on a number of factors including solar angle, latitude, elevation, stratospheric ozone concentration, atmospheric turbidity, and degree of cloud cover. Additionally, the earth-sun distance and minor solar fluctuations also contribute to seasonal and annual variations in irradiance (6). Since stratospheric ozone is a primary attenuator of solar UV-B radiation (UV-B between 280 and 320 nm), a reduction in the ozone column would increase the transmission of UV-B

*Address for correspondence

NATO ASI Series, Vol. I 18
Stratospheric Ozone Depletion/
UV-B Radiation in the Biosphere
Edited by R. H. Biggs and M. E. B. Joyner

radiation at wavelengths greater than approximately 290 nm. Though representing only a fraction of the total solar electromagnetic spectrum, UV-B radiation has a disproportionately large photobiological effect due to its absorption by important macromolecules such as proteins and nucleic acids (10). Therefore, it is not surprising that both plant and animal life are greatly affected by UV-B radiation. While we have some information on the responses of plants to UV-B radiation, we still can not make an adequate assessment of the potential consequences of an increase in solar UV-B radiation to plants.

Much uncertainty results from the experimental protocols that have been previously utilized. Controlled environment studies, which account for a large majority of previous UV-B studies, do not adequately simulate field conditions and most field studies have utilized artificial sunlamps to simulate increases in solar UV-B radiation. The artificial irradiation does not precisely match solar spectral characteristics. Consequently, because of the strong wavelength dependency of photon absorption (e.g., UV-B effectiveness), weighting functions, based on action spectra for specific responses, must be utilized to assess the biological effectiveness of both the irradiation sources and predicted ozone depletion (9).

In addition to qualitative differences in irradiation (i.e., spectral shifts) between natural and artificial UV-B radiation sources, difficulties may also arise in the quantitative delivery of the desired irradiance that would simulate an anticipated increase in UV-B radiation due to a given ozone depletion scenario. Generally, an empirical model is utilized to predict UV-B irradiance for a particular location, time of year, and ozone thickness (2, 11). In this manner, an "experimental dose" of supplemental UV-B can be calculated on an instantaneous or daily basis. However, since solar UV-B irradiance varies diurnally, seasonally, and due to local weather and atmospheric conditions, the selection and supplementation of a single "experimental dose" of UV-B radiation may not be appropriate for studies that are designed to realistically assess the likely consequences of ozone depletion. Furthermore, neither of the previously mentioned empirical models appear to predict total UV-B fluences very accurately. For example, published spectral data (7) indicate that the Björn and Murphy (2) model substantially underestimates UV-B irradiance in many locations. However, Booker et al. (3) found a closer agreement between measured values (using a broadband RB meter) and those predicted by Björn and Murphy (2). Although neither model very accurately estimates total UV-B fluence, both models give similar estimates for projected changes in solar UV-B that may be associated with a particular ozone depletion.

An alternative to supplying a constant daily "experimental dose" of UV-B radiation provided by a square-wave (SQ) delivery system is to supply UV-B as a proportional supplement to ambient conditions. Such a modulated (MOD) system was first developed and employed by Caldwell et al. (8). Modulated irradiation systems account for daily and seasonal changes in UV-B fluence, maintain a constant supplemental to ambient UV-B ratio and place much less emphasis on modeled prediction of UV-B fluxes since ambient UV-B continuously controls the incremental UV-B supplement. In spite of these advantages, out of practical considerations many studies continue to use SQ irradiation systems. Therefore, we undertook this study in an effort to compare the biological responses of plants subjected to supplemental UV-B irradiance from these two contrasting UV-B delivery systems. The first system used a seasonally based supplemental UV-B irradiance provided in a square-wave fashion (SQ) and the second system employed state-of-the-art modulated (MOD) UV-B supplementation. Figures 1A and 1B demonstrate theoretical differences in the irradiance provided by the two systems under clear sky and overcast conditions.

This study represents the first comparison between SQ and MOD irradiation systems. We framed two hypotheses: (1) the square-wave system would provide both a higher total irradiance (seasonal basis) and a higher supplemental to natural UV-B ratio during parts of the day, especially on cloudy days, than the modulated system, and (2) any exposure-dependent UV-B effects such as yield, growth, photosynthesis, and flavonoid production would be greater in SQ than MOD systems. In this paper we summarize the results of a single growing season in terms of differences in UV-B irradiation received by the plants and some of the biological responses. Greater detail in terms of physiological and biochemical measurements and responses will be published at a later date.

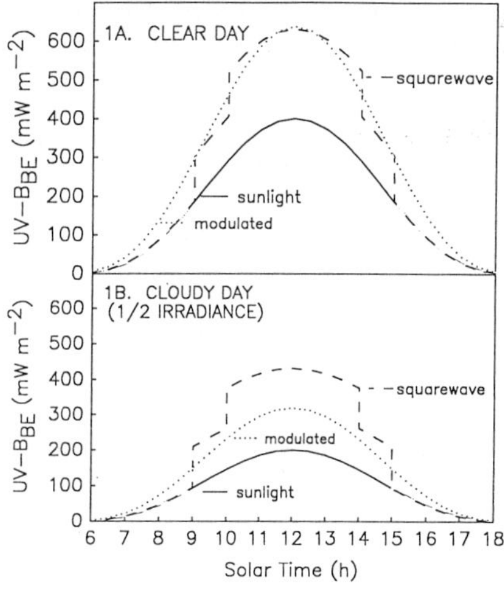


Figure 1. Calculated exposure to UV-B radiation under ambient conditions and with two UV-B delivery systems. The top panel represents modelled (11) calculations for a clear day and the lower panel is for a uniformly overcast day where ambient UV-B radiation is one-half of the previous day.

## METHODOLOGY

### PLANT GROWTH AND PHYSIOLOGICAL MEASUREMENTS

Soybean [*Glycine max* (L.) Merr.] cv CNS was grown in the field from June through October 1991 at the Beltsville Agricultural Research Center, USDA, Beltsville, MD. Seeds were sown in rows spaced 0.4 m apart at a seeding density of 33 seeds $m^{-1}$ in 8 plots of 4.5 m x 2.5 m each. Following initial seedling establishment the plants were thinned to 15 plants $m^{-1}$ in accordance with commonly used practices for the mid-Atlantic region. The overall experimental design was a randomized complete block with 2 blocks and 3 UV treatments. One treatment, MOD, contained 2 replicates within each block.

During the growing season a number of biochemical, physiological, and growth measurements were made on plants from all treatments. These included analysis of foliar photosynthetic and nonphotosynthetic pigments, polyamines, starch and sugar accumulation, and photosynthetic gas exchange characteristics. Specific measurement procedures followed previously published methodology and may be found in Britz et al. (5) for carbohydrate analysis, Krizek et al. (12) for pigment and enzyme analysis, and Sullivan and Teramura (15) for photosynthetic oxygen evolution measurements.

During pod development in late August, ten plants were harvested from each treatment plot and total leaf area, plant height, and number of nodes were measured before oven-dry biomass was determined on leaves, stems, and pods. SLW was calculated on a whole plant basis (leaf dry biomass/ leaf area). In addition to the above measurements, the area of single leaflets and SLW was measured on leaflets harvested for assessments of sugar and starch accumulation. In this case SLW was expressed using residual dry matter (i.e., DW—starch and soluble sugars).

All plants were harvested at maturity from the two treatment rows beneath the lamp racks according to the procedures outlined in Teramura et al. (16). Seeds were counted, dried, and weighed to determine total seed yield per plant.

### IRRADIATION PROTOCOLS

In all treatments supplemental UV-B radiation was provided by filtered Q-Panel UVB-313 sunlamps following the construction design outlined in Lydon et al. (13). Lamps were suspended above and perpendicular to the plant rows and filtered with either presolarized 0.13 mm-thick cellulose diacetate (transmission down to 290 nm) for supplemental UV-B irradiation or 0.13 mm polyester (spectrally equivalent to Dupont, Mylar Type S) films (absorption of almost all radiation below 318 nm) as a control. The spectral irradiance at plant height under the lamps was measured with an Optronic Laboratories Model 742 Spectroradiometer (Optronics Laboratories Inc., Orlando, FL, USA). The spectroradiometer was equipped with a dual holographic grating and modified to maintain constant temperature by the addition of Peltier heat exchange units. The spectroradiometer was calibrated against a NIST-traceable 1000 W tungsten filament quartz halogen lamp and wavelength alignment was checked against mercury vapor emission lines from a Hg arc lamp. The absolute spectral irradiance was weighted with the generalized plant response action spectrum (6) and normalized at 300 nm to obtain the daily biologically effective irradiance (UV-$B_{BE}$). Plants beneath the polyester filtered lamps received only ambient UV-B irradiance and were designated as controls (CO).

Two methods of UV-B radiation supplementation were employed with the diacetate filtered lamps. In one system the irradiation was provided in a seasonally adjusted two-step square-wave increment for 6 h daily centered around solar noon. Peak daily UV-$B_{BE}$ supplementation was 5 kJ $m^{-2}$ for a cloudless day near the summer solstice as approximated by the model of Green et al. (11). The peak supplement was adjusted at approximately three-week intervals in accordance with modelled estimations so that the maximum daily supplement was 3 kJ $m^{-2}$ at the end of the growing season. Daily adjustments were also made in the SQ system by turning off either one-half or all lamps in those plots to compensate for overcast days.

The second exposure system was the YMT-6 Ultraviolet Irradiation Modulation System (YMT-UIMS), developed by William Yu and Susan Mai in the laboratory of Alan Teramura at the University of Maryland. This system utilizes a new broadband UV-B detector (spectral characteristics verified by NIST) and a true condition closed-loop feedback control system that employs two detectors beneath each lamp bank (Figure 2). The true-condition feedback system continuously monitors solar UV-B irradiances and those present beneath the lamp banks (solar plus supplemental UV-B) and maintains a constant proportional supplement of UV-B. Detailed description of the electronics and the YMT broadband detector may be found in Yu et al. (18).

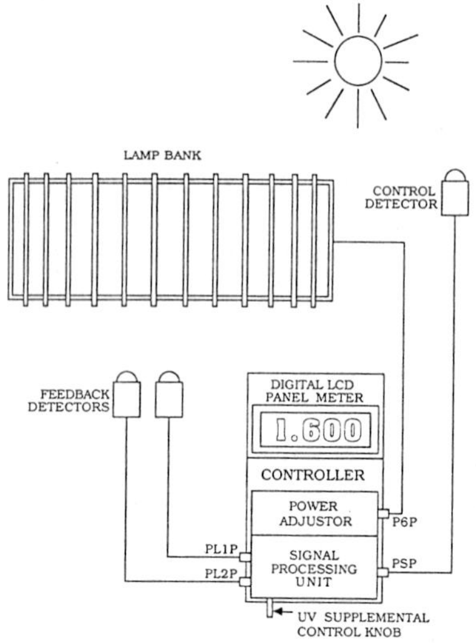


Figure 2. Diagramatic representation of the YMT modulated UV-B irradiation system.

## RESULTS AND DISCUSSION

### UV-B SUPPLEMENTATION

Daily and seasonal UV-B fluxes were obtained by a combination of spectral measurements, YMT broadband data and, on days where measurements were not made, by conversion of data obtained at the site with a Li-Cor pyranometer (LI 200SA). Since irradiance was estimated by broadband meters calibrated to UV-$B_{BE}$ and with pyranometric data, Figure 3 represents only a relative comparison between the two irradiation systems and ambient rather than actual UV-B monitoring data. Nonetheless, a clear difference between CO, SQ, and MOD was observed throughout the season. Daily ambient UV-$B_{BE}$ ranged from less than 2 to about 7 kJ $m^{-2}$ and supplemental UV-B varied accordingly. As ex-

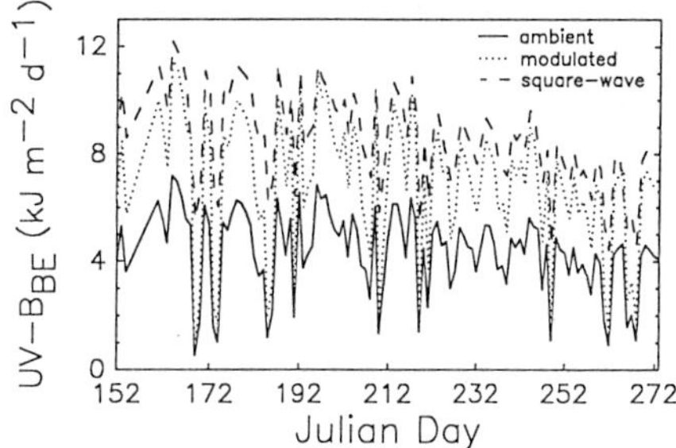


Figure 3. Seasonal UV-B irradiance estimated under ambient conditions and with two UV-B delivery systems. Both systems were designed to simulate a 25% depletion of stratospheric ozone over Beltsville, MD according to an empirical model (11).

pected, the SQ system provided more supplemental UV-B radiation on all days and the difference between SQ and MOD was greatest on days with lower ambient irradiance (Figure 3). These data demonstrate the range of daily difference that would likely occur between SQ and MOD irradiation systems. The smallest differences would occur on clear days while larger differences would be observed on overcast days and magnified further as shown in Figure 3, if no daily correction for sky conditions is made in the SQ system. The agreement between a MOD and a SQ system would also vary on an annual basis. In other words, the SQ and MOD systems would be in closer agreement during a growing season characterized by low precipitation and few cloudy days than in a wet or cloudy season such as 1991. Therefore, due to the nature of the growing season in which this study was conducted and to the fact that we did not account for daily modification of the SQ system in our calculations, these data probably represent a worst-case scenario. In this case, the average daily UV-B supplement for the season with MOD was about 60% above ambient, as predicted for a 25% ozone depletion by Green et al. (11) and Björn and Murphy (2). In the SQ system, the UV-B supplement was up to 90% greater than ambient. On a average daily basis the UV-B$_{BE}$ received by the plants was about 4.7, 7.5, and 9.0 kJ m$^{-2}$ for the CO, MOD, and SQ systems, respectively. Although this represents an extreme case, clearly, if the goal of a specific study is to realistically simulate a specific ozone depletion or UV-B enhancement scenario, then the use of MOD systems would be preferred over SQ irradiation systems.

PHOTOSYNTHETIC CAPACITY AND EFFICIENCY

During a one-week period in mid-August the influence of UV-B radiation on photosynthetic capacity and quantum efficiency was assessed by measuring photosynthetic oxygen evolution under conditions of $CO_2$ saturation (i.e. stomatal limitations removed). Photochemical efficiency and direct effects on photosystem II were also evaluated by assessment of chlorophyll fluorescence in dark-adapted leaves. Photosynthetic pigments were quantified in order to determine whether indirect effects on photosynthesis may have resulted from changes in pigment concentration.

Reductions in photosynthetic capacity tended to correlate with exposure to UV-B radiation but the reductions were not statistically significant at the 0.05 level of probability (Table 1). This parameter was more strongly influenced by the total solar irradiance (e.g., degree of cloudiness) present the day before and the day of measurements. Solar irradiance explained 57% of the variation observed in measurements of maximum photosynthetic capacity while the average daily UV-B$_{BE}$ explained only about 17% of the observed variation. Therefore it is possible that changes in ambient irradiance (including UV-B) due to local weather and atmospheric conditions could obscure more subtle effects of supplemental UV-B radiation.

Measurements of photosynthetic efficiency were significantly affected by the UV-B delivery system. Plants receiving MOD UV-B radiation showed reduced dark respiration, lower light compensation points (LCP), and increased oxygen evolution at very low light values (Table 1). Measurements of chlorophyll fluorescence suggest that plants grown

Table 1. Effects of UV-B radiation supplied by either square-wave or modulated delivery systems on some photosynthetic characteristics of CNS soybean as measured by oxygen evolution. Data are means of 9-12 plants per treatment %SE and values in each columns followed by the same letter are not significantly different according to LSD. $A_{max}$ is the light and $CO_2$ saturated rate of oxygen evolution, $A_{78}$ is oxygen evolution at a PPF of 78, $R_d$ is the rate of oxygen consumption in the dark, LCP[1] is the estimated light compensation point and AQE is apparent quantum efficiency.

| TRT | $A_{max}$ | $A_{78}$ (mmol m$^{-2}$ s$^{-1}$) | $R_d$ | LCP[1] | AQE (mmol/mmol) |
|---|---|---|---|---|---|
| 1 | 36.5 ±1.3 a | 1.4 ±0.5 b | -2.8 ±0.6 ab | 50 | 0.054 ±0.003 a |
| 2 | 34.1 ±2.2 a | 2.7 ±0.3 a | -1.6 ±0.3 b | 25 | 0.055 ±0.004 a |
| 3 | 31.8 ±1.9 a (P=0.06) | 0.5 ±0.8 b | -3.1 ±1.1 a | 62 | 0.046 ±0.007 a |

[1]Point estimated from regression analysis. No statistics available.

TRT 1 = CONTROL (PE)
TRT 2 = MODULATED UV
TRT 3 = SQUARE-WAVE

under the MOD UV-B systems had an increase in the size of the pool of electron acceptors on the reducing side of PSII (increase in T 1/2) and the observed reductions in $F_o$ and $F_m$ would also be consistent with this (Figure 4). The mechanisms responsible for these changes are presently unknown but the plants showed responses that apparently were independent of total UV-B exposure. It is possible that changes in UV-B to UV-A or visible irradiance during early mornings or evenings in the

Table 2. Effects of UV-B radiation supplied by either square-wave or modulated delivery systems on some photosynthetic pigments of CNS soybean. Data are means of 9-12 plants per treatment %SE and values in each row followed by the same letter are not significantly different according to LSD.

| Pigment Measured (mg mgDW$^{-1}$) | UV-B Treatment: Control | Modulated | Square-wave |
|---|---|---|---|
| Chl *a* | 1.58 ±0.23 b | 2.38 ±0.17 a | 1.95 ±0.20 ab |
| Chl *b* | 0.80 ±0.09 a | 1.08 ±0.11 a | 0.86 ±0.10 a |
| Total Chl | 2.38 ±0.32 b | 3.45 ±0.28 a | 2.80 ±0.30 ab |
| Chl *a/b* (ratio) | 1.95 ±0.09 a | 2.33 ±0.13 a | 2.30 ±0.08 a |
| Carotenoids | 113.0 %14.6 a | 158.3 %12.7 a | 139.8 %5.3 a |

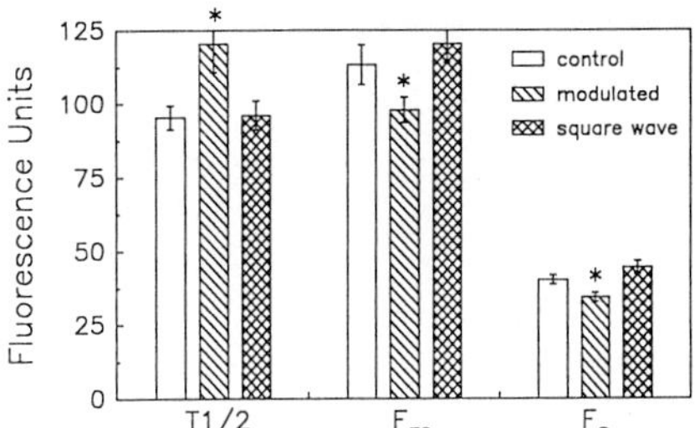


Figure 4. Chlorophyll fluorescence data for CNS soybean grown under ambient conditions or exposed to supplemental UV-B irradiation supplied by either square-wave or modulated delivery systems. Each bar on the histogram is the mean of 12 plants per treatment %SE. An asterisk (*) denotes a statistically significant difference from controls according to LSD at P < 0.05.

MOD systems, which would not have occurred with the SQ system, may have contributed to these effects.

The trend toward lower photosynthetic capacity in the SQ irradiated plants was probably due to direct effects since the concentrations of photosynthetic pigments were not reduced. The concentration of chlorophylls, especially chl *a*, actually increased under MOD UV-B exposure (Table 2). In fact, the expression of photosynthesis on a chlorophyll basis showed reductions of 25 and 35% in the SQ and MOD systems, respectively. Carotenoids also tended to increase under supplemental UV-B and this was especially true in the MOD treatment (Table 2). These changes in pigments and photosynthetic rates in the MOD system suggest that the UV-B portion of the solar spectrum was important to the development of photosynthetic capacity and efficiency and that this importance was not simply quantitative. Perhaps the most striking example of a greater plant response to MOD than to SQ irradiation was in carbohydrate metabolism. Plants in three of four MOD systems demonstrated significantly lower starch levels at the end of the night (Figure 5A) in spite of a higher daily starch accumulation rate (Figure 5B). It is presently unclear whether the slight increases in UV-B radiation in the morning and, more importantly, in the evening before sunset, supplied with the MOD but not the SQ system was involved in these responses or why plants in one of the MOD systems responded in an opposite manner. Possibly increases in UV-B radiation early or late in the photoperiod may affect the daylength response of photosynthate partitioning (4). Clearly additional research is needed in order to further understand these responses to small and perhaps qualitative changes in UV-B radiation.

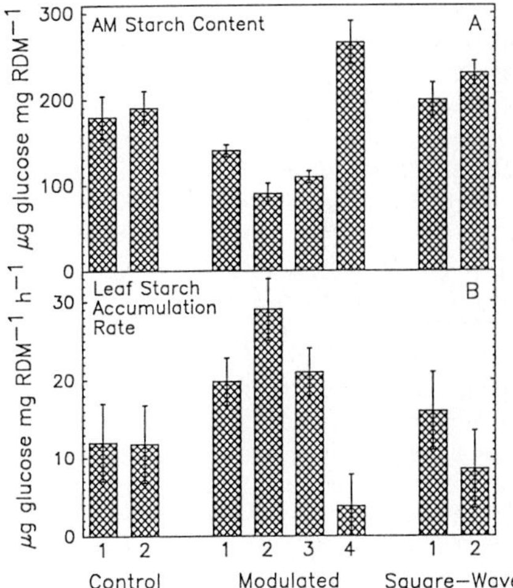


Figure 5. Morning starch accumulation (A) and accumulation rate during the day (B) for CNS soybean grown under ambient conditions or exposed to supplemental UV-B irradiation supplied by either square-wave or modulated delivery systems. Each bar on the histograms is the mean of 4 plants per treatment replication %SE. There were 2 replicated in the CO and SQ systems and 4 in the MOD systems.

Table 3. Effects of UV-B radiation supplied by either square-wave or modulated delivery systems on some protective responses to UV-B radiation. Data are means of at least 9 plants per treatment %SE and values in each row followed by the same letter are not significantly different according to LSD.

| PARAMETER MEASURED | UV-B Treatment | | |
|---|---|---|---|
| | Control | Modulated | Square-wave |
| LEAF MOPHO-LOGY/ANATOMY | | | |
| Leaflet area ($cm^{-2}$) | 26.4 ±1.3 a | 26.6 ±0.9 a | 23.7 ±0.8 b |
| SLW ($g\ m^{-2}$) | 36.2 ±2 ab | 31.1 ±1 b | 38.4 ±3 a |
| UV-ABSORBING COMPOUNDS ($A\ mg^{-1}$ RDM) | 1.25 ±0.04 b | 1.30 ±0.04 b | 1.44 ±0.03 a |
| ENZYMES ($A\ min^{-1}\ g^{-1}$ FW) | | | |
| Catalase | 23.3 ±1.1 a | 19.9 ±1.0 a | 18.9 ±1.3 a |
| SOD | 14.7 ±0.8 a | 17.5 ±1.0 a | 18.3 ±1.0 a |
| TOTAL AMINES ($nmol\ g^{-1}$ FW)Y | 243 ±13 a | 253 ±14 a | 213 ±19 a |

## UV-B PROTECTION RESPONSES

A range of protective responses including changes in anatomy, morphology and foliar biochemistry and pigmentation have been reported with respect to UV-B radiation (e.g. 1,17). In this study, leaf area was reduced by UV-B radiation only in the SQ system. SLW, a correlate of leaf thickness, was less in the MOD system compared to the SQ system, and UV-absorbing compounds increased only in the SQ system (Table 3). No changes were observed in polyamines or in free radical scavenging enzymes such as catalase or SOD which have also been associated with UV protection (12). In general, few protective responses were observed that appeared to be a function of total UV-B exposure. This is not too surprising, however, since CNS soybean was found to be very sensitive to deleterious effects of UV-B radiation in a greenhouse study (14). In that study, like this field study, CNS soybean did not appear to exhibit well-developed protective responses such as increasing leaf thickness or accumulation of UV-absorbing compounds. The reduction in SLW upon exposure to MOD UV-B irradiation was surprising and again suggests the subtle importance and our lack of understanding of the role of the UV-B component of sunlight in leaf development.

## PLANT GROWTH AND YIELD

Vegetative growth of these plants was not significantly altered by UV-B radiation (Table 4). The only variables showing a statistically significant effect were leaf size which was reduced with the SQ system and SLW as previously discussed. Supplemental UV-B radiation did not alter seed weight, seed number, or total seed yield per plant compared to controls (Table 4). Both seed number and yield, however, were larger in MOD compared to SQ plots. Reduction in seed number in soybean as a result of UV-B exposure has been reported for Essex soybean in a previous field study (16). Since seed number rather than seed size was affected, one possible explanation for the differences might be that flowering was affected by the contrasting treatment regimes. Unfortunately, no flower counts were made and no field observations demonstrated any obvious differences in flowering phenology between treatments.

The effects of UV-B radiation on plant growth

Table 4. Effects of UV-B radiation supplied by either square-wave or modulated delivery systems on vegetative growth and yield characteristics of CNS soybean. Data are means of 20 plants for vegetative growth and 89-168 plants for yield %SE. Values in each row followed by the same letter are not significantly different according to LSD.

| PARAMETER MEASURED | UV-B Treatment Control | Modulated | Square-wave |
|---|---|---|---|
| Plant Height (cm) | 121.0 ±3 a | 132.0 ±5 a | 121.0 ±4 a |
| Total DW (g) | 57.0 ±5 a | 55.0 ±6 a | 57.0 ±5 a |
| Seed Number | 131.0 ±6 ab | 140.0 ±7 a | 121.0 ±7 b |
| Yield (g plant$^{-1}$) | 15.6 ±1 ab | 16.3 ±1 a | 13.1 ±1 b |

and productivity suggest that the plants responded differently to the two UV-B delivery systems but these data do not simply fall into an expected exposure-response pattern such as CO < MOD < SQ. Flowering, for example, is a complex process in which UV-B may have some importance. Ziska et al. (19) observed that flowering was enhanced in some high-elevation plant species grown under high levels of UV-B radiation, while other studies have demonstrated reductions in flowering in response to UV-B radiation. This study suggests that subtle changes in UV-B may be important to some aspects of the floral induction, fertilization, or seed development process.

## CONCLUSIONS

Undoubtedly, the plants grown under the SQ system were exposed to greater UV-B fluences on a seasonal basis. Our approximations suggest that the seasonal average UV-$B_{BE}$ exposures were about 4.7, 7.5, and 9.0 kJ m$^{-2}$ for the CO, MOD, and SQ systems, respectively. The type of UV-B irradiation system used is important and will greatly affect the exposure of organisms to UV-B radiation and the interpretive ability of the study. The manner in which UV-B radiation is supplied can modify biological responses and affect our overall assessment of potential UV-B damage.

In addition to total UV-B exposure, some developmental and biochemical responses may be associated with short-term changes in UV-B radiation and possibly with diurnal changes in the UV-B to UV-A or total irradiance ratio as observed in this study. In this study, responses that did not follow an anticipated dose response included changes in photosynthetic pigments, carbohydrate metabolism, and possibly flowering phenology. Our hypothesis is that the UV-B enhancement with the MOD system in the early morning and evening, before and after the operation of the SQ system, was an important factor for the plants. Unlike seasonal total irradiance where the SQ system would result in over-irradiation of the plants, the SQ system actually provides less UV-B radiation than expected with ozone depletion during parts of the day and may underestimate some responses to UV-B radiation. In either case, a MOD system will provide a more realistic assessment and should be used where economically feasible. Further research will be required in order to better understand these effects which may be due to subtle differences, other than total UV-B exposure, in the irradiation systems.

## REFERENCES

1. Beggs, C.J., U. Schneider-Ziebert, and E. Wellman. 1986. UV-B radiation and adaptive mechanisms in plants. *In* R.C. Worrest and M.M. Caldwell (eds.), Stratospheric Ozone Reduction, Solar Ultraviolet Radiation and Plant Life. Springer-Verlag, Berlin Heidelberg New York. pp.235-250.

2. Björn, L.O., and T.M. Murphy. 1985. Computer calculation of solar ultraviolet radiation at ground level. Physiologia Vegetatio 23:555-561.

3. Booker, F.L., E.L. Fiscus, R.E. Philbeck, A.S. Heagle, J.E. Miller, and W.W. Heck. 1992. A supplemental ultraviolet-B radiation system for open-top field chambers. J. Environ. Qual. 21:56-61.

4. Britz, S.J. 1991. Regulation of photosynthate partitioning into starch in soybean leaves. Plant Physiol. 94:350-356.

5. Britz, S.J., W.E. Hungerford, and D.R. Lee. 1987. Rhythms during extended dark periods determine rates of net photosynthesis and accumulation of starch and soluble sugars in subsequent light periods in leaves of sorghum. Planta 171:339-345.

6. Caldwell, M.M. 1971. Solar UV irradiation and the growth and development of higher plants. *In* A.C. Giese (ed.), Photophysiology, vol 6. Academic Press, New York. pp. 131-177.

7. Caldwell, M.M., R. Robberecht, and W.D. Billings. 1980. A steep latitudinal gradient of solar ultraviolet-B radiation in the arctic-alpine life zone. Ecology 61:600-611.

8. Caldwell, M.M., W.G. Gold, G. Harris, and C.W. Ashurst. 1983. A modulated lamp system for solar UV-B (280-320 nm) supplementation studies in the field. Photochem. and Photobiol. 37:479-485.

9. Caldwell, M.M., L.B. Camp, C.W. Warner, and S.B. Flint. 1986. Action spectra and their key role in assessing biological consequences of solar ultraviolet radiation change. *In* R.C. Worrest and M.M.Caldwell (eds.), Stratospheric Ozone Reduction, Solar Ultraviolet Radiation and Plant Life. Springer-Verlag, Berlin Heidelberg New York. pp.87-112.

10. Giese, A.C. 1964. Studies on ultraviolet radiation action upon animal cells. *In* A.C. Giese (ed.), Photo-physiology, vol 2. Academic Press, New York. pp.203-245.

11. Green, A.E.S., K.R. Cross, and L.A. Smith. 1980. Improved analytical characterization of ultraviolet skylight. Photochem. and Photobiol. 31:59-65.

12. Krizek, D.T., G.F. Kramer, A. Upadhyaya, and R.M. Mirecki, R.M. 1993. UV-B response of cucumber seedlings grown under metal halide and high pressure sodium/deluxe lamps. Physiol. Plant. 88:350-358.

13. Lydon, J., A.H. Teramura, and E.G. Summers. 1986. Effects of ultraviolet-B radiation on the growth and productivity of field-grown soybean. *In* R.C. Worrest and M.M. Caldwell (eds.), Stratospheric Ozone Reduction, Solar Ultraviolet Radiation and Plant Life. Springer-Verlag, Berlin Heidelberg New York. pp.313-325.

14. Reed, H.E, A.H. Teramura, and W.J. Kenworthy. 1992. Ancestral U.S. soybean cultivars characterized for tolerance to ultraviolet-B radiation. Crop Sci. 32:1214-1219.

15. Sullivan, J.H., and A.H. Teramura. 1993. The effects of ultraviolet-B radiation on loblolly pines. 3: Interaction with $CO_2$ enhancement. Plant Cell and Environ. (in press).

16. Teramura, A.H., J.H. Sullivan, and J. Lydon. 1990. The effectiveness of UV-B radiation in altering soybean yield: A six-year field study. Physiologia Plant. 80:5-11.

17. Tevini, M., and A.H. Teramura. UV-B effects on terrestrial plants. Photochem. and Photobiol. 40:479-487.

18. Yu, W., J.H. Sullivan, and A.H. Teramura. 1991. The YMT Ultraviolet-B irradiation system: Manual of operation. Final Report to U.S. E.P.A., Environmental Research Laboratory, Corvallis OR.

19. Ziska, L.E., and A.H. Teramura. 1992. $CO_2$ enhancement of growth and photosynthesis in rice (*Oryza sativa*): Modification by increased ultraviolet-B radiation. Plant Physiol. 99:473-481.

# THE NON-SCANNING UV-B SPECTROPHOTOMETER

John E. Rives

Department of Physics and Astronomy
University of Georgia, Athens, GA 30602

**ABSTRACT**

A UV-B spectrophotometer is described which uses an integrating sphere flux collector, a non-scanning single grating monochrometer, and an intensified photo diode array detector. Simultaneous rapid recording of the entire spectral data with good wavelength resolution, while maintaining a fixed grating position, offers certain advantages over scanning instruments for full-sky irradiance measurements. Poor stray light rejection represents a serious limitation of a single grating instrument in the UV-B cutoff region of the spectrum. The use of an interference filter to partially solve this problem is discussed.

NATO ASI Series, Vol. I 18
Stratospheric Ozone Depletion/
UV-B Radiation in the Biosphere
Edited by R. H. Biggs and M. E. B. Joyner

# STRATOSPHERIC OZONE DEPLETION/UV RADIATION: A DATA AND INFORMATION ACCESS SYSTEM

Vincent Abreu and Ronald Emaus

EPA-Corvallis, Oregon

**ABSTRACT**

The CIESIN Information Gateway provides access to a family of information guides, applications, and services, which are accessible via the Internet as well as by modem dial-in access. The current version is implemented using Internet Gopher and WAIS, and future enhancements will include a World Wide Web (WWW) server. The Information Gateway will provide users with easy access to on-line information pertaining to global environmental change, with an emphasis on the human dimensions. The Information Gateway also provides to users the ability to link to other relevant servers and services This global environmental web of networked servers provides "one-stop shopping" for data, information, and science applications.

The core contents of the Information Gateway are built around a series of topics. Topics include issues such as the effects of ozone depletion; information related to regional areas, such as sub-Saharan Africa or the Saginaw Bay Watershed; or categories of data such as industrial pollution. Within a topic, bibliographies, papers, reports, and directories of data sets are collected about that topic. Guides are written that describe each topic and the information related to it. In many cases, multiple guides will be written for different categories of users.

Here we will preview the guides being developed for the topic of ozone depletion and effects of enhanced UV-B radiation.

NATO ASI Series, Vol. I 18
Stratospheric Ozone Depletion/
UV-B Radiation in the Biosphere
Edited by R. H. Biggs and M. E. B. Joyner

# HOW UV-B MEASUREMENTS CAN BE COORDINATED

# MEASUREMENT AND INSTRUMENT COORDINATION: PROBLEMS AND SOLUTIONS

Bruce Baker

National Climatic Data Center, Federal Building
Asheville, NC 28801-2723

**SUMMARY**

## INTRODUCTION

It is evident that there needs to be a level of coordination on a global scale to understand the spatial and temporal changes in the UV-B radiation field at the surface. There are scientific and social issues that need to be addressed when discussing the UV-B in the biosphere. To put this into perspective, Figure 1 illustrates the implications enhanced UV-B will have on a variety of disciplines. As we can see, the premise on which the implied effects are based is an increase in the UV-B at the surface of the earth. The understanding of the global spatial and temporal trends will require making accurate measurements coupled with model validation to comprehend the ramifications of the impacts. The quantification of the potential impacts can be seen in Figure 2. This figure illustrates the mechanics of quantifying the required measurements and models once the measurements have been made. It is essential to have these feedback mechanisms. Fundamentally, a suite of measurements are needed and modelling is necessary to obtain adequate global coverage.

## STATEMENT OF THE PROBLEM

It is evident that the part of the scientific community involved in understanding the UV-B radiation field at the surface of the earth has a number of scientific issues to address to define and refine the problems of determining the magnitude as a function of wavelength of UV-B radiation at the surface of the earth and to understand the fluctuations in this electromagnetic radiation field spatially and temporally. There is agreement that there are four basic scientific issues.

1) Spectral measurements need to be quantified in terms of their accuracy from static tests; their representativeness of capturing the electromagnetic field in the atmosphere needs to be determined.

2) Long-term trends are important and warrant an effort to initiate and implement a global historical database to complement any new measurement programs.

3) All of the respective programs need to ensure that all instruments chosen be intercompared for data continuity.

4) Ancillary measurements that have either a first- or second-order effect on the UV-B radiation field at the surface of the earth need to be characterized in conjunction with the UV-B radiation measurements.

NATO ASI Series, Vol. I 18
Stratospheric Ozone Depletion/
UV-B Radiation in the Biosphere
Edited by R. H. Biggs and M. E. B. Joyner

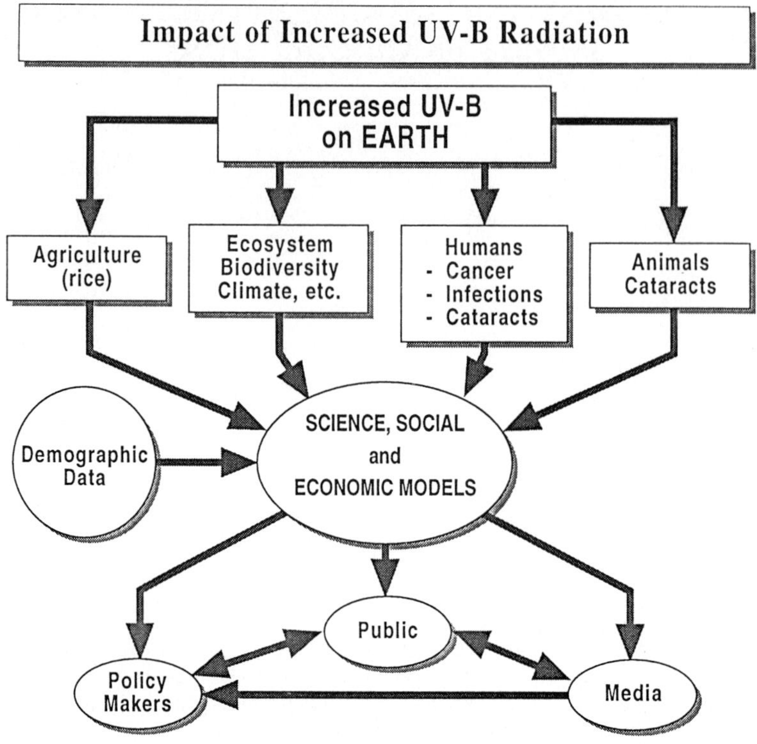


Figure 1.

## PROPOSED SOLUTIONS

Paramount for a UV-B monitoring network will be high-quality spectral and broadband UV-B measurements. Static tests of all instruments must be standardized and traceable to NIST. Static tests in this instance refer to the physical characterization of all of the UV-B instruments. All instruments should have a cosine response test and a linearity test over the dynamic range of the instrument. Quantification is necessary in the slit function of the spectral instruments and the filter function of the broadband instruments. If the static tests are standardized then the view of the UV-B radiation field each instrument "sees" in the atmosphere will be representative.

An effort needs to be undertaken to assimilate a global historical UV-B database. To identify long-term trends scientists need this historical database. This database would complement the current programs by providing historical data in the US and also provide the avenue to examine the question "Is there

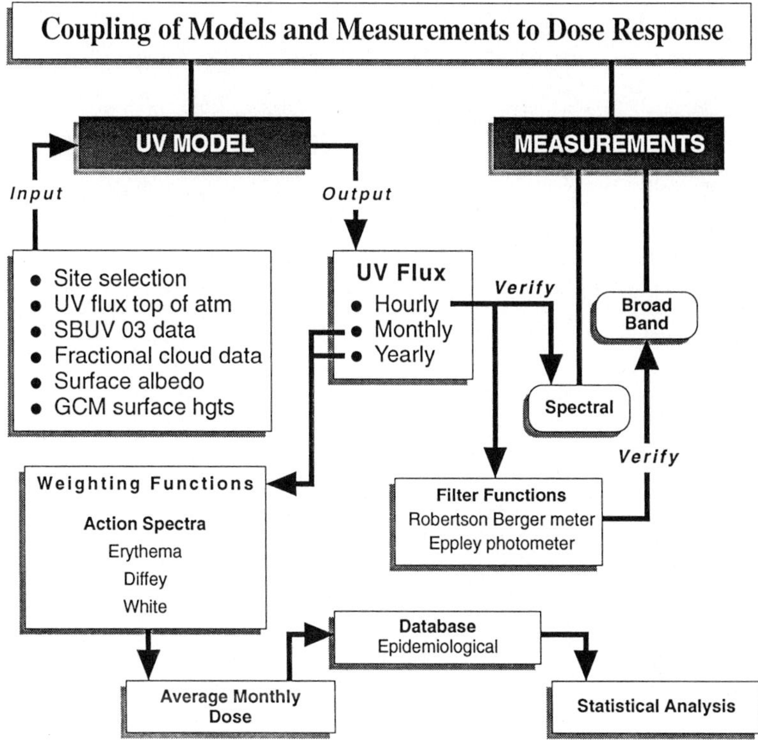


Figure 2.

a trend in the global UV-B radiation at the surface of the earth?"

A field program that would provide a site to continually compare both broadband instruments and spectral instruments is needed to provide data continuity for all measurement programs. Due to different funding agencies having different foci (thus a different suite of instrumentation), an effort needs to be made to make measurements at a single location for data continuity between programs. This is indeed necessary to detect long-term trends spatially and temporally between disparate measurements from different programs.

The ancillary measurements coupled with the UV-B measurements will help minimize the variance in the natural variability, thus providing a higher degree of confidence in the long-term trend. These measurements will also provide for model validation and perhaps a better understanding of the influence of clouds and aerosols on this spectral regime.

# NATO'S PROGRAM ON THE SCIENCE OF GLOBAL ENVIRONMENTAL CHANGE

Norton D. Strommen

USDA World Agricultural Outlook Board
14th & Independence Avenue, SW, Washington, DC 20250

The objectives of NATO's special programs are to promote research dealing with the potential for global changes within the earth's environmental system. In particular, the programs' aims are to describe and understand the interactive, physical, chemical, and biological processes that regulate the total earth system. Their primary goal is to advance our capability to predict changes in the global environment, especially those which result from human activities and their resulting impacts on climate. This requires understanding change that occurs from both natural and human causes.

Today much of the research activity related to global environmental change is being organized at international and national levels. However, inter-disciplinary research is not developing at a rate required to meet our needs to better understand the complex changes occurring in the environmental system. Most of the effort continues to be confined to specific disciplines. This special NATO program fosters the development of strong interdisciplinary collaboration to investigate in depth the interactions and feedback between different components of the global earth system.

One specific purpose is to elaborate the role of the biosphere, how it affects the chemistry and composition of the atmosphere and how, through biogeochemical and hydrologic cycles, it interacts with the physical processes to produce the climate of the earth.

NATO ASI Series, Vol. I 18
Stratospheric Ozone Depletion/
UV-B Radiation in the Biosphere
Edited by R. H. Biggs and M. E. B. Joyner

# MONITORING SURFACE UV-B RADIATION FROM SPACE


Richard McPeters

Laboratory for Atmospheres, NASA Goddard Space Flight Center
Code 916, Greenbelt, Maryland 20771 USA



Key words: NOAA, NASA, ozone measurement, SBUV/2 instrument, TOMS, UV radiation


## INTRODUCTION

While the problem of estimating expected global changes in surface UV-B is complex, the UV-B input into the troposphere can be estimated quite accurately because of the high quality record of global ozone measurements from the TOMS instrument on Nimbus 7. A continuous record of daily total column ozone measurements extends from November 1978 through May 1993. The relative accuracy of the ozone record from TOMS has been maintained to within ±2% (2 sigma error) over this period, so the measured changes in global ozone over this period can reliably be used to compute UV transmission through the atmosphere. The uncertainties in the calculation of surface UV-B will then be the uncertainties involved in the transmission of UV through the troposphere--uncertainties in assumed cloud cover and in tropospheric aerosol loading.

## THE OZONE MEASUREMENT

The Total Ozone Mapping Spectrometer (TOMS), which was launched on Nimbus-7 in 1978, measures total column ozone at high spatial resolution (50 km field of view) while scanning across the orbital track. The Solar Backscattered Ultraviolet instrument (SBUV), which accompanied TOMS on Nimbus 7, measures the vertical distribution of ozone along the orbital track with about 10 km vertical resolution. Because Nimbus-7 is in a sun-synchronous polar orbit, TOMS measures ozone over the entire globe (except in areas of darkness) on a daily basis. This measurement program is being continued by the National Oceanic and Atmospheric Administration (NOAA) in cooperation with NASA with a series of SBUV/2 instruments which are improved versions of SBUV. The first was launched on NOAA-9 in 1984 and is still operating, although because of orbital drift the measurements have been compromised by high solar zenith angles since 1987. The second SBUV/2 was launched on NOAA-11 in late 1988 and also continues to operate. The third instrument in the series was launched on NOAA-13 in 1993 but was lost when the spacecraft power systems failed.

Ozone can be calculated from the backscattered radiance because of differential absorption by ozone in the ultraviolet. In a very simple approximation, the albedo, the ratio of the radiance backscattered from

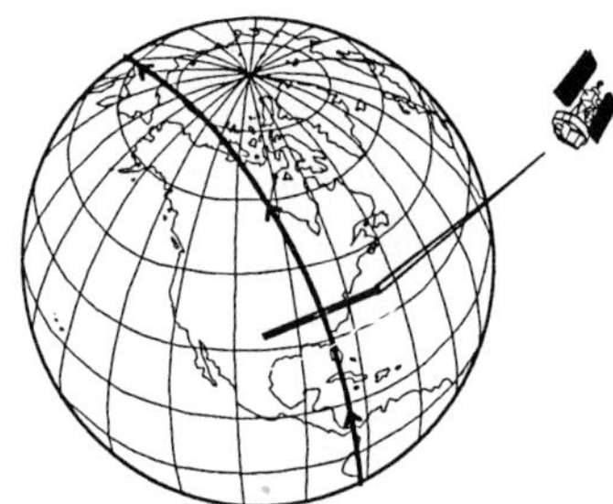

Fig. 1 Nimbus 7 with TOMS and SBUV.


NATO ASI Series, Vol. I 18
Stratospheric Ozone Depletion/
UV-B Radiation in the Biosphere
Edited by R. H. Biggs and M. E. B. Joyner

the atmosphere to the extraterrestrial solar irradiance, is related to ozone by:

$$I_{ground} = F_0 \times e^{-(1/\cos)\,\alpha \times \Omega} \tag{1}$$

$$I_{sat} = I_{ground} \times R \times e^{-\alpha \times \Omega} \tag{2}$$

$$albedo = I_{sat}/F_0 = R \times e^{-(1+1/\cos)\,\alpha \times \Omega} \tag{3}$$

The ground reflectivity R is measured at a wavelength for which ozone absorption is zero. Knowing R, Equation 3 can then be solved for total ozone Ω. The actual retrieval must of course account for the fact that most of the backscattered light comes from multiple Rayleigh scattering in the atmosphere. Because of the computational complexity, a set of precomputed standard tables are used. The tables are computed for 23 climatological ozone profiles (low latitude 225-325 DU, mid latitude 125-575 DU, and high latitude 125-575 DU) for solar zenith angles from 0° to 88°. Details of the ozone retrieval are described by Klenk et al. [3]. Comparison of the derived ozone with ground measurements shows that ozone is generally derived with 1-2% accuracy. But because only a fraction of the scattered UV penetrates to the ground, ozone in the troposphere is sensed with only about 50% sensitivity, so errors of a few percent can be made if tropospheric ozone varies. Also, at large solar zenith angles (greater than 80°) errors as large as 10% can be made if the profile shape deviates significantly from our assumed climatology. This error affects only high latitudes under polar winter conditions. Finally, when large amounts of volcanic aerosol are present at altitudes above 20 km, as was the case in 1991 and 1992 after the eruption of Mount Pinatubo, the ozone retrieval will be affected depending on the TOMS view angle. Scans looking to the left and right have a small positive error, while scans looking down have a small negative error, so the average error is near zero.

The calibration of the instrument is maintained by the "pair justification" technique. While single wavelengths can be used to infer total ozone, it is much more accurate to use wavelength pairs to eliminate the effect of wavelength independent errors such as calibration bias or aerosol scattering. The key to our calibration technique is that the different wavelengths used to measure total ozone also have different sensitivity to wavelength dependent calibration error. Time dependent changes in the albedo calibration (usually traceable to the diffuser plate since changes in the spectrometer itself cancel in the albedo ratio) result in relative drifts between pair ozone values. An internal calibration can be done based on the requirement that total ozone measured by different wavelength pairs agree (pair justification) [2].

## OZONE TRENDS

The entire TOMS total ozone data set of 13+ years has been re-processed using pair justification to maintain the calibration from the beginning to the end of the data record to an estimated relative accuracy of ±1.5%. The data for the period 1979-1990 have been used to derive latitude dependent trends. A statistical model including terms for seasonal variation, linear trend, quasi-biennial oscillation, solar cycle, and second-order autore-gressive noise has been fit to TOMS data [4]. Figure 2 shows the relative variation in TOMS data for the 30° to 40°N zone after the average annual variation has been removed. The solar cycle dependence, which amounts to only a percent or so, has also been removed.

We find that global ozone (65°N-65°S) declined by -2.6 ±1.4% per decade between 1979 and 1990. Almost no decline is seen in the equatorial region, while, as shown in Figure 2, at 35°N a 4.5% decline is observed over this period. The strongest trend in the northern hemisphere is seen near 50°N in winter and early spring, amounting to 8% per decade.

## LOW OZONE IN 1992-1993

The global average ozone measured by TOMS in 1992-1993 was 2-3% lower than in any previous year (1979-1991). While ozone at any given location is so highly variable that changes of a few percent might not be obvious, the global average ozone (area weighted) is quite stable. As shown in Figure 3, beginning in early 1992 global ozone dropped well below the 1979-1990 average, and in 1993 was more than 3% lower. When the latitude dependence of the decrease is analyzed, the largest decreases occur between 10°S and 20°S and between 10°N and 60°N. While the cause of the low ozone is not certain, a first guess might be that the decrease is related to the continuing presence of aerosol in the stratosphere from the June 1991 eruption of Mount Pinatubo.

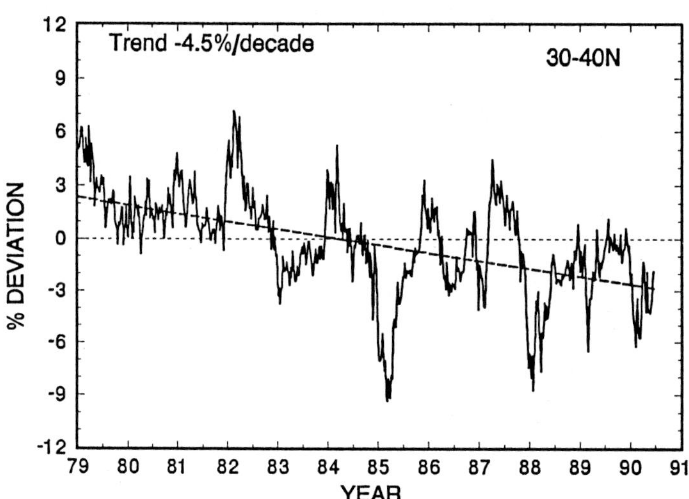


Figure 2. De-seasonalized ozone measured in the 30°–40°N zone showing linear trend of -4.5% per decade.

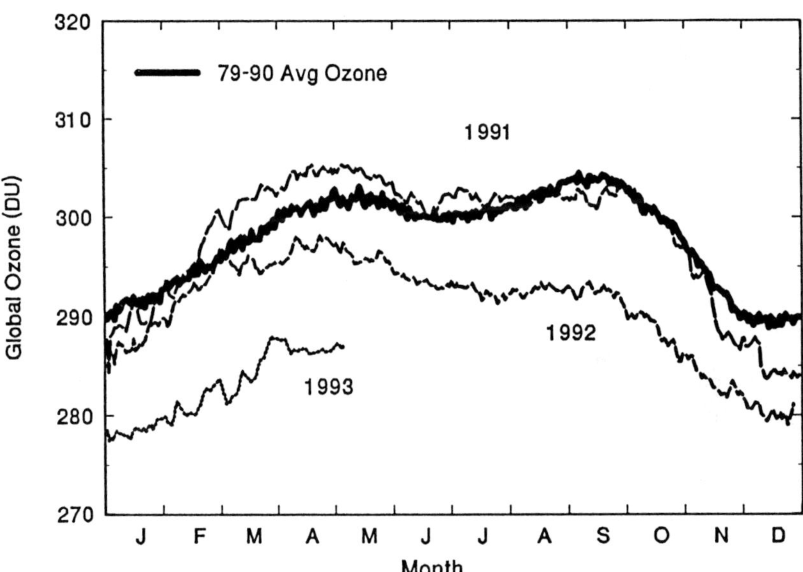


Figure 3. Daily global average ozone (70°S–70°N) measured by Nimbus 7 TOMS.

The initial drop in ozone appears in late 1991, after the aerosol has had time to spread throughout both hemispheres. What must be explained is why the ozone decrease was largest in 1993, almost two years after the eruption. While the combination of solar cycle decrease plus a continuing long term decrease in ozone would predict a downturn in ozone in 1992, the observed decrease in 1992-1993 is much larger than would be predicted from previous ozone observations. The observed decreases have been confirmed by data from other satellites (the NOAA 11 SBUV/2 and the Meteor 3 TOMS saw similar decreases) and by data from ground-based Dobson instruments.

## REFERENCES


1. Gleason, J.F., P.K. Bhartia, J.R. Herman, R. McPeters, P. Newman, R. Stolarski, L. Flynn, G. Labow, D. Larko, C. Seftor, C. Wellemeyer, W. Komhyr, A.J. Miller, and W. Planet. 1993. Record Low Global Ozone in 1992. Science 260:523-526.

2. Herman, J.R., R. Hudson, R. McPeters, R. Stolarski, Z. Ahmad, S. Taylor, and C. Wellemeyer. 1991. A method for self calibration of TOMS to measure long term global ozone change. J. Geophys. Res., 96:7531-7546.

3. Klenk, K.F., P.K. Bhartia, A.J. Fleig, V.G. Kaveeshwar, R.D. McPeters, and P.M. Smith. 1982. Total ozone determination from the backscattered ultraviolet (BUV) instrument. J. Appl. Meteor. 21:1672-1684.

4. Stolarski, R.S., P. Bloomfield, R. McPeters, and J. Herman. 1991. Total ozone trends deduced from Nimbus 7 TOMS data. Geophys. Res. Lett. 18:1015-1018.

# UV SPECTRORADIOMETRIC MONITORING IN POLAR REGIONS

Charles R. Booth and Timothy B. Lucas

Biospherical Instruments, Inc.
5340 Riley Street, San Diego, CA 92110-2621

**ABSTRACT**

In 1987, the United States National Science Foundation, Office of Polar Programs, called for the development of a monitoring network for UV irradiance in Antarctica. This network was established in 1988 and presently consists of six monitoring stations: three in Antarctica, one at Ushuaia, Argentina, one at Barrow, Alaska, and a system at San Diego, California. High-resolution (0.7 nm) spectral scans are made hourly and the data is routinely distributed to scientists around the world on CD-ROM and by other means. Details of the design and implementation of the network along with examples of the data are presented.

NATO ASI Series, Vol. I 18
Stratospheric Ozone Depletion/
UV-B Radiation in the Biosphere
Edited by R. H. Biggs and M. E. B. Joyner

# UV RADIATION MONITORING IN NEW ZEALAND

R.L. McKenzie

National Institute of Water and Atmospheric Research
NIWA-Atmosphere, Lauder, Central Otago, New Zealand

Key words: actinometer, erythemally weighted, Lauder, New Zealand, ozone, Robertson-Berger meter spectroradiometer, UV radiation

## ABSTRACT

When it was realized that global ozone reductions were occurring, a program to measure the spectral distribution of UV was initiated in New Zealand. The aims of this program are to understand the reasons for variability in UV, to build up a climatology of UV and relate it to UV levels at other locations, and to monitor future long-term trends.

A purpose-built instrument was developed to enable routine measurements of cosine weighted spectral UV irradiances incident on a horizontal surface at the ground. The spectral range covered is 290-450 nm, and the instrument bandpass is approximately 1 nm. Measurements have been made routinely at Lauder Central Otago (45°S, 170°E, elevation 370m) since December 1989. Scans are made each day at 5° intervals in solar zenith angle (SZA, for SZA <75°) whenever weather permits. Near local solar noon, extra scans are logged, including measurements of the diffuse component during clear sky conditions.

The observation period is still too short to allow determination of trends in UV. However, the sensitivity of UV spectra to changes in ozone, cloud, sun angle, and aerosols is demonstrated. Preliminary results from a recent intercalibration of spectrometers are discussed. These enable a comparison to be made between UV spectra measured in New Zealand and those as measured in Australia and Europe.

The spectral measurements at Lauder are complemented by measurements with broad-band instruments there and at other sites. The performance of these instruments is discussed, and recent results showing upward trends in UV from 10 years of R-B meter observations from Invercargill (46.3°S, 168.3°E) are reviewed.

## INTRODUCTION

Ozone in the atmosphere effectively filters damaging solar ultraviolet radiation in the UV-B region (290-320 nm) from the Earth's surface. A major concern about reductions in stratospheric ozone is they will lead to increases in UVB radiation. Ozone reductions are expected to continue well into next century as chlorine levels continue to increase. Therefore there is a need to monitor UV, and to understand the factors that influence it.

### REGIONAL ISSUES

In the New Zealand/Australia region there is statistical and anecdotal evidence that UV levels were already high compared with corresponding latitudes in the Northern Hemisphere before ozone levels became perturbed. Model calculations indicate that in 1980 UVB levels in NZ were at least 10-15% greater than at corresponding latitudes in the Northern hemisphere [10], and this has been suggested as a contributing factor to higher rates of incidence of skin cancer in NZ [13]. However, before the 1980s, no quantitative data was available. From the early 1980s data became available in NZ from 2 broadband filter instruments called Robertson-Berger (R-B) meters. Results from these instruments suggested that during clear sky conditions, UV levels in NZ were relatively high compared with comparable latitudes in the Northern Hemisphere [18]. A recent study has drawn

NATO ASI Series, Vol. I 18
Stratospheric Ozone Depletion/
UV-B Radiation in the Biosphere
Edited by R. H. Biggs and M. E. B. Joyner

attention to the higher levels of UV experienced in New Zealand compared with similar latitudes in Europe [21].

The discovery of the Antarctic "Ozone Hole" in 1985 stimulated public interest and debate in New Zealand about the incidence levels of UV irradiance at the surface, and future changes in it. Very large enhancements in UV have been documented during the Antarctic spring in recent years [27]. By the end of the 1980s it was apparent that ozone losses were not confined to the Antarctic, and global ozone reductions had already occurred. With our close proximity to Antarctica, the risks of future enhanced UV levels are greater. An episode of enhanced UV has already been documented in Melbourne, following the transport of ozone-poor air from Antarctica, in the summer of 1987-88 [20].

## SPECTRAL MEASUREMENTS

Absolute UV spectral irradiance measurements are potentially more useful than broad band measurements because a) greater confidence in calibration can be achieved, b) the reasons for variabilities in UV can be investigated, and (c) the wavelength dependence of damage to the biosphere is not yet well known. However, these spectral measurements are demanding, and good quality data has become available only in recent years. In Australia, measurements have been made for several years [19] but prior to 1989, the only spectroradiometric UV data available in New Zealand was from two short measurement campaigns in 1980 and 1988 [6]. This study showed that in 1988, the UV levels for clear sky conditions and for similar ozone amounts were similar to those in 1980. In 1988 the DSIR initiated a programme to characterize the spectrum of solar UV radiation reaching the ground in New Zealand. A purpose-built instrument was developed to enable routine measurements of cosine weighted spectral UV irradiances incident on a horizontal surface at the ground. The aims are to determine the climatology of UV, to identify regional differences, to understand the reasons for its variability, and to detect long term trends. The spectral range covered is 290-450 nm, and the instrument bandpass is approximately 1 nm. Measurements have been made routinely with this instrument in Lauder Central Otago (45°S, 170°E, elevation 370m) since December 1989. Scans are made each day at 5° intervals in solar

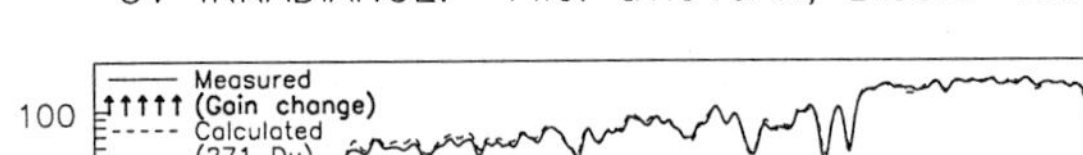


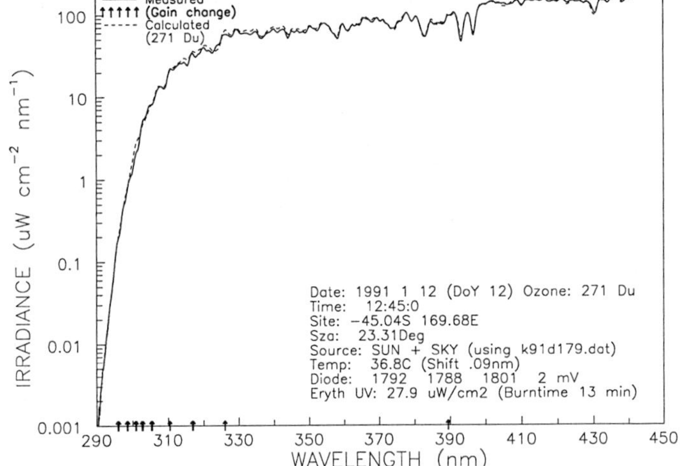


Figure 1. Clear sky spectrum measured and calculated at Lauder, NZ

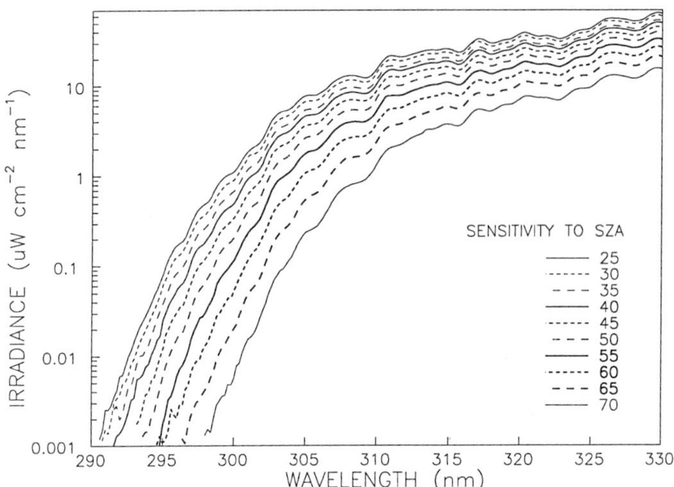


Figure 2. Sensitivity to changes in solar zenith angle.

zenith angle (SZA, for SZA <75°) whenever weather permits. Near local solar noon, extra scans are logged, including measurements of the diffuse component during clear sky conditions. A description of the instrument, the calibration procedures, and samples of data is available [14].

An application of the results used concurrent ozone measurements to deduce the percentage increase in erythemal radiation [17] resulting from a 1% reduction in ozone [15]. The study showed that this Radiative Amplification Factor (RAF) for erythema is 1.25±0.20, in approximate agreement with calculated RAFs of 1.1 [25]. An algorithm to derive total column ozone amounts directly from UV spectral irradiance data [22] has been implemented and found to give good agreement with Dobson measurements. For high sun conditions (Figure 1) the measured clear sky spectra agree well with spectra calculated with a simple model [10].

Change in solar zenith angle is a more important factor than changes in ozone in controlling UVB (Figure 2 and Figure 3). Changes in cloud are also important, and frequently attenuate UVB by more than 80%. During partly cloudy conditions when the sun is not obscured, UV levels can also be enhanced by up to 20% due to scattering from clouds (Figure 4). There are still uncertainties about the relative importance to UV of changes in aerosols, or changes in surface albedo. Given the wide range of factors that influence UV, and the uncertainties in measuring it, we find that the observation period at Lauder is still too short to be able to detect trends in erythemally weighted UV (Figure 4). The recent eruption of Mt Pinatubo provided an opportunity to investigate the effects of volcanic aerosols on UV irradiances. Global irradiances were found to be insensitive to the volcanic aerosols (Figure 4), but there were marked changes in the distribution between diffuse and direct radiation [12]. At Lauder after the eruption, diffuse/direct ratios increased by more than 50% in the UVA region, but changes were smaller in the UVB region (Figure 5).

An intercomparison of UV spectroradiometers from New Zealand, Australia, and Germany was made at Lauder in February 1993. The results showed that the three groups agreed at the 10% level for high sun conditions, and enabled systematic differences to be identified so that a comparison of spectra measured

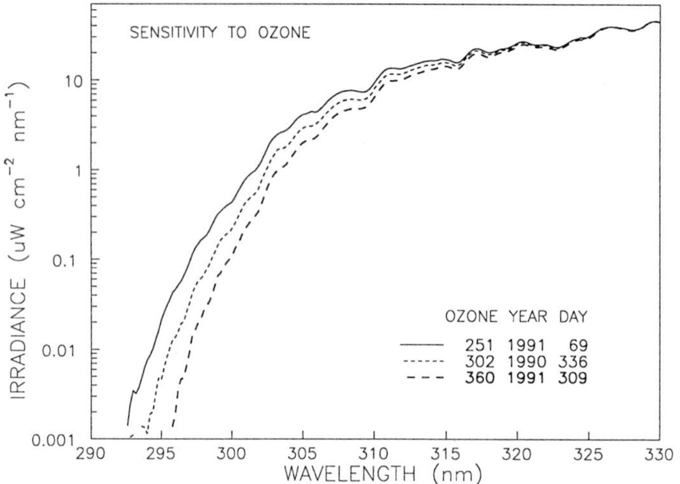


Figure 3. Sensitivity to changes in ozone column.

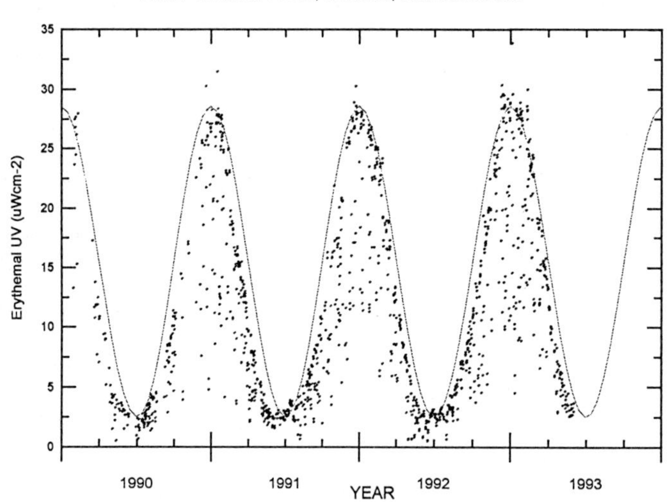


Figure 4. Time series of erythemally weighted UV at Lauder, NZ.

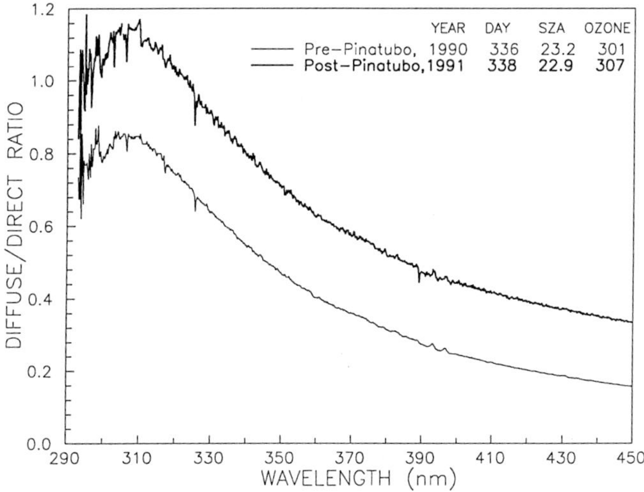


Figure 5. Sensitivity of diffuse/direct ratios to Pinatubo aerosol.

in the three countries could be made. The maximum *clear sky* spectra were found to be greatest in Australia and least in Germany. The DNA-weighted clear sky maxima for Melbourne (Australia), Lauder (NZ), and Neuherberg (Germany) were 242 (100%), 191 (79%), and 130 (54%) mW m$^{-2}$ respectively. Differences were smaller for erythemally weighted UV [16].

## BROAD-BAND MEASUREMENTS

The longest record of UV data in New Zealand is the 10 year time series of UVB from the R-B meter at Invercargill [18]. A recent study has identified an upward trend in UVB consistent with the downward trend in ozone observed in New Zealand [2].

Previous attempts to study long term trends in UV with other R-B instruments had shown negative trends. This may have been because other atmospheric changes were taking place [25,27], or it may have been due to instrument problems or calibration uncertainties. The R-B meter's sensitivity to ozone is not high [25], and there may have been problems due to its strong temperature co-efficient of 1%/°C [8]. However, the Invercargill R-B instrument was recently recalibrated both in the laboratory and in the field against the Lauder spectrometer. It was found to have remained stable over the observation period [9]. Studies have found that other R-B instruments to be relatively stable over the long term [7], although there is a non-linear relationship between R-B weighted UV and biologically weighted (e.g., erythemal) UV [5].

In the late 1980s, a NZ-wide network of filter instruments (International Light Inc. 1730 series actinic UV radiometer type SED/ACTS270/W) was established [23,26]. A similar network is maintained in Australia [19,20]. Data from the NZ instruments are disseminated to local radio stations, and are used to inform the public about current UV levels in the summer. These results are expressed as "burn-times," defined as the time taken for erythemally-weighted irradiance to reach the minimum erythemal dose (MED) of 210 J m$^{-2}$. The noon summer burn-time is typically 15 to 20 minutes in NZ.

More modern temperature-stabilised "R-B meters" have recently become available (eg 501

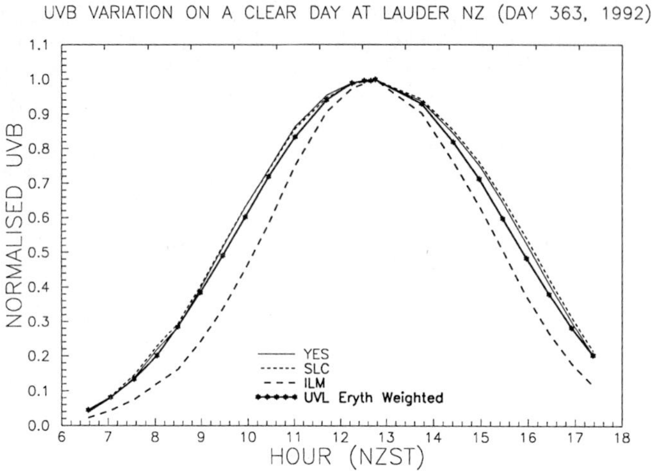

Figure 6. Diurnal variation of UVB measured by two "R-B meters" supplied by Yankee Environmental Systems (YES) and Solar Light Co (SLC), and by an actinometer supplied by International Light (ILM). The variation in erythemally weighted UV measured by the Lauder spectroradiometer is shown for comparison. The range of solar zenith angles is 75° to 22° in the morning, with an afternoon maximum of 60°.

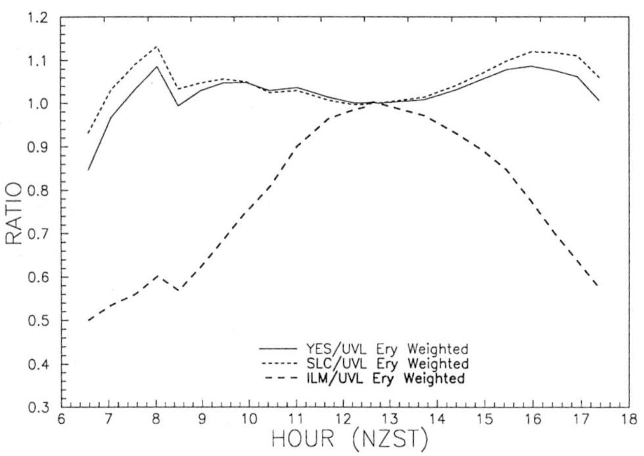


Figure 7. Ratios of measurements from the instruments in Figure 6 compared with erythemally weighted UV spectra.

biometer from Solar Light Co., and UVB-1 pyranometer from Yankee Environmental Systems). Compared with the original R-B meters, these have an improved cosine response and a greater sensitivity to ozone changes. Examples have been evaluated by intercomparing the data with suitably weighted spectra which were measured simultaneously by the spectroradiometer at Lauder (Figures 6 and 7). The results indicate that the new "R-B meters" may be a suitable inexpensive alternative to monitor future long-term changes in erythemal UV. However, the issue of long-term stability still has to be adequately addressed. The comparisons in Figures 6 and 7 show that the diurnal changes measured by the "R-B meters" have a much closer match to erythemal changes than the International Light Monitor, which has a relatively narrow field of view, and which is also designed to measure radiation from shorter wavelengths (giving greater sensitivity to ozone and SZA changes).

## THE FUTURE

Efforts are continuing to extend and improve the quality, and geographical coverage of UV measurements throughout NZ and the surrounding Pacific region through the deployment of new instrumentation and through participation in intercomparisons with other instruments. Because of instrumental differences, it is unlikely that an extensive global network of scanning instruments is practical. A register of data availability may however be appropriate, so that the data is available to the scientific community through direct application to the provider of the data.

A second spectroradiometer with improved cosine response, throughput, temperature stability, and automation has recently been developed at Lauder. This has the capability of correcting for intensity changes during scans, so will be more suitable for investigating cloud effects. An instrument will be available for campaign studies such as field intercomparisons, investigation of solar simulators used in growth chambers, and other experimental investigations.

More broad-band instruments are required to investigate geographic differences, and trends in UV levels. These should be calibrated regularly: (a) in the laboratory by measuring absolute responsivity, and (b) in the field by intercomparisons with properly maintained and calibrated UV spectro-radiometers over a wide range of sun angles observing conditions. It is essential that future UVB monitoring networks in NZ are comparable with global networks, so that our UV levels can be related to those in other regions [1]. Ideally these networks would use standard instruments, standard calibration procedures, and the data should be centrally archived.

Current UVB research in New Zealand, future needs, and priorities for future research have been summarised recently [11]. A bibliography of recent UV research relevant to the New Zealand region, including recent work in Antarctica [3,4,24] is given in the extended reference list below.

## REFERENCES

1. Basher, R.E. 1989. Perspectives on monitoring ozone and solar ultraviolet radiation. Trans. Menzies Foundation, Melbourne, pp.81-88.

2. Basher, R.E., X. Zheng, and S. Nichol. Upward trend detected in UV-B series. Geophys. Res. Lett. submitted June 1993.

3. Beaglehole, D., and G.G. Carter. 1992. Antarctic skies 1: Diurnal variations of the sky irradiance, and UV effects of the ozone hole, spring 1990. J. Geophys. Res. 97(D2):2589-2596.

4. Beaglehole, D., and G.G. Carter. 1992. Antarctic skies 2: Characterization of the intensity and polarization of skylight in a high albedo environment. J. Geophys. Res. 97(D2):2597-2600.

5. Bittar, A., and J.D. Hamlin. 1991. A perspective on solar ultraviolet calibrations and filter instrument performance. Proc CIE, Vienna, v.1, 1, II, pp.9-10.

6. Bittar, A., and R.L. McKenzie. 1990. Spectral UV intensity measurements at 45°S: 1980 and 1988. J. Geophys. Res. 95(D5):5597-5600.

7. De Luisi, J., J. Wendell, and F. Kreiner. 1992. An examination of the spectral response characteristics of seven Robertson-Berger meters after long-term field use. Photochem. Photobiol. 56:115-122.

8. Dichter, B.K., D.J. Beaubien, and A.F. Beaubien. 1993. Effects of ambient temperature changes on

UV-B measurements using thermally regulated and unregulated fluorescent phosphor ultraviolet pyranometers. Proc. Quad Ozone Symposium, Charlottesville, Va., June 1992. In press.

9. Grainger, R.G., R.E. Basher, and R.L. McKenzie. 1993. UV-B Robertson-Berger Meter characterization and field calibration. Applied Optics 32:343-349.

10. McKenzie, R.L. 1991. Application of a simple model to calculate latitudinal and hemispheric differences in ultraviolet radiation. Weather and Climate 11:3-14.

11. McKenzie, R.L. 1992. Current status of UVB research in New Zealand. National Science Strategy Committee for Climate Change, Royal Society of New Zealand Information Series 2.

12. McKenzie, R.L. 1993. UV spectral irradiance measurements in New Zealand: effects of Pinatubo volcanic aerosol. Proc. Quad Ozone Symposium, Charlottesville, Va., June 1992.

13. McKenzie, R.L., and J.M. Elwood. 1990. Intensity of solar ultraviolet radiation and its implications for skin cancer. NZ Medical J. 103:152-154.

14. McKenzie, R.L., P.V. Johnston, M. Kotkamp, A. Bittar, and J.D. Hamlin. 1992. Solar ultraviolet spectro-radiometry in New Zealand: Instrumentation and sample results from 1990. Applied Optics 31:6501-6509.

15. McKenzie, R.L., W.A. Matthews, and P.V. Johnston. 1991. The relationship between erythemal UV and ozone derived from spectral irradiance measurements. Geophys. Res. Lett. 18:2269-2272.

16. McKenzie, R.L., M. Kotkamp, G. Seckmeyer, R. Erb, C. Roy, and P. Gies. 1993. First Southern Hemisphere intercomparison of spectroradio-meters. Geophys. Res. Lett. In preparation.

17. McKinlay, A.F., and B.L. Diffey. 1987. A reference action spectrum for ultraviolet induced erythema in human skin. CIE J. 6:17-22.

18. Nichol, S.E., and R.E. Basher. 1987. Sunburning ultraviolet radiation at Invercargill, New Zealand. Weather and Climate 7:21-25.

19. Roy, C.R., H.P. Gies, and G. Elliot. 1988. Solar ultraviolet radiation: Personal exposure and protection. J. Occup. Health Safety - Aust NZ 4:133-139.

20. Roy, C.R., H.P. Gies, and G. Elliot. 1990. Ozone depletion. Nature 347:235-236.

21. Seckmeyer, G., and R.L. McKenzie. 1992. Elevated ultraviolet radiation in New Zealand (45°S) contrasted with Germany (48°N). Nature 359:135-137.

22. Stamnes, K., J. Slusser, and M. Bowen. 1991. Derivation of total ozone abundance and cloud effects from spectral irradiance measurements. Appl. Optics 30:4418-4426.

23. Smith, G.J., K. Ryan, and M.G. White. 1992. Trends in erythemal and carcinogenic ultraviolet radiation at mid-southern latitudes. Photochem. Photobiol. In press.

24. Trodahl, H.J., and R.G. Buckley. 1990. Enhanced ultraviolet transmission of Antarctic sea ice during the Austral spring. Geophys. Res. Lett. 12:2177-2179.

25. UNEP. 1991. Environmental effects of ozone depletion: 1991 update. United Nations Environment Program, Nairobi, Kenya.

26. White, M.G., G.J. Smith, K.G. Ryan, R.B. Coppell, D. Backshall, C.J. Edmunds, and R.L. McKenzie. 1992. Erythemal and carcinogenic ultraviolet radiation at three New Zealand cities. Weather and Climate 12:83-88.

27. WMO. 1992. Scientific assessment of ozone depletion: 1991. World Meteorological Organization Ozone Research and Monitoring Project, No. 25.

# THE EC's STEP UV-B PROGRAM

Ann R. Webb

University of Reading, Department of Meteorology
PO Box 239, Reading RG6 2AU United Kingdom

**ABSTRACT**

Spectral measurements of solar UV-B radiation are made at an increasing number of sites in Europe with a range of different instruments and various levels of supporting data. Standard lamps are predominantly used for calibration, but lamps originate from several National Standards laboratories and facilities for their use vary from site to site. Since 1990, scientists from a number of countries have been working together to standardize the perform-ance of their UV spectroradiometers as part of the European Community's STEP (Science and Technology for Environmental Protection) Program. The work has involved two intercomparisons of UV spectroradiometers. Instrument performance was tested both in the laboratory during calibration procedures and outdoors during the measurement of solar radiation. Results have shown that it is possible to achieve consistent agreement between some instruments in both situations if the instrument characteristics are well known and there is rigorous attention to detail at all stages of operation. To maintain consistency of calibration between instruments at different sites, a traveling lamp unit has been designed to provide periodic confirmation of relative stability (not absolute calibration) of each spectroradiometer. A prototype unit is currently being tested.

## INTRODUCTION

The European Community's Science and Technology for Environmental Protection (STEP) UV-B Program began in 1991 with the working title "Determination of Standards for a UV Monitoring Network." The intention of the program is not to establish a routine operational network for UV monitoring, but to enable national groups involved in UV measurements to work together to provide a series of recommendations and guidelines for UV data collection based on practical experience. The program is targeted at high-resolution spectral measurements and does not consider broadband UV monitors. There is no standard instrument for solar spectral UV-B radiation measurement, nor is there an international reference standard for calibration purposes. Much of the work to date therefore has been an assessment of the performance of instruments currently used for spectral UV-B measurements and their relative calibrations which are traceable to different national standards laboratories. With a better understanding of the technical limitations associated with the instrumentation, the program is now beginning to assess protocols for measurement and calibration at individual sites, methods of quality control and quality assurance, the maintenance of relative calibration between sites, and site specification. The contractors to the program are given in Appendix A and the home sites of the instruments involved are shown in Figure 1.

## INSTRUMENT CHARACTERISTICS

Initial demands on instrument performance were kept to a minimum, resulting in a diverse set of spectroradiometers. The instruments were required to measure incident solar UV radiation from a hemi-

NATO ASI Series, Vol. I 18
Stratospheric Ozone Depletion/
UV-B Radiation in the Biosphere
Edited by R. H. Biggs and M. E. B. Joyner

sphere above the horizontal plane at a resolution of 1 nm across the entire UV-B waveband. Many of the instruments have longer wavelength ranges (up to 600nm) and usual scanning resolutions of as little as 0.1 nm.

The instruments range from commercially available units with varying degrees of modification to custom-built equipment (Appendix A, Table 1). Double- and single-grating monochrometers are both represented, with input optics of cosine diffusers or integrating spheres. Photomultiplier tube detectors are used in all but one of the instruments which has a diode array. Further details of all instruments can be found in Figures 1–3.

A first intercomparison of six spectroradiometers was held in Thessaloniki, Greece, during July 1991 [1,2]. It became clear from this campaign that interpreting the results from a diverse set of instruments is only possible if each instrument is well characterised so that anomalous results may be attributed to a particular aspect of instrument design or performance. The steep slope of the solar spectrum in the UV-B makes the slit function of the instrument, and its stray light performance, vital, especially at the shortest wavelengths in the region of 300nm. Detailed knowledge of the cosine response is also required, especially for comparisons of global solar radiation when results may be significantly different to investigations in the laboratory with lamps at normal incidence to the input optics. The individual aspects of each instrument's performance have, in general, been explored at the home sites. Results below show some of the characteristics of two of the more commercial instruments, the Bentham DM150 and Optronic 742. It should be noted in this instance that both instruments have undergone some degree of modification. Tests were made at the University of Innsbruck using the same experimental procedures for both instruments [4].

### SLIT FUNCTION

The slit functions of the two instruments measured with a helium cadmium laser [4] are shown in Figure 2. An indication of the slit function of an instrument is usually given by quoting the bandwidth at full-width half-maximum (FWHM), but the shape of the shoulders of the slit function also become important when measuring solar UV-B because the spectral intensity changes so rapidly. Broad-shouldered slit functions can lead to significant near-field stray light from longer wavelengths. Its significance becomes evident when comparing measured solar UV-B spectra (Figure 5).

### STRAY LIGHT

In addition to near-field stray light, radiation may also reach the detector from wavelengths well outside the waveband covered by the slit function. This is far-field stray light and can lead to deceptively high measured irradiances. As 0.1% of the radiation at 400 nm or 1% of radiation at 320 nm is equivalent to the solar radiation at 300 nm, far-field stray light must be rigorously suppressed. The far-field stray light for the Optronic 742 was $10^{-4}$ of the signal at long wavelengths, and $10^{-5}$ for the DM150 [5]. Some of the other instruments in the campaign, particularly the single-grating monochrometers, showed evidence of stray light problems at ~300 nm or lacked the sensitivity to detect either true signal or stray light at the shortest wavelengths.

### COSINE RESPONSE

Imperfect cosine response can lead to diurnally-varying discrepancies between instruments making solar irradiation measurements. As a large proportion (more than 50%) of global UV-B irradiance is diffuse at temperature latitudes, zenith angle-dependent errors are less obvious than at large wavelengths. However, cosine errors of 15% at 60° and 30% at 75° (Optronic 742, Figure 3) can still introduce errors of a few percent in individual observations when the sun is low in the sky or the sky is overcast. Good agreement between two instruments over a range of zenith angles (Figure 5) may just indicate equally poor cosine responses (Figure 3). There is room for improvement in the angular response of many of the spectroradiometers tested.

## CALIBRATION

Absolute calibration of the spectroradiometers is maintained by reference to standard lamps. Irradiance scales of standard lamps used by participating groups are derived from different national standards laboratories, allowing for two sources of discrepancy in absolute calibration: differences between nationally derived standards, and imperfect calibration pro-

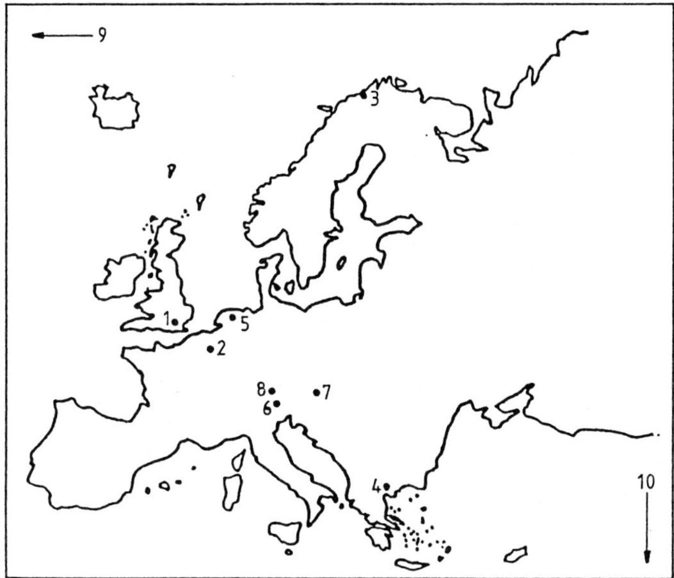


Figure 1. Home sites of instruments involved in 1992 intercomparison (see Appendix A).

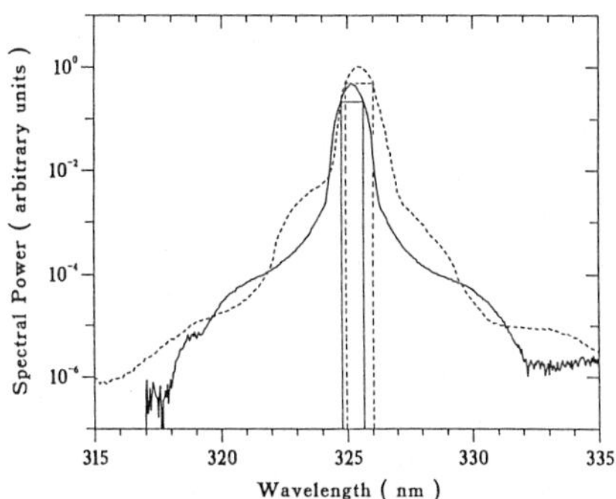


Figure 2. Slit functions of the Bentham DM150 (——) and Optronic 742(- - -). The FWHM of each instrument is marked. A fiber-optic light guide used with the Optronic 742 in Thessalonicki broadens the function.

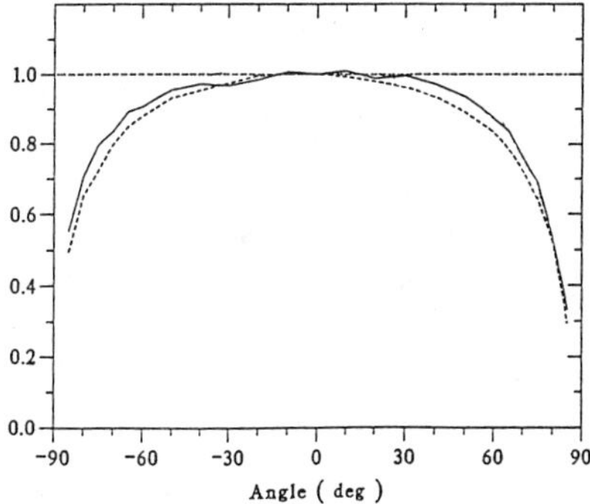


Figure 3. Cosine responses of the Bentham DM150 (——) and Optronic 742(- - -) expressed as (actual response / true cosine response).

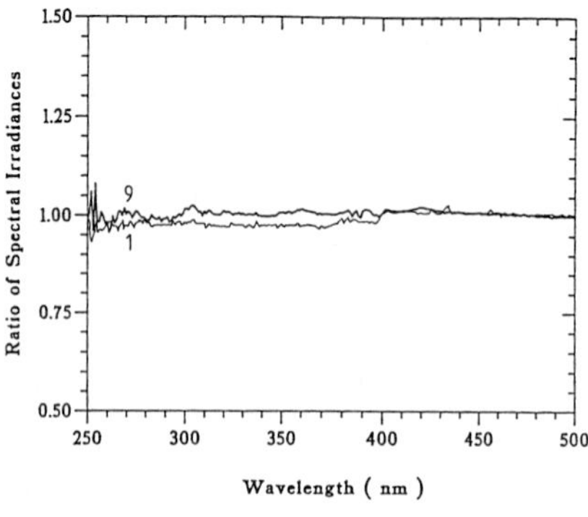


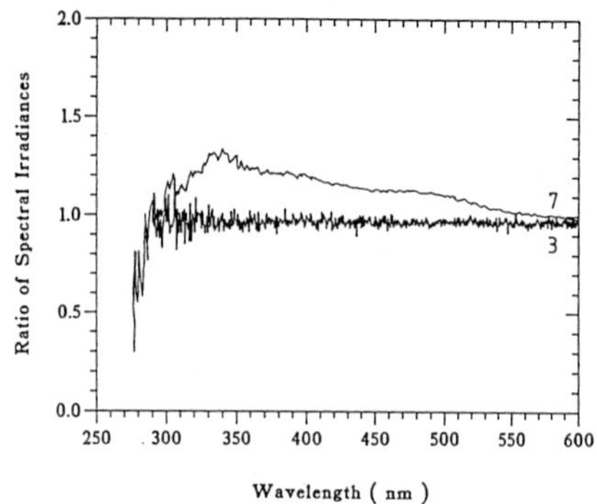


Figure 4. (Measured irradiance / Certificate) for lamp F-300 as measured by instruments 9,1,7,3 (Appendix A).

cedures. Standard lamp certificates of irradiances have typical uncertainties in the UV-B of 2-3% so agreement to better than a few percent between differently sourced calibrations cannot be expected. Lamps used by contractors within the program are referred back to NIST (National Institute of Standards and Technology), PTB (Physikilsch–Technishe Bundesanstalt), and NPL (National Physical Laboratory).

Investigation of calibration procedures in Innsbruck [4] indicated that, provided the lamp is correctly oriented and supplied with its designated precision current, the major sources of calibration error are the long-axis distancing between lamp and instrument and stray light from the surroundings. Reducing stray light and careful alignment procedures yielded agreement of better then 3% between instruments, and 1% between independent lamps [4]

Maintaining such agreement becomes more difficult when working in the field or when moving instruments from calibration room to solar observation site. During intercomparisons [1,2,3] contractors were faced with the latter situation, and these less than ideal conditions introduced an additional test of each instrument's consistency and robustness of calibration as measurements of an independent lamp and then the solar radiation were compared.

The following results are from the 1992 intercomparison [3] attended by ten spectroradiometers. Each instrument calibration was self-sufficient, that is, each group used only their own equipment for calibration purposes. An independent NIST certificate 1000W lamp was then measured by each spectroradiometer and the measured spectra compared with the certificate. Results from four of the instruments measuring over a wide spectral range are shown in Figure 4.

Three of the spectroradiometers showed agreement with the certificated lamp irradiances to within a few percent across the entire spectral range, and other spectroradiometers gave equally satisfying results when operating within their normal wavelength range [3]. The wavelength dependent inaccuracies of the fourth instrument illustrated in Figure 4 show some of the problems which may in part be caused by mechanical instability (instruments had to be turned on their side for calibration) and temperature-dependent behavior (temperature of the enclosed calibration room could not be controlled).

Results from the field calibration were generally very encouraging, proving that a disparate group of instruments using standards of different origin can reproduce a single independent lamp spectrum to within the intrinsic uncertainties associated with national reference standards.

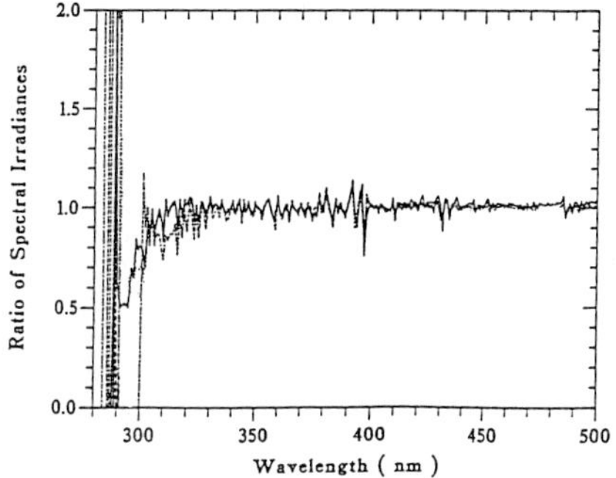


Figure 5. Ratio of sky scans (Bentham DM15 / Optronic 742) at zenith angles of 30° (—— ) and 83° (— • —).

## INTERCOMPARISON OF SOLAR IRRADIANCE MEASUREMENTS

The ultimate test of the UV spectroradiometers came in trying to reproduce the success of the calibration room when measuring global solar radiation in a range of conditions. Figure 5 shows the ratio of scans (Bentham DM150/Optronic 742) for zenith angles of 30° and 83°. Scans at intermediate zenith angles throughout a cloudless day were very close to the line for the zenith angle 30°. The consistancy between the instruments over a range of zenith angles indicates similar (if imperfect) cosine responses and good temperature stability (both instruments were housed in temperature controlled units). The decreasing ratio seen at 305 nm (z=30°) illustrates the effect of near-field stray light from the broader-slit function of the Optronic 742. At < 295 nm both instruments are recording only noise. When the zenith angle is large (83°) the slit function effect and noise equivalent irradiances are both observed at longer wavelengths. The structure in the ratios is a result of the different resolutions of the two instruments, convoluted with the structure of the solar spectrum. Wavelength alignment was not a problem with these, or any of the other instruments.

The agreement indicated by Figure 5 is the best that was achieved during the intercomparison. Other instruments had aspects of their performance that were better than these of the Optronic or Bentham instruments. Consistently good agreement in laboratory and beneath the sky, however, is best represented by the information in Figures 4 and 5. Inconsistencies between lamp and sky scans became apparent in a number of cases (Figures 4 and 6) when a good performance in the calibration room was not repeated outdoors. In the case of the Norwegian spectroradiometer (number 3) an anomalous cosine response is suspected. A similar, smaller zenith angle-dependent discrepancy was observed with a second Bentham (DM300) and is probably attributable to differences

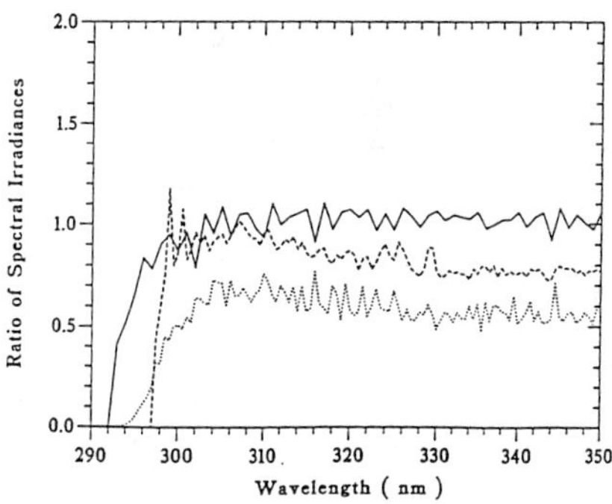


Figure 6. Ratio of simultaneous sky scans for instruments 9/1 (——), 9/3 (•••), and 9/7 (- - - ) Appendix, Table 1).

in cosine response of the input optics. The wavelength-dependent discrepancy in the calibration of instrument 7 (Figure 4) is also evident in the sky scans (Figure 6). Full details of the 1992 intercomparison of spectroradiometers can be found in Gardner et al. (unpublished). The limited results above have been chosen to illustrate important aspects of instrument performance and operation and the level of agreement that it is possible to achieve between independently maintained spectoradiometers of different design. While aiming to retain and improve this compatibility and understand some of the characteristic differences between measured spectra, it is possible to begin exploring the practicalities of using spectroradiometers at their isolated home sites as an operational network.

## NETWORKING

The networking of independently-operated spectroradiometers requires a number of considerations, such as:

How much data is to be collected and when? This will depend not only upon the intended use of the data but also the ability of the instruments and operators to perform measurements at specified times and under specific conditions.

How should the overall data set be handled when data formats differ from site to site? A central facility that can reduce all raw formats to a form suitable for combined analysis is required.

What quality control and quality assurance measures should be instigated by each site? Poor data should not be submitted to the central facility.

How is relative consistency of calibration to be maintained between sites? A frequent independent check is needed for each site's internal standard.

Finally, what additional information about the site, or co-located instruments, should be required? Indications of sky state and albedo become particularly important when comparing UV-B from sites with different climates. The EC STEP Program is now approaching some of these problems and searching for practical solutions.

### CALIBRATION STANDARD

A portable calibration unit has been designed as a traveling reference for relative calibration checks. The unit is based upon a rotating carousel with six 200W lamps which, through internal referencing, will ensure a stable output from a single lamp to the spectoradiometer. The unit will move from site to site and will act as an independent relative reference for the absolute calibration procedures of each contractor. A prototype unit has been built and after initial tests is now being modified before beginning operational evaluation in circulation.

### SITE DEFINITION

The siting of the individual instruments is outside the control of the program but to assist in data analysis, details of each site should be provided with the data. Information should include the details of the site horizon with direction and elevation of any obstructions and any times when the instrument may be shaded from the direct beam solar radiation. An estimate of the albedo over surrounding areas of increasing radius is requested, especially if and when albedo changes dramatically, such as with the fall of snow. Finally a co-located standard pyranometer is required to provide estimates of sky state and a connection to more widespread radiation climatology.

### DATA COLLECTION AND MANAGEMENT

Contractors are currently being asked to make spectral scans at least once per hour on a minimum of five clear, or partially clear, days in each three-month period of the year. During scan days total solar radiaion from the co-located pyranometer should be recorded at least every 60 seconds.

Data is sent to the British Antarctic Survey in basic site format and is then converted into a more uniform style for analysis. Once submitted to the central database, data should not be revised. This places the onus of quality control on the individual operators. It is suggested that a record is kept of frequent calibration checks both for individual reference lamps and the program reference standard when in circulation to enable drift in calibration to be spotted and amended rapidly.

None of the spectroradiometers has been in operation more than a few years, therefore there is no experience of long-term stability of calibration against which to assess the necessary frequency of calibration. This knowledge will only come with time.

## SUMMARY

The EC STEP UV-B Program has held two intercomparisons of spectroradiometers normally capable of solar spectral UV-B measurements[1,2,3]. The technical aspects of the instruments' optical performances have been investigated, and the operational procedures which determine consistency of results at an isolated site inspected. Areas for improvement in both these facets of data collection often become apparent only when instruments come together in this way; all instruments have undergone improvements between the two campaigns. Further progress towards greater uniformity of performance will depend upon the work of future intercomparisons, but agreement is now sufficiently good between some instruments to begin gathering data which may be viewed as a single cohesive dataset. The tasks of data management and continuous quality control must now be tackled.

### Acknowledgments

This work was funded by the European Community STEP Program 0076. The contribution of all participants (Table 1) to the results presented are gratefully acknowledged.

## REFERENCES

1. Gardiner, B.G., and P.J. Kirsch (eds.). 1992. European intercomparison of ultraviolet spectrometers. CEC Air Pollution Research Report 38.

2. Gardiner, B.G., A.R. Webb, A.F. Bais, et al. 1993. European intercomparison of ultraviolet spectrometers. Environ. Technol. 14:25-43.

3. Gardiner, B.G., P.J. Kirsch, and A.R. Webb. Second European intercomparison of ultraviolet spectrometers. CEC (in preparation).

4. Webb, A.R., B.G. Gardiner, M. Blumthaler, P. Forster, M. Huber, and P.J. Kirsch. Laboratory investigation of two commercially available ultraviolet spectroradiometers. Photochem. Photobiol. (submitted).

## Appendix A

Table 1. European Community Step Program 0076 participants and instruments.

| Contractors | Spectrometer | Gratings | Sites |
|---|---|---|---|
| A.R. Webb<br>Department of Meteorology<br>United Kingdom | Optronics 742 | 2 | 1 |
| P. Simon<br>Institut d'Aeronomie Spatiale de Belgique<br>Avenue Circularie 3<br>Brussels | JY DHIO | 2 | 2 |
| K. Henrikson<br>Auroral Observatory<br>University of Tromso<br>Norway | JY HR320 | 1 | 3 |

(cont.)

Table 1 (continued). European Community Step Program 0076 participants and instruments.

| Contractors | Spectrometer | Gratings | Sites |
|---|---|---|---|
| B. Gardiner<br>British Antartic Survey<br>Cambridge<br>United Kingdom | No Instrument | - | - |
| C. Zerefos<br>University of Thessaloniki<br>Laboratory of Atmospheric Physics<br>Greece | Brewer MK 11 | 1 | 4 |
| G. Holmes<br>Botany School<br>University of Cambridge<br>United Kingdom | No Instrumnet | - | - |
| J. Lenoble<br>Universite Science et Techniques<br>de Lille, Villeneuve d'Ascq<br>France | No instrument | - | - |
| H. Kelder<br>Royal Netherlands Meteorological Institute<br>Div. Physical Meteorology<br>de Bilt, Netherlands | EG&G 1453A<br>Diode Array | 1 | 5 |
| M. Blumthaler<br>Institut fur Medizinische Physik<br>Universitat Innsbruck<br>Austria | Bentham DM150 | 2 | 6 |
| I. Dirmhirn<br>Universitat fur Bodenkulter<br>Institut fur Meteorologie und Physik<br>Wien, Austria | JY DH10 | 2 | 7 |
| **Additional Participants in Intercomparison** | | | |
| G. Seckmeyer<br>Fraunhofer-Institut fur Atmospharische<br>Garmisch-Partenkirchen<br>Germany | Bentham DM300 | 2 | 8 |
| G. Smith<br>Industrial Research Ltd<br>Lower Hutt<br>New Zealand | Kratos GM200-2 | 1 | 10 |
| K. Lamb<br>SCI-TECH Instruments Ltd<br>Saskatoon<br>Canada | Brewer MK III | 2 | 9 |

# TOTAL OZONE AND UV-B BEHAVIOR IN 1992-1993 AT A MID-LATITUDE STATION

A.M. Siani,[1] N.J. Muthama,[2] and S.Palmieri[1]

[1]University of Rome "La Sapienza," Physics Department
Piazzale Aldo Moro 2, 00185, Rome, Italy
[2]University of Nairobi, Meteorology Department, PO Box 30197, Nairobi, Kenya

Key words: Rome (Italy), total ozone, ultraviolet irradiance

**ABSTRACT**

Daily total ozone data derived by means of a Brewer spectrophotometer at Rome (Italy) during the European Arctic Stratospheric Ozone Experiment (EASOE) were subjected to time series analyses to investigate any possible variations.

Graphical and statistical analyses revealed the existence of below-normal occurrences of total ozone for the months of January and February in both winters. In an effort to filter out the influence of natural fluctuations on the observed below-normal ozone occurrences, long-term Dobson total ozone records (from 1957 to 1986) of Vigna di Valle station (Rome) were analyzed to ascertain the existence of the respective high and low frequency variations. The removal of short-term fluctuations shows that mean ozone values are still below normal. UV-B data collected within the same period are also considered.

A trend analysis of the ozone time series of Vigna di Valle (Rome) station shows a decreasing tendency in the last ten years, particularly during the winter season.

## INTRODUCTION

The temporal behavior of total ozone amounts at various stations around the world have been frequently investigated, revealing various fluctuations and trends [17,3,13,10]. The reasons attributed to these variations have been associated mainly with two factors: influence on atmospheric chemistry brought about by nuclear tests, sun spots, volcanoes, and injections of chlorofluorocarbons; and changes in atmospheric large-scale circulations causing anomalies in the horizontal and vertical motions [13]. It is well known that ozone variations influence the UV radiation directly, especially at shorter wavelengths where ozone absorption increases with decreasing wavelength [1].

Total ozone has been observed to be decreasing over the Antarctic during their spring. The reductions have been thought to be due to chemical processes dependent on the injection of anthropogenic substances from the troposphere into the stratosphere [7,12,15,]. A crucial role is seen for stratospheric clouds which disappear in the spring, releasing trace constituents in gaseous form. Solar cycle-induced changes in nitrogen species have also been suggested to account for the Antarctic decrease [3,11].

Similar but smaller decreases in ozone has been observed in the Arctic region during late winter [8]. This observation resulted in the launch of the European Arctic Stratospheric Ozone experiment (EASOE) program between November 1991 and April 1992. The experiment involved the monitoring of total ozone

NATO ASI Series, Vol. I 18
Stratospheric Ozone Depletion/
UV-B Radiation in the Biosphere
Edited by R. H. Biggs and M. E. B. Joyner

and its vertical profile by use of ground-based and airborne instruments. As part of the experiment, the Brewer station at Rome participated in the collection of daily total ozone and UV-B data during the experimental period. The analysis of the collected total ozone and UV-B data since the beginning of the EASOE campaign up to winter 1992/1993 forms the basis of this paper.

## DATA PREPARATION

The total ozone observations were performed using a Brewer MKIV spectrophotometer No. 067 located at the Physics Department of University of Rome "La Sapienza" (Latitude 41.9°N, Longitude 12.5°E). Daily total ozone measurements were collected for the period January 1992 to March 1993 to investigate their temporal characteristics during the 1991-1992 and 1992-1993 winter seasons. However, in view of the short Brewer ozone data records, a climatological frame of reference had to be established. This was achieved by utilizing the long-term Dobson total ozone data records of Vigna di Valle (Rome) station. The two stations are controlled by the same meso-scale circulation patterns, therefore their ozone data are considered comparable. The Dobson total ozone data from Vigna di Valle were collected for the period 1957-1991. The Brewer and Dobson data are comparable to better than 1% [9].

The Brewer instrument is also able to measure incident solar radiation in the entire UV-B spectral range (290nm-325nm). It provides a measure of damaging ultraviolet radiation (DUV), defined as the incident radiation weighed against an action spectrum that relates the sensitivity of the human body to UV radiation [18]. The DUV values are calculated using the erythemal weighting curve recommended by the American Conference of Governmental Industrial Hygienists, National Institute for Occupational Safety and Health (ACGIH-NIOSH). These DUV measurements were collected for the period January 1992 to March 1993. The stability of the instrument is ensured by controls involving calibration tests of internal mercury and halogen lamps and an external tungsten-halogen lamp.

Long-term mean surface pressure values, together with daily surface pressures for the period May 1991 to April 1992 from Ufficio Centrale di Ecologie Agrarie in Rome were also used in investigating the reasons associated to the total ozone fluctuations in Rome.

## ANALYSIS AND DISCUSSION

### TOTAL OZONE

Graphical and statistical methods were used for the time series analyses of the Brewer total ozone data at Rome. Figure 1 shows the mean distribution of daily total ozone measurements with respect to smoothed long-term mean ozone pattern (line c) at Rome during 1992 and winter 1992-1993. The graph shows a notable decrease of total ozone for the months of January and February in both years.

Below-normal ozone decreases of 12% and 8% were observed in January and February 1992, respectively. A similar below-normal ozone behavior was evident in January and February 1993, with respective ozone decrease of 12% and 10%. In other months of 1992, no ozone anomaly was observed.

To better understand the below-normal total ozone occurrence, an attempt was made to filter out the natural variations to see whether or not the observed decrease was due to natural factors. The data were then subjected to autocorrelation analysis and periodogram analysis. The natural fluctuations evident from these tests, namely the twelve month and the twenty-four month cycles were then filtered out of the series. The variation patterns of the smoothed series do not explain the below-normal values of January and February of both years.

A significant correlation between total ozone and surface pressure at Rome was obtained, in agreement with the well known relationship between these two parameters [16]. This would suggest that the intense anti-cyclonic circulation that brought about the observed above normal surface pressure in Rome (Figures 2a and 2b) either caused the total ozone decrease, or both the pressure and ozone were influenced by the same factor(s).

A further investigation based on another statistical methodology was performed on the long-term total ozone records of Vigna di Valle. The analysis was based on three different nonparametric tests, namely Mann-Kendall Sneyers (MKS) to detect trends, and Pettitt and Lombard to estimate the number and the location of the change points in the series [14].

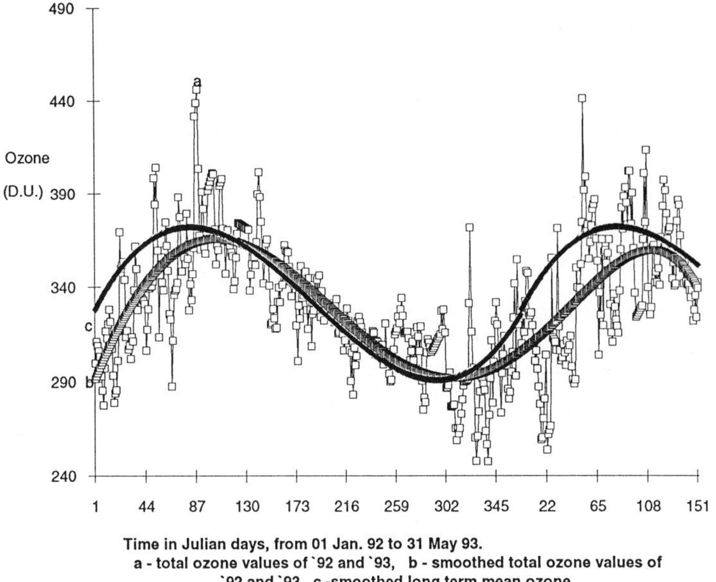


Figure 1. Time sequence plot of daily total ozone at Rome from January 1992 through May 1993.

Applying these tests to ozone series selected on a monthly basis, a statistically significant decreasing trend was noted for the winter months of 1973-1992, (Figure 3). This depletion is graphically estimated at 0.2% per year. This is in good agreement with many recent studies in the mid-latitudes [4,19].

A recent study on the Mount Pinatubo eruption of June 1991 indicated that the large amounts of aerosols that were ejected into the stratosphere have increased the stratospheric concentrations of aerosols over Europe by about 2 orders of magnitude. It has been suggested that the aerosols effects on ozone at Rome are expected to last for about two years after the eruption [5,6]. In view of these results, therefore, it is hypothesised that the Mount Pinatubo eruption may have influenced to a large extent the chemistry and the dynamics of ozone at Rome.

UV-B IRRADIANCE

Total ozone variations influence UV-B measurements directly. However, several other atmospheric parameters (like cloudiness, aerosols, various gases) and geometrical factors (variation of sun-earth distance, incident solar angle) have also been known to influence the amount of UV-B irradiance reaching the earth's surface [1]. Additionally, many recent studies have documented that the ozone-layer depletion leads to increased damage to human skin due to an increase in the erythemal solar UV-B dosage. In view of these facts a preliminary investigation was performed on a selected set of DUV measurements obtained during clear sky days. Figure 4 shows an example of the diurnal behavior of DUV during two clear-sky days in April 1992 with total ozone difference of 26 D.U. It is evident from the graph that the peak DUV differ-

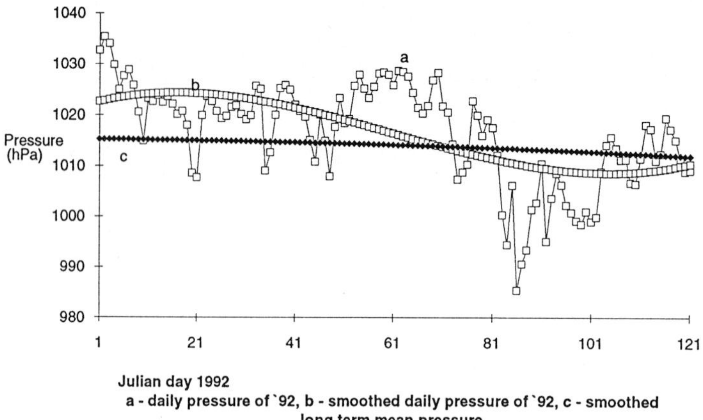


Figure 2a. Time sequence plot of daily surface pressure at Rome from January through April 1992.

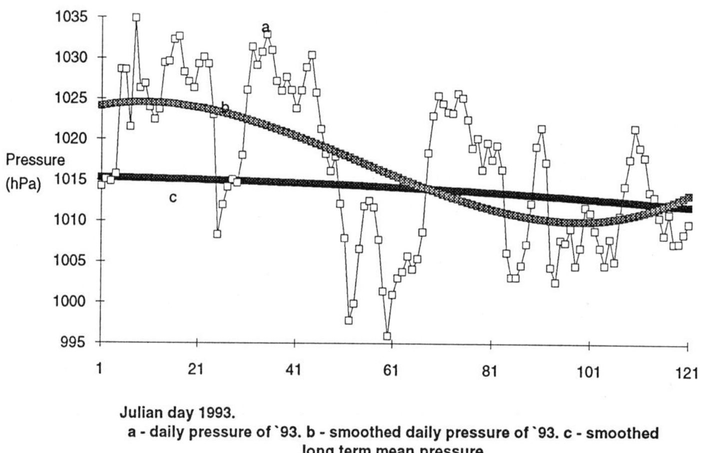


Figure 2b. Time sequence plot of daily surface pressure at Rome from January through April 1993.

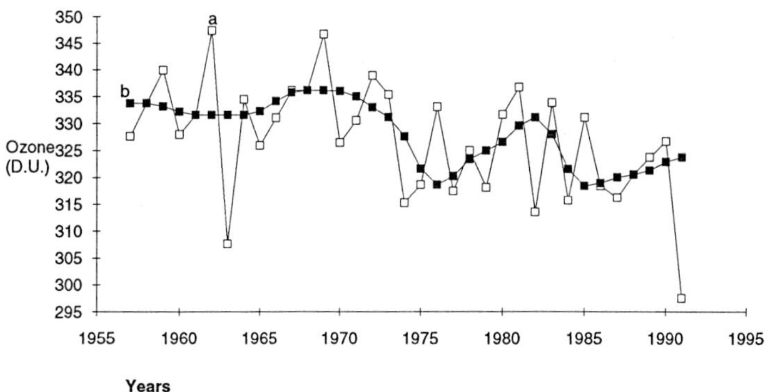


Figure 3. Time series plot of mean winter total ozone at Vigna di Valle.

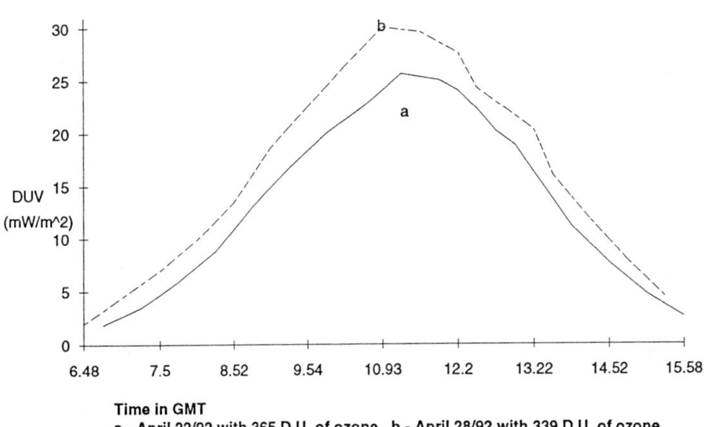


Figure 4. Diurnal course of DUV during 2 clear days with ozone difference of 26 DU, April 1992.

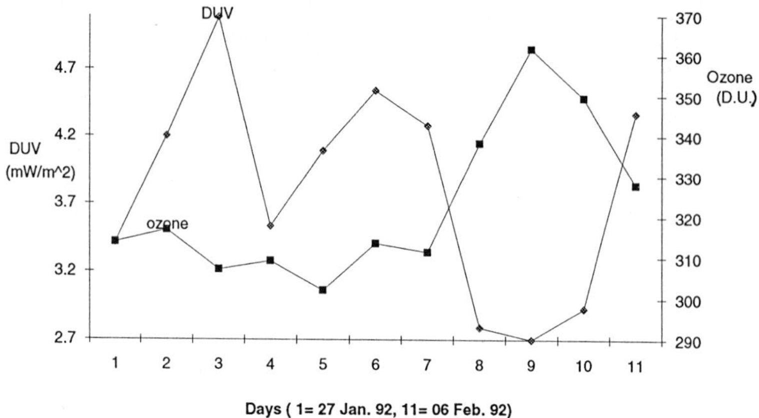


Figure 5a. DUV and total ozone values observed within solar zeneth angles 61.7° and 63.0°.

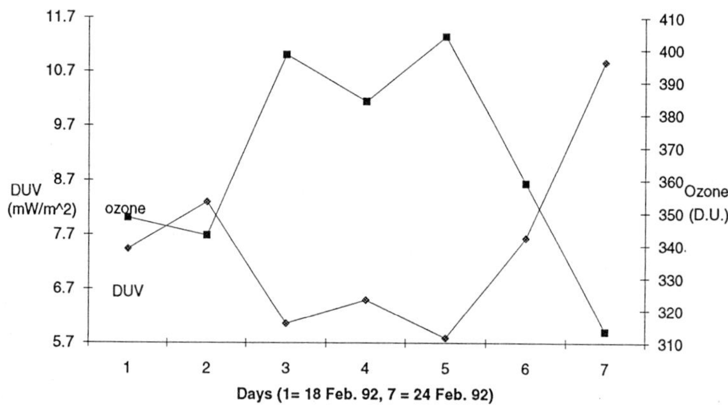


Figure 5b. DUV and total ozone values observed within solar zeneth angles -51.5° and -53.8°.

ence is at around local noon (10:00 GMT). Taking this into consideration, some days with DUV values measured at around noon and the corresponding total ozone observations were considered for graphic analysis.

Figures 5a, 5b, 5c, and 5d are examples of the temporal patterns of the DUV and total ozone measurements in 1992 and 1993. Despite the very limited data sets available, it is clear from the figures that, in general, a decrease in ozone is associated with an

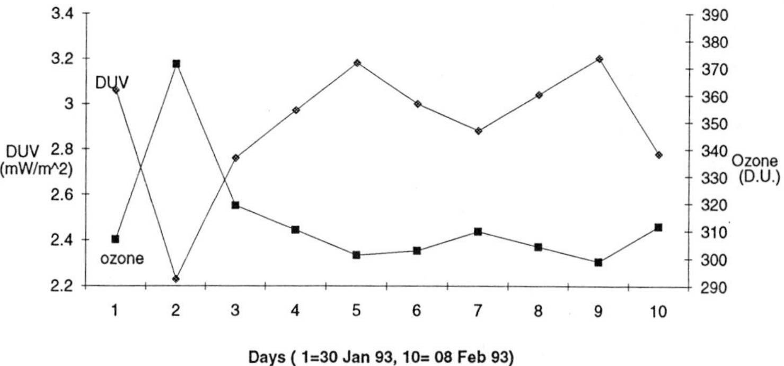


Figure 5c. DUV and total ozone values observed within solar zeneth angles -64.4° and -64.5°.

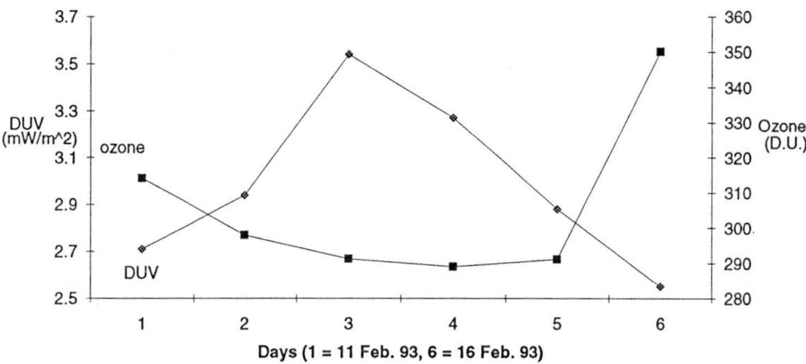


Figure 5d. DUV and total ozone values observed within solar zeneth angles -64.9° and -65.3°.

increase in DUV, as has been documented in many other places in the world. However, this pattern is of notable importance due to the following consideration: DUV and ozone values are observed at the center of Rome, where the contribution of pollutants (such as $SO_2$) may compensate for the increase of UV-B radiation due to the ozone depletion. In this case, however, the impact of the compensation on the UV radiation is weak.

## CONCLUDING REMARKS

The filtration of the seasonal fluctuations and the quasi-biannual cycle were not enough to explain the total ozone decreases during January and February 1992 and 1993 at Rome. However, the significant correlation between surface pressure and total ozone at Rome may suggest that the Mount Pinatubo eruption may have played some role in the observed ozone patterns. More studies about this aspect may throw some light onto the influence on the ozone layer of the eruption and any anthropogenic processes during the winter seasons under consideration. Some examples of temporal patterns of DUV and total ozone during the winters of 1992 and 1993 are presented. Increase in DUV is noted as total ozone decreases.

In an effort to further address the origins these problems, additional investigations on total ozone, $SO_2$, some meteorological parameters, and DUV are needed to provide a more complete picture of the situation, as more DUV data become available.

## REFERENCES


1. Bais, A.F., C.H. Zerefos, C. Meleti, I.C. Ziomas, and K.Tourpali. 1993. Spectral measurements of solar UV-B radiation and its relations to total ozone, $SO_2$ and clouds. J. Geophy. Res. 98:5199-5204.
2. Callis, L.B., and M. Natarajan. 1986. Ozone and nitrogen dioxide changes in the stratosphere during 1979-84. Nature 323:772-777.
3. Cariolle, D., and M. De'que'. 1986. Southern hemisphere medium scale waves and total ozone disturbances in a spectral general circulation model. J. Geophy. Res. 91:10825-10846.
4. Cervini, M., and G. Giovanelli. 1991. Spectral analysis and trend of total ozone observations at Vigna di Valle. Il Nuovo Cimento 14C(6):575-585.
5. De Benedetti, G. 1993. The evolution of aerosol load in the stratosphere following the Mount Pinatubo eruption. Graduation Thesis, Department of Physics, University of Rome "La Sapienza," 1993.
6. Di Sarra, A., M. Cacciani, G. Fiocco, D. Fua, B. Knudsen, N. Larsen, and T.S. Joergesen. 1992. Observation of correlated behavior of stratospheric ozone and aerosol at Thule during winter 1991-1992, Geopys. Res. Lett. 19:1823-1826.
7. Farman, J.C., B.G. Gardiner, and J.D. Shanklin. 1985. Large losses of total ozone in Antarctica reveal seasonal $ClO_x/NO_x$ interaction. Nature 315:207-210.
8. IPCC. 1991. Intergovernmental panel on climate change.
9. Kerr, J.B., I.A. Asbridge, and W.F.J. Evans. 1988. Intercomparison of total ozone measured by Brewer and Dobson spectrophotometers at Toronto. J. Geophys. Res. 93:11129-11140.
10. Lacis, A., D.J. Wuebbles, and J.A. Logan. 1990. Radiative forcing of climate changes in the vertical distribution of ozone. J. Geophy. Res. 95:9971-9981.
11. Lefe'vre, F., L.P. Riishojgaard, D. Cariolle, and P. Simon. 1991. Modeling the February 1990 polar stratospheric cloud event and its potential impact on the Northern Hemisphere ozone content. J. Geophy. Res. 96:22509-22534.
12. McElroy, M.B., R.J. Salawitch, S.C. Wofsy, and J.A. Logan. 1986. Reduction of antarctic ozone due to synergistic interactions of chlorine and bromine. Nature 321:759-762.
13. Newell, R.E., and H.B. Selkirk. 1988. Recent large fluctuations in total ozone. Q.J.R. Meteorol. Soc. 114:595-617.
14. Siani, A.M., S. Bruni, N.J. Muthama, S. Giannoccolo, E. Veccia, and S. Palmieri. 1993. The total ozone time series at Rome: Temporal fluctuations and trend analysis. To be presented at the International Conference on Applications of Time Series Analysis in Astronomy and Meteorology, University of Padua, Italy, September 6-10, 1993.
15. Solomon, S., R.R. Garcia, F.S. Rowland, and D.J. Wuebbles. 1986. On the depletion of Antarctic ozone. Nature 321:755-758.
16. Taba, H. 1961. Ozone observations and their meteorological applications. WMO Tech. Note No. 36.
17. WMO/NASA. 1988. Report of the international ozone trends panel, 1988. WMO Rep. No. 18, Geneva.
18. WMO. 1990. Consultation on Brewer ozone spectrophotometer operation, calibration and data reporting. WMO Rep. No. 22, Arosa.
19. WMO. 1991. Report of the first meeting of the ozone managers of the parties to the Vienna Convention for the protection of the ozone layer. WMO Rep. No. 23, Geneva.

# THE ATHENS STATION FOR ATMOSPHERIC OZONE AND SOLAR RADIATION MONITORING

C. Varotsos

University Athens, Dept. of Applied Physics,
33, Ippokratous Street, 106 80 Athens, Greece

**ABSTRACT**

Measurements of atmospheric ozone at the surface level as obtained in Athens, Greece, from the beginning to the end of the twentieth century are used to estimate the UV-B level and determine changes in the ozone-producing potential.

Total ozone measurements with a Dobson spectrophotometer (No. 118) were instituted in Athens in 1988 and they now form part of the data published by the WMO, World Ozone Data Center. We have used this dataset to examine the consistency of TOMS data with the corresponding Dobson data on a daily basis. The results show that Athens Station may be used as a ground truth for satellite-based total ozone measurements and also provide reasonably accurate total ozone values for southeastern Europe.

The reprocessed daily total ozone observations made by the TOMS on Nimbus-7 over Athens (38°N, 23°E), Dundee, Scotland (56.5°N, 3°W), and St. Petersburg, Russia (60°N, 30°E) have been analyzed from November 1978 to January 1992 to investigate the temporal and spatial variations of both total ozone and UV-B radiation over this geographical area.

A number of vertical ozone profiles up to 35 km in height have been measured using the method of free balloon launch (ozonesondes), at the Athens Station from 1989 as part of various EC projects (TOASTE, TOR-EUROTRAC, EASOE, OCTA) to investigate among others the stratosphere-troposphere ozone exchange, the lamination in ozone profiles, and the horizontal transport of ozone. Further, the dependence of the Radiative Amplification Factor of the erythemally active UV on the changes of the vertical ozone profile is attempted.

Broad-band SUVR measurements have been obtained at Athens Station since 1986 by using three Eppley radiometers and the recently acquired YES UV-B1 radiometer. In collaboration with the Photobiology Units of Dundee University and Athens University, the indoor and outdoor MED have been also measured using the polysulphone film method.

## BACKGROUND

### SURFACE OZONE IN ATHENS IN THE TWENTIETH CENTURY

For the first four decades of the twentieth century, atmospheric ozone at the surface level was monitored in Athens using De James colometric papers. The location of measurements enhances the importance of this record as it is the only surface ozone record for the southeastern Mediterranean for the beginning of this century. Considering the severe air quality problems that large cities around the world (Athens being a pronounced example) experience at the present, knowledge of the background ozone concentration give invaluable information for the proper treatment of photochemical modelling and for the theoretical approach of the UV-B radiation reaching the ground.

A method was developed to allow the conversion

NATO ASI Series, Vol. I 18
Stratospheric Ozone Depletion/
UV-B Radiation in the Biosphere
Edited by R. H. Biggs and M. E. B. Joyner

of the colometric papers to actual ozone mixing ratios. To be compatible with recent experimental findings, the method accounted for the role of a number of factors such as relative humidity and exposure time on measurements. Application of this method resulted in the corrected surface ozone record for Athens for the period 1901-1940 which differs substantially from the original uncorrected record. This large discrepancy stresses the need for a standardized approach for the treatment of the historic surface ozone records available at several sites around the world.

The ozone data recorded at the National Observatory of Athens during the first four decades of this century are significantly lower compared to more recent ozone data, both for the area of Athens and abroad. This is not surprising as the increased rates of ozone production in both urban and rural areas are today well documented. On the contrary, the annual cycle for Athens (1901-1940) is in phase with the annual cycle in Montsouris (1877-1907), a result which strongly demonstrates the uniformity of the physical and chemical mechanisms prevailing at different sites of the early past.

This historic record is analyzed in conjunction with present ozone measurements in Athens (1987-1990) to show an increase of the ozone mixing ratio from 28.2 to 57.4 for daytime, and from 26.6 to 48.5 for night-time. The annual cycle of surface ozone contains one maximum for the daytime data and two maxima for the night-time data, the situation being reversed when a subperiod (1901-1907) of the historic record is considered.

## TOTAL OZONE DEPLETION OVER ATHENS AS DEDUCED FROM SATELLITE AND GROUND-BASED INSTRUMENTATION

Recently Varotsos and Cracknell [17] suggested that the total ozone depletion over Athens from January 1979 to January 1992 shows a strong seasonal variation from more than 7% per decade in winter and early spring to about 2.5% per decade in summer. They have adopted as a plausible reason for these large negative trends the fact that heterogeneous reactions modify the pure gas-phase chemistry within the Arctic vortex leading to the observed large ozone decrease with time [15]. They have also suggested that this effect is evident at mid-latitudes when transport of polar air occurs as the vortex is distorted or as it breaks up. It is at that time that a negative gradient in the amount of active chlorine away from the pole takes place. Due to the fact that as one moves from the pole the available sunlight becomes stronger, the ozone destruction through the chlorine process increases, causing large negative ozone trends at mid-latitudes.

Very recently Varotsos and Cracknell [18] made an intercomparison of the ground-based total ozone observations which they derived from the Dobson No.118 spectrophotometer with the satellite total ozone measurements they deduced from TOMS instrument during its passes above Athens. They have concluded that TOMS is in good agreement with the Dobson spectrophotometer (+4% in 2).

A major impact of decreased total ozone is that greater amounts of biologically active UV-B radiation (280-315 nm) may reach the earth's surface. A 1% decrease in total ozone can cause about a 2% in UV-B radiation doses weighted by the DNA action spectrum [16,11].

Considering the above-mentioned results, for the rate of total ozone loss over Athens from 1979 to 1991, it is easily concluded that an increase of the erythematically active irradiance reaching the ground of the order of 14% per decade in winter and 5% per decade in summer should be expected for the same time period (UNEP, 1991).

## SUVR REACHING THE GROUND AS DEDUCED FROM MEASUREMENTS AND A PARAMETRIC MODEL

Two years ago, Katsambas et al. [10] proposed a simple parametric model for the calculation of direct and diffuse SUVR reaching the earth's surface for different atmospheric conditions (as ozone abudance in the atmosphere) and varying time periods (as the time of the day, the season) and geographic locations.

In particular, the direct SUVR propagating through the atmosphere and reaching the ground, was calculated following an expression by Bird and Riordan [2]:

$$I(d\lambda)= (o\lambda)*D*\tau*(r\lambda)*t(a\lambda)*\tau(w\lambda)*\tau(o\lambda)*\tau(u\lambda) \quad (1)$$

where:

I(dτ(wλ) stands for the direct normal irradiance

H(oτ(wλ) the extraterrestrial solar radiation at the mean distance Earth-Sun at a givenwavelength

D the correction factor for the change of the Earth-Sun distance along the year.

The parameters τ(rλ), τ(aλ), τ(oλ), τ(wλ) and τ(uλ) denote the transmittance functions of the atmosphere at wavelength for molecular scattering, absorption by aerosols, ozone absorption, water vapor absorption, and uniformly mixed gas absorption, respectively.

The total diffuse irradiance was calculated using the expression:

$$I(s) = I(r\lambda)+I(a\lambda)+I(g\lambda) \qquad (2)$$

where:

I(rλ) is the Rayleigh scattering component,
I(aλ) is the aerosol scattering components, and
I(gλ) is the component accounding for the multiply reflected irradiance between the air and the ground.

Following is a list of the various transmission functions used for calculating the components of formulation (1):

Rayleigh scattering:

$$\tau(r\lambda) = \exp\{-M'/[\lambda\lambda(115.6406 - 1.335/)]\}$$

Aerosol transmission function:

$$\tau(a\lambda) = \exp[-\beta\lambda^{a}M)$$

Ozone transmission function:

$$\tau(o\lambda) = \exp[-ao\lambda O_3 M0]$$

Transmission function for uniformly mixed gases:

$$(u\lambda)=\exp[-1.41a(u\lambda)M'/(1+118.3\alpha M')^{0.45}]$$

Water vapor absorption function:

τ(wλ) = 1 for the wavelengths from 0.28 μm to 0.75 m.

M is the relative air mass,
M' is the pressure corrected air mass,
λ the wavelength,
α is 1.0274 if $\lambda<0.5$ m and is 1.2060 if $\lambda>0.5$ m,
$\beta =$ $[0.025 + 0.100\cos(C_pT)] \exp(-0.7h)$,
LAT is the geographic latitude,
h is the hour of the day,
a(oλ) is the absorption coefficient given by Bird and Riordan [2]
$O_3$ is the column ozone in atm-cm,
Mo is the ozone mass, and
a(uτ) is a combination of an absorption coefficient and gaseous amount given by Neckel and Labs [13] revised extraterrestrial spectrum and atmospheric absorption coefficients at 122 wavelengths.

The model modified by Bird and Riordan [2] was used for calculating the diffuse irradiance with special care given to the treatment of clouds and in the use of the appropriate albedo value. In particular clouds are treated in an approximate manner and in trems of sky coverage (%). Three different cases are specified:

| | |
|---|---|
| clear sky: | SK=1 |
| partly cloudy sky: | SK=0.7 |
| overcast sky: | SK=0.2 |

where SK a parameter used in the calculation of total diffused irradiance. The values of SK are from Diffey [4].

The following cases are specified for the albedo values:

| | |
|---|---|
| fresh snow | 85 |
| ice | 0.35 |
| sand | 0.25 |
| grassy field | 0.2 |
| dry, plowed field | 0.125 |
| water | 0.1 |
| forest | 0.75 |
| thick clouds | 0.75 |

Finally, the combined use of formulae (1) and (2) yielded the amount of total irradiance to the earth's surface, with respect to the solar zenith angle z:

$$I(total) = I(d\lambda)\cos z + I(s\lambda) \qquad (3)$$

An inspection of the mean monthly values of the measured and modeled SUVR reaching the Athens basin as deduced from 1989 to 1991 at local noon shows that there is close agreement between the similated SUVR values and the measured ones. The comparable value of SUVR in April with those in June and July is due to the high atmospheric turbidity.

The aforementioned UV model was run for $\lambda$=305 nm, $\lambda$=315 nm, and 290<$\lambda$<320 nm to detect the trend of the most harmful part or the SUVR through the time period 1979-1991. It is clearly seen that the positive trend of the annual course of the broad-band UV-B radiation correlates well with the UV-B at $\lambda$=315 nm while in the case of $\lambda$=305 nm, a larger sensitivity is noted. For the larger sensitivity at wavelength $\lambda$=305 nm, a plausible explanation might be the fact that at this wavelength the ozone cross section is larger. In particular, there is a factor of about 3 between (290-320 nm) and 305 nm in all months.

## VERTICAL OZONE STRUCTURE OVER ATHENS

The existence of a laminar structure (layers of enhanced and depleted ozone) in ozone profiles was first reported about thirty years ago [7,8], based on ozone soundings performed over North America and Switzerland. Dobson [5] made a systematic analysis of the laminar structure in ozone profiles in the lower stratosphere over a wide longitudinal and latitudinal range. As criteria for the detection of laminae he adopted ozone partial pressure change >30 nanobars. For the statistical analysis of the laminar structure he divided the ozone profiles into three groups:

Group 0: profiles containing almost no lamina,
Group I: profiles exhibiting moderate lamination,
Group II: profiles extensively laminated.

Dobson discovered that the lamination phenomenon is most frequently found between January and April and that the features of the laminated structure vary with latitude. At latitudes below about 20°N in the spring and below about 30°N in the autumn, a laminated structure of the ozone is very seldom found. About 35% of all profiles were Group II at high latitudes in spring, while Group I days reached a peak frequency of 70% in June. Group II days were rarely observed at any latitude during the summer, while Group I was found in Polar regions throughout the year. Dobson [5] could not find an explanation for the incidence of these layers at a preferred height of 15 km or for the constancy of the preferred height with latitude. He showed, though, that there is a strong correlation between the existence of a characteristic ozone minimum at 15 km and the occurence of a double tropopause, especially in latitudes around 40°N. He also made the assumption that the laminated structure and the ozone minima at 15 km are of the same origin since the variations in the frequency of their appearance with both season and latitude are very similar.

Another statistical study of the lamination phenomenon by Reid and Vaughan [14] was published in 1991. In this study, only the laminar structures found in the altitude range of 9.5-21.5 km were examined. Laminae were categorized according to their vertical extent (called depth) as well as the change in ozone partial pressure inside them (called magnitude). They also used the same criteria to separate genuine laminar features from instrumental noise or large scale features in the ozone profile. They chose an acceptable depth for laminae of between 0.2 and 2.5 km while the minimum magnitude was 20 nbars. The above study showed that laminae are most abundant below 18 km at high latitudes during winter and spring, which is in accordance with the findings of the Dobson study. But Reid and Vaughan disagree with Dobson about the origin of laminae. Based on the observation that the magnitude is greatest and the depth least during winter and spring, they rule out the suggestion by Dobson that laminae all originate near the subtropical jetstream, although the possibility remains that a small number of laminae may be generated by stratosphere-troposphere exchange around these features [1]. Instead, they suggest that laminae may represent evidence for a process which can cause exchange of air into and out of the polar vortex. This suggestion is supported by recent polar ozone research where laminae have been detected that are particularly sharp and deep near the winter polar vortex [12]. Given the ozone distruction that is known to take place within the vortex, the possibility that such exchange can spread ozone depletion to lower latitudes should be taken into consideration.

The first important finding of our study is that the lamination phenomenon has been observed very frequently over our region. These laminar events were significantly more frequent than those observed in previous measurements in our latitudes. This difference is much more pronounced in February-March where the lamination frequency over Athens is about twice that observed over Cagliari, Italy (39.1°N). We should mention that the data of Cagliari were taken between 1972-1975. This fact, in combination with the observation that the laminar events are much more pronounced at the edge of the polar vortex, leads to a possible explanation that the polar vortex has moved southward during the last twenty years, a possibility mentioned in the on-going discussions about ozone depletion in the middle latitudes over the last fifteen years [15]. It is worth noting that the nearest ozone sounding station to Athens (in Hopenpeissenberg, Germany, 47.5°N) is quite far-reaching in order to get more information about the situation over our region.

If we examine the weather maps provided by the German Weather Service, we observe that there is a significant correlation between the appearance of laminar events and the establishment of a north-northwest circulation in the lower stratosphere over our site. In this case, we could also say that the site is under the influence of the polar vortex. This observation is in agreement with the findings mentioned above about the enhancement of laminar phenomena at the vortex boundary.

The occurence of the laminated ozone structure as well as the appearance of a characteristic ozone minimum at 14-15 km, correlated with the general circulation, leads to the following conclusions: a) The laminated ozone profiles are associated with the north-northwest circulation in the lower stratosphere and b) the characteristic ozone minimum is related to the influence of the subtropical jet stream circulation.

## REFERENCES


1. Begun, D.A., 1989. Studies of the total ozone field around the sub-tropical jet stream. Ph.D. Thesis, University of Wales.

2. Bird, R., and C. Riordan. 1986. Simple solar spectral model for direct and diffuse irradiance on horizontal and tilted planes at the earth's durface for cloudless atmosphere. J.Clim. Appl. Meteor. 25:87-97.

3. Cartalis C., C. Varotsos, H. Feidas, and A. Catsambas. 1992. The impact of air pollution in an urban area on the amount of solar ultraviolet radiation at the surface. Tox. Envir. Chem. (in press).

4. Diffey, B. 1982. The consistency of studies of ultraviolet erythema in normal human skin. Phys. Med. Biol. 37:715-720.

5. Dobson, G.M.B. 1973. The laminated structure of the ozone in the atmosphere. Quart. J. R. Met. Soc. 99:599-607.

6. Drummund, A.J., and A.H. Wade. 1969. Instrumentation for the measurement of solar ultraviolet radiation. *In* F. Urback (ed.), The Biological Effect of Ultraviolet Radiation. Pergamon, Elmsford, NY.

7. Hering, W.S. 1964. Ozone observations over N. America. Vol. 1, A.F.C.R.F.

8. Hering, W.S. 1964. Ozone observations over N. America. Vol. 2, A.F.C.R.F.

9. Justus, C.G., and M.V. Paris. 1985. A model for solar spectral irradiance at the bottom and top of a cloudless atmosphere. J. Clim. Appl. Meteor. 24:193-205.

10. Katsambas, A., C. Antoniou, J. Stratigos, I. Arvanitis, F. Zolota, C. Varotsos, C. Cartalis, and D.N. Asimakopoulos. 1991. A simple algorithm for simulating the solar ultraviolet radiation at the earth's surface: An application in detering the minimum erythema dose. Earth, Moon and Planets 53:191-204.

11. Liu, S.C., A.A. McKeen, and S. Madronich. 1991. Effect of anthropogenic aerosols on biologically active ultraviolet radiation. Geophys. Res. Lett. 18:2265-2268.

12. McKenna, D.S., R.L. Jones, J. Austin, E.V. Browell, M.P. McCormick, A.J. Krueger, and A.F. Tuck. 1989. Diagnostic studies of the Antarctic vortex during the 1987 airborne Antarctic ozone experiment: Ozone miniholes. J. Geophys. Res. 94:11641-11669.

13. Neckel, H., and D. Labs. 1981. Improved data of solar spectral irradiance from 0.33 to 1.25 m. Solar Physics 74:231-249.

14. Reid, S.J., and G. Vaughan. 1991. Lamination in

ozone profiles in the lower stratosphere. Quart. J. R. Met. Soc. 117:825-844.

15. Stolarski, R.S., P. Bloomfield, R.D. McPeters, and J.R. Herman. 1991. Total ozone trends deduced from Nimbus 7 TOMS data. J. Res. Lett. 18:1015-1018.

16. UNEP. 1989. Environment Effects Panel Report. United Nations Environmental Programme, Nairobi, Kenya.

17. Varotsos, C., and A.P. Cracknell. 1992a. Ozone depletion over Greece as deduced from Nimbus 7 TOMS measurement. Int. J. Rem. Sensing (in press).

18. Varotsos, C., and A.P. Cracknell. 1992b. Three years of total ozone over Athens-Greece using the remote sensing technique of Dobson spectrophotometer. Int. J. Rem. Sensing. (in press).

19. Varotsos, C., and C. Cartalis. 1993. Surface ozone in Athens, Greece at the beginning and at the end of the twentieth century. Atmospheric Environ. (in press).

# HIGH-RESOLUTION SPECTRAL UV MONITORING AND ATMOSPHERIC TRANSFER MODELING IN THE NETHERLANDS

Harry Slaper, Henk Reinen, and Johan Bordewijk

Laboratory for Radiation Research
National Institute of Public Health and Environmental Protection (RIVM)
Bilthoven, The Netherlands

**ABSTRACT**

During the past decade a decrease in stratospheric ozone has been observed over large parts of the globe. Stratospheric ozone is the primary absorber of solar UV-B and UV-C and serves as a partially-protective shield against harmful UV irradiance in the biosphere. Decreased levels of ozone appear to increase the levels of harmful UV-B (280-315 nm) in the biosphere, leading to a variety of adverse effects on human health and on aquatic and terrestrial ecosystems. Changes in other atmospheric parameters, like aerosol content, cloud cover, and tropospheric ozone could also influence UV-B irradiance at ground level. High resolution spectral monitoring data are necessary to detect the present biologically-effective UV irradiance and to investigate the possible trends in effective UV. The advantage of spectral measurements is that the data can be used for evaluations with various biological action spectra.

A project was started in the Netherlands to gain knowledge of the present biologically-relevant UV climate and its dependence on atmospheric conditions. The measurements will be used to validate atmospheric UV transfer models; long-term monitoring should provide information on trends in the biologically-relevant UV. To meet these aims, a UV spectrometer system consisting of a highly accurate scanning double monochromator especially for the biologically-relevant UV-B region with a focal length of 50 cm, grating with 2400 l/mm, linear dispersion 0.375 nm/mm, and a multichannel detection system (diode array) for UV-A radiation measurements. Spectral coverage of the combined system ranges from 290-450 nm. UV monitoring needs will be outlined from the perspective of effect evaluations; preliminary results from the RIVM measurement system will be shown and the necessary international collaboration will be discussed.

NATO ASI Series, Vol. I 18
Stratospheric Ozone Depletion/
UV-B Radiation in the Biosphere
Edited by R. H. Biggs and M. E. B. Joyner

# GROUND-BASED MONITORING OF UV-B RADIATION IN CANADA

C.T. McElroy, J.B. Kerr, L.J.B. McArthur, and D.I Wardle

Atmospheric Environment Service of Canada , 4905 Dufferin Street
Downsview, Ontario M3H 5T4 Canada

Key words: Brewer spectrophotometer, Canadian Ozone Watch, ozone, UV radiation

**ABSTRACT**

A program of ultraviolet-B radiation monitoring using specially equipped Brewer Ozone Spectrophotometers has been in place in Canada since 1989. The network has recently been expanded to 12 stations which are distributed throughout the country and now report ozone and UV-B values automatically in near-real-time to the Canadian meteorological data collection network. These data are used to prepare operational forecasts of ozone and UV-B radiation levels for distribution to the Canadian public. Some results of the monitoring program, the performance of the instrumentation and the methods used for the calibration of the field instruments will be discussed.

## WHAT INFORMATION DO WE DERIVE FROM UV-B OBSERVATIONS?

UV-B data which are collected by the global UV observing network have a number of applications. These include the determination of UVB levels at particular sites, the development of a global climatology of ultraviolet light, the determination of the trend in UV at selected sites and the elucidation of the relationship between measured UV-B levels and the independent variables on which they depend. Some of these goals can be realized using broadband measurements, particularly those related to the public dissemination of information about UV-B and education.

In the long term, an accurate estimate of ultraviolet radiation levels at all points on the globe will be required, together with the appropriate biological response functions, to determine the impact of atmospheric change on the Earth's biosystems. Similar assessments will be needed to determine the effects (for example: economic impact) that such changes would have on human life. To accomplish this goal, an atmospheric radiative transfer model which is capable of determining the ground-level UV-B levels given an estimate of the composition and structure of the atmosphere through which the radiation is transmitted must be employed.

The development and testing of the model needed for this analysis will require measurements of surface UV-B levels as well as the many atmospheric properties on which the level of radiation depends. These include cloud, ozone amount (both troposheric and stratospheric), sulphur dioxide and nitrogen dioxide levels, haze, surface albedo, and other factors. Many sites in different climate and latitude zones must be instrumented to make observations of these elements to provide the information needed to generalize locally determined relationships between UV-B levels and atmospheric properties to the global scale.

The research approach taken in Canada is to make spectroradiometric measurements of UV-B so that the process of measurement can be decoupled as much as possible from the application of the collected data. This ensures that the data will be useful in the assessment of all types of ultraviolet effects, regardless of the action spectra associated with the process

NATO ASI Series, Vol. I 18
Stratospheric Ozone Depletion/
UV-B Radiation in the Biosphere
Edited by R. H. Biggs and M. E. B. Joyner

under consideration. The spectroradiometric data will, in particular, support materials testing and the application of the data to biological systems with response spectra which have not yet been measured, if they turn out to be significantly different from response spectra currently quantified.

## THE BREWER OZONE SPECTROPHOTOMETER

The Brewer Ozone Spectrophotometer was developed by Environment Canada primarily to address the need for an ozone measuring instrument to replace the venerable Dobson Ozone Spectrophotometer in the Global Ozone Observing System operated by the World Meteorological Organization. The instrument was put into commercial production in the early 1980s by Sci-Tec Instruments Limited of Saskatoon, Saskatchewan. The device was specifically designed to meet the stringent spectroscopic performance criteria associated with making accurate ultraviolet measurements for the calculation of ozone column amounts using direct-solar-beam absorption observations.

In its ozone measuring mode, the Brewer instrument makes nearly simultaneous observations at a set of 5 wavelengths in the nearultraviolet. These wavelengths are sampled rapidly using a steppingmotor chopper which opens and closes five exits slits placed at the focal plane of a modified Ebert spectrometer. The spectrograph incorporates a correction lens and a slightly asymmetric geometry which produce a flat focal plane and a high-quality spectral image. The instrument resolution (slit width) is about 0.6 nm, and the five slits are placed at extreme points in the solar spectrum at 306.3, 310.0, 313.4, 316.7, 319.9 nm. The instrument passband is carefully controlled in production and accurately characterized by spectral response tests.

The most challenging aspect of making ozone and UV-B measurements through a large slant path of ozone is that of controlling stray light in the spectrometer. The solar spectrum, as observed from the Earth's surface, changes by more than 4 orders of magnitude between 295 and 325 nm, so light from the visible portion of the solar spectrum must be strongly rejected at the short wavelength end of the measurement region. The Brewer achieves the required rejection ratio, in spite of being a single monochromator, by virtue of being a reflection instrument (unlike the refractive Dobson spectrophotometer) and through the use of a nickel-sulphate/cobalt glass band-limiting filter to reduce the amount of visible radiation which can reach the detector.

The Brewer system includes automatic wavelength calibration using an internal mercury discharge lamp, and a relative spectral intensity source which based on a quartz-halogen lamp. It also has automatic solar pointing which is accomplished using a computer-controlled azimuth mount, together with a pointing prism drive which can change the elevation angle of the instrument field-of-view.

All functions of the instrument are under the control of an IBMPC compatible computer which communicates with the instrument via an RS-232 serial communications link. The instrument wavelength setting is monitored using an automatic calibration algorithm which makes measurements at a number of diffraction grating angles (wavelength settings) and computes the optimal setting of the grating from the analysis of a scan of a line in the spectrum of the mercury discharge lamp. The precision with which the Brewer can calculate the wavelengths at which observations are made is of the order of 0.0004 nm. The resolution of the stepping motor drive is 1/170 nm.

During the instrument design phase a horizontal diffuser was added so that the Brewer could be used as a UV-B monitor. Most of the spectrometer performance requirements for the accurate measurement of ozone are the same as those needed for good UV-B measurements. A fixed directing prism is used to allow the foreoptics pointing prism to view the diffuser. The diffuser is mounted on the outside of the instrument case and is protected from the elements by a clear quartz dome.

The design of the Brewer was optimized to make the best possible measurements in the UV cutoff region of the solar spectrum where ozone is the principal absorber of solar radiation. When used for monitoring of UV-B radiation the instrument is scanned using the one of the ozone observing slits across a range of wavelengths using the diffraction grating motor drive. The scans take about 5 minutes. A scan in each direction (short to long and long to short wavelength) is made between 290 and 325 nm and the two scans are co-added to produce one spectral measurement. An observation is made every 0.5 nm for a total of 71 spectral data points per scan. These data are numerically convolved in near-real-time with the Diffey erythemal action spectrum (or the 71-point vector representing the action spectrum of some other effect) and printed out. The collected data are archived on disk for permanent storage in the ozone and UV-B data base.

There is an additional contribution of the Diffey erythemal action spectrum which comes from the UV-A region. In the current operating mode, there is a correction of about 15% added to the measured UV-B contribution using the brightness of the 325 nm spectral element as an estimator. The uncertainty in this flux does not contribute a significant error to the total UV-B value calculated.

Brewer instruments are calibrated for making UV-B measurements using a source of known spectral irradiance which is traceable to NIST. The responsivity data derived are stored and used by the analysis software to translate measured photon counts into absolute pectral irradiance values. The calibration source is a certified DXW-type quartz-halogen lamp provided by Optronics Corporation. The lamp is operated on direct current from a high-precision switching regulator which is adjusted at each use to put a fixed voltage across a precision resistance connected in series with the lamp. The power dissipated in the lamp is determined from measurements made using a high-quality digital voltmeter and the specified resistance value of the shunt.

Brewer UV-B measurements are corrected for scattered light by subtracting the mean photon count value between 290 and 292.5 nm from the measured counts at all other wavelengths. This is based on a model of instrumental stray light which has a constant contribution at all wavelengths. Because of ozone absorption there is a negligible amount of liqht short of 293 nm in the solar spectrum at the qround.

## THE CANADIAN OZONE MONITORING NETWORK

The Canadian monitoring network began operating in 1957 and by 1964 there were five sites (Toronto, Edmonton, Goose Bay, Churchill, and Resolute) from which measurements of total ozone were made using the Dobson spectrophotometer. During the 1980s the Dobson instruments were replaced by automated Brewer spectrophotometers at four stations. In Toronto, both Dobson and Brewer instrument types are now operating [5,6]. The Brewer instrument is capable of making measurements more frequently becuse of its high level of automation.

Ground-based column ozone data are supplemented by weekly ozonesonde measurements. These measurements began in the late 1960s and early 1970s at four sites (Edmonton, Goose Bay, Churchill, and Resolute) using the Brewer-Mast ozonesonde. Around 1980, the ozonesonde type was changed to the Electrochemical Cell (ECC) sonde which produced an improvement in the quality of the ozone profile data [2].

The monitoring network was expanded and improved during the 1980s and 1990s in response to increased concern regarding ozone depletion after the discovery of the Antarctic ozone hole [3]. There are now twelve stations in the network; the new ones being Saturna, Saskatoon, Winnipeg, Montreal, and Halifax (at mid-latitudes) as well as Alert and Eureka in the Arctic. Eureka is the location of the new Canadian Arctic stratospheric ozone research observatory which is being outfitted with ground-based instrumentation for monitoring stratospheric constituents as part of the World Meteorological Organization (WMO) Network for the Detection of Stratospheric Change (NDSC). Lidar ozone and aerosol measurements began at this site in early 1993 [1]. Figure 1 shows the location of the Canadian observing stations and Table 1 summarizes their operational history.

## OBSERVED TREND IN OZONE

Data from the Canadian ozone stations have been used in studies to evaluate global trends in total ozone [10,12,13]. These data make an important contribution toward the findinqs at middle and hiqh latitudes in the Northern Hemisphere.

Ground-based direct sun total ozone measurements at Toronto between 1960 and 1991 have been analyzed [4]. A downward trend of 4.2% per decade was observed for the annual average during the 1980s. This negative trend has a seasonal dependence with a maximum decrease of about 7% per decade during the winter/spring season, a time when total

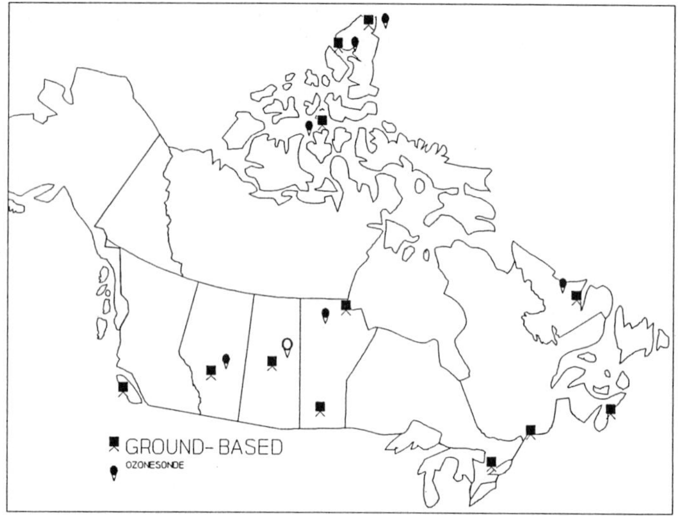


Figure 1. Map of Canada showing the locations of stations in the ozone monitoring network.

ozone is at its annual maximum. Despite this loss during winter and spring, ozone values remain higher than those in fall, a time of year which has shown little or no loss. These results are in good agreement with the Total Ozone Mapping Spectrometer (TOMS) satellite data which show a similar decrease at mid-latitudes averaged over the entire Northern Hemisphere 9]. The Toronto study also shows that the seasonal behavior of the 1980s trend is quite distinct from smaller fluctuations before 1980 which were presumably due to natural variations. Figure 2 shows the decrease in ozone seen at Toronto until 1992 and Figure 3 shows the seasonal dependence of the decrease during the 1980s.

Table 1. Listing of stations in the Canadian ozone monitoring network showing the locations and start of continuous records.

| | | | |
|---|---|---|---|
| Edmonton | 54N,114W | 1957 | 1972 |
| Resolute | 75N, 95W | 1957 | 1966 |
| Toronto | 44N, 79W | 1960 | |
| Goose Bay | 53N, 60W | 1962 | 1969 |
| Churchill | 59N, 94W | 1964 | 1973 |
| Saskatoon | 52N,107W | 1988 | |
| Alert | 82N, 62W | 1987 | 1987 |
| Saturna | 49N,123W | 1990 | |
| Winnipeg | 50N, 97W | 1992 | |
| Halifax | 45N, 64W | 1992 | |
| Montreal | 46N, 74W | 1993 | |
| Eureka | 80N, 86W | 1993 | 1993 |

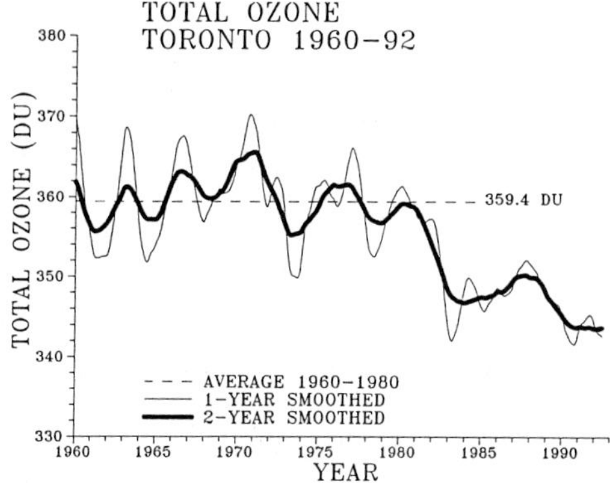


Figure 2. This plot shows one-year and two-year, smoothed direct sun total ozone measurements made at Toronto between 1960 and 1992. The value of 359.4 Dobson Units (DU) is the average prior to 1980.

## UV-B MEASUREMENTS

Measurements of ultraviolet-B (UV-B) radiation began at Toronto in March, 1989 using the Brewer spectrophotometer. Absolute intensity measurements of global (diffuse plus direct) radiation are made at wavelengths between 290 and 325 nm at 0.5 nm wavelength intervals with a resolution of 0.5 nm full width at half intensity (FWHI). Between 1990 and 1993 the UV-B measurements started operationally at the other sites. Figure 4 shows examples of spectral irradiance measurements made on a clear day in Toronto.

Erythemally weighted ultraviolet radiation has received considerable attention primarily because of concern over sunburn and skin cancer. The generally

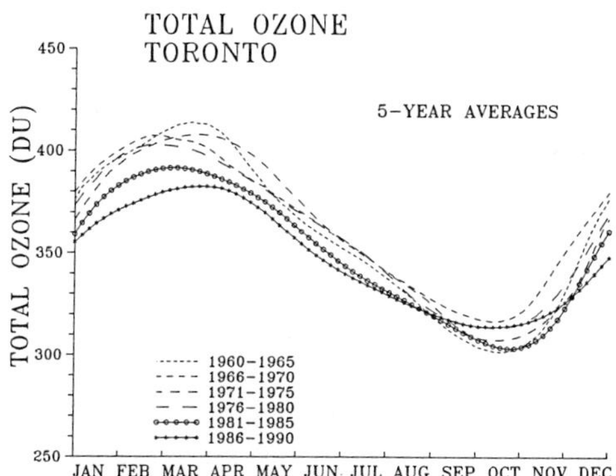


Figure 3. Averages of 60-day smoothed ozone values for 5-year periods showing a decrease in total ozone during the 1980s primarily during the winter and spring seasons.

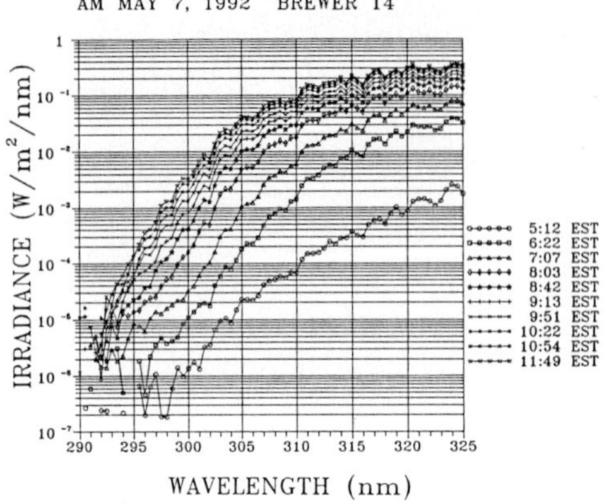


Figure 4. Spectral irradiance measurements made at different times on a clear day.

accepted "action spectrum" for erythema (skin reddening) is that of McKinley and Diffey [8], shown in Figure 5. Multiplying this spectrum by the spectrum of observed ultraviolet radiation gives an irradiance at 298 nm which would, in principle, cause the same skin-reddening as the radiation with the observed spectrum. Figure 6 shows the result of applying the Diffey erythemal weighting to the spectral measurements of Fiqure 4.

The length of the UV-B data record is not long

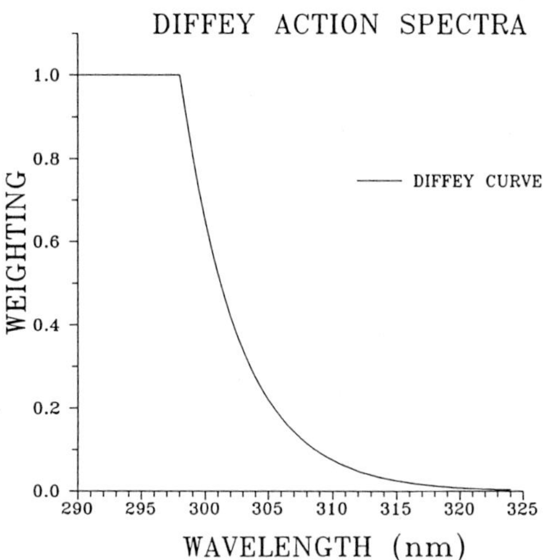


Figure 5. The Diffey erythemal action spectrum. This curve shows the relative effectiveness of radiation at different UV-B wavelengths for causing skin reddenning.

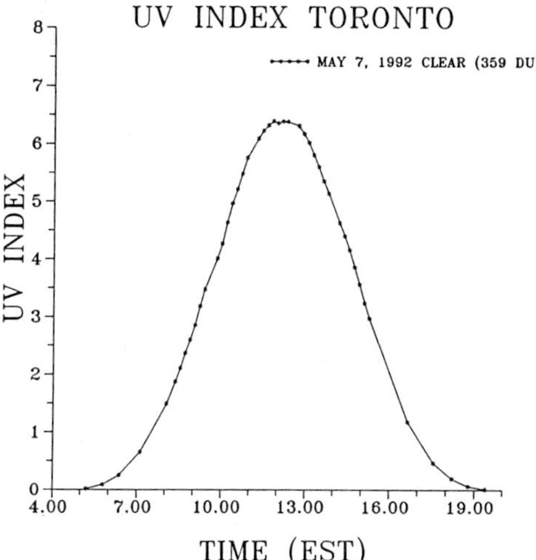


Figure 6. Effective irradiance values determined by integrating the spectral irradiance measurements (Figure 4) weighted by the erythemal action spectrum (Figure 5).

enough to determine whether long-term trends of UV-B radiation have occurred. However, it is possible to determine the dependence of UV-B irradiance on total ozone by looking at day-to-day variations. Results of this dependence are summarized in Figure 7. Measurements of total ozone and UV-B radiation made over a three year period were used. The UV-B spectral measurements were weighted with the Diffey action curve. Data were selected from days when the sky was completely cloudless and when the total ozone varied

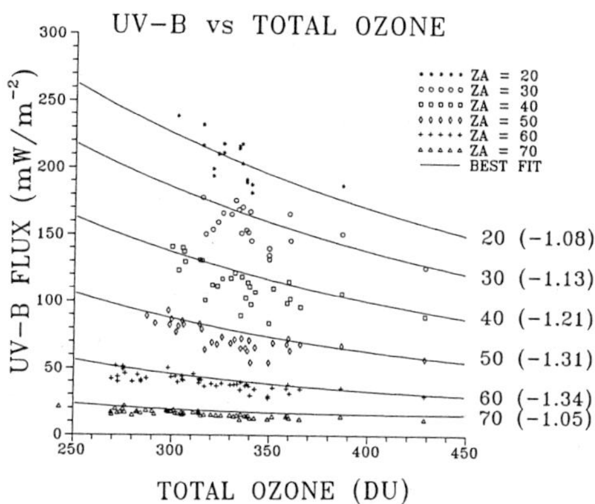


Figure 7. UV-B flux measurements as a function of total ozone. The data are grouped according to solar zenith angle. Values in parentheses indicate the per cent change in UV-B for a 1% change in total ozone.

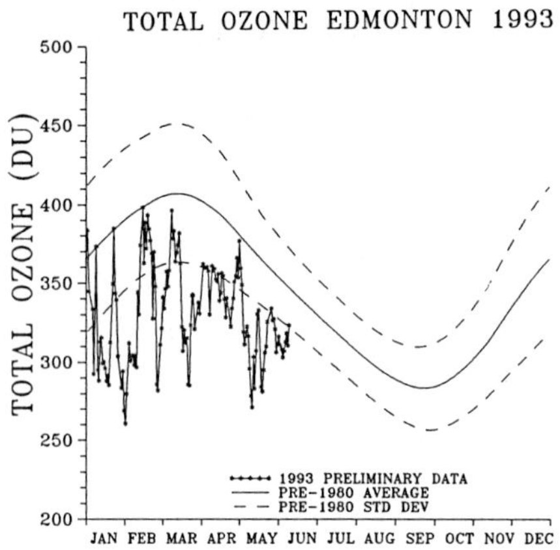


Figure 8. Daily measurements of ozone at Edmonton, Alberta during 1993 compared with a pre1980 reference curve.

by less than 2%. The data points in Figure 7 depict near simultaneous measurements of ozone and UV-B flux. The different symbols indicate that the data points were measured when the solar zenith angle was within ±2.5° of the value indicated in the legend. The solid lines through the data points are the best fit of the data points for the indicated solar zenith angles.

## PUBLIC AWARENESS PROGRAMS IN CANADA

Early in 1992, NASA reported new results from satellite and aircraft measurements indicating high levels of reactive chlorine over a wide area at middle and high latitudes in the Northern Hemisphere and

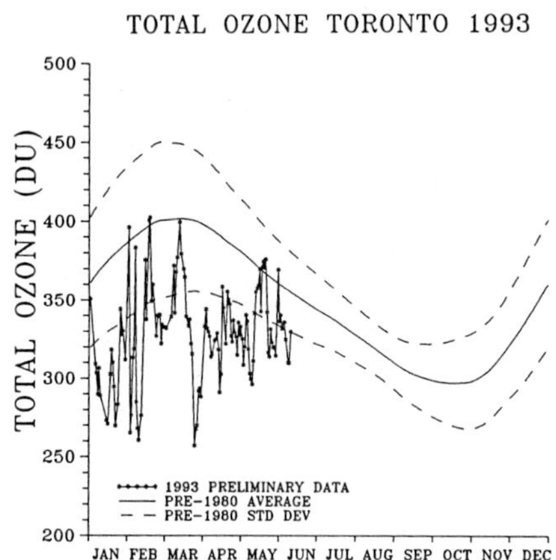


Figure 9. Daily measurements of ozone at Toronto during 1993 compared with a pre-1980 reference curve.

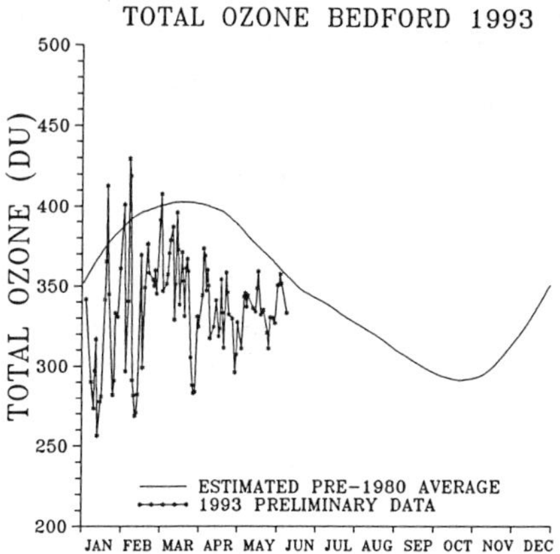


Figure 10. Daily measurements of ozone at Bedford, Nova Scotia during 1993 compared with a pre-1980 reference curve.

suggested that these large values could potentially cause large ozone reductions. In response to public concern regarding these findings, Environment Canada introduced the Ozone Watch program in March, 1992 and UV Index advisory program that May [7]. Both are part of Canada's Green Plan, a national agenda of research, regulation, provision of public information and other activities aimed at protecting the environment. The programs depend on measurements made by the Canadian ozone monitoring network.

## THE OZONE WATCH PROGRAM

The Ozone Watch program is based on the comparison of recent measurements of total ozone with historical "climatological" reference values. Examples of the daily direct sun measurements made during 1993 compared with reference curves are shown in Figure 8 for Edmonton, Figure 9 for Toronto, and Figure 10 for Bedford. Ozone Watch values are the percentage difference of a two-week average of current daily measurements compared with the corresponding two-week average for the reference curve.

The percentage differences for all ground-based stations are issued weekly and are based on data from the previous two weeks. The information has been reported to the public as a time series (Figure 11) and a mapped display (Figure 12). These weekly reports have given the Canadian public an up-to-date and accurate assessment of the current state of the ozone layer both from the large scale (all of Canada) and regional perspectives.

The reference curves are determined from direct sun measurements made prior to 1980 at the five older sites where these data are available. The curves are based on approximately 20 years of data smoothed with a 60-day triangular filter. At the more recently established sites, the reference curves were determined with the aid of early TOMS Version 6 satellite data. The TOMS data between 1978 (start of record) and 1982 (prior to the El Chichon volcanic eruption) for all sites were smoothed by the same 60-day filter.

Annual ratio curves (TOMS/ground-based) were made for the four midlatitude sites with long records. The average of these four ratio curves were then used to adjust the TOMS annual curves at the new stations. Based on the spread of the ratio curves at the long-term sites, the reference curves at the new sites are estimated to be within ±2% of the required long-term pre-1980 values at all times of the year.

## THE UV INDEX PROGRAM

The aim of the UV Index program is to provide a forecast of the intensity of ultraviolet radiation for the next day throughout Canada. The UV Index value is

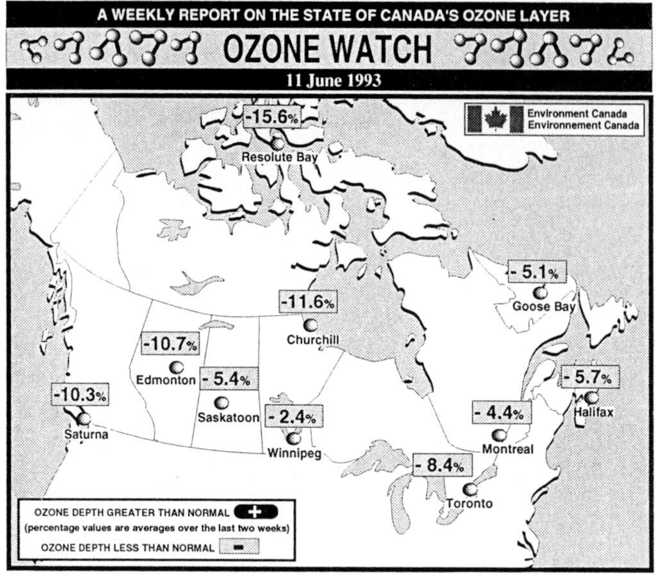


Figure 11. Ozone Watch comparisons at Canadian stations during the spring of 1993.

defined as the downward spectral irradiance weighted according to the Diffey erythemal action spectrum [8] and divided by 25 $mWm^{-2}$. It is non-dimensional and was chosen so that typical sea-level tropical values would be in the 10 to 12 range. The forecast index value is valid at local solar noon (daily maximum) under typical clear sky conditions (which usually include the effects of some attenuation due to haze and atmospheric aerosols). Index values as a function of time of day are determined from the forecast noontime value and the solar elevation throughout the day.

The forecast of the UV Index is a two-stage process. The first step is the prediction of a total ozone field and the second step is to derive a noontime UV Index field from the ozone field.

The first stage of the forecast is based on the statistical correlation of past total ozone to meteorological data. The correlation is applied to the appropriate meteorological values predicted by the Canadian Meteorological Center (CMC) forecast to produce a total ozone field. Real time ozone data (current day's measurements) are used to correct the ozone field. This is done by comparing the forecast values (of the current day) made the previous day with the actual measurements made on the current day. Real-time reporting of ozone data to the AES wide area network (WAN), which carries meteorological data throughout Canada, is now being carried out automatically by the field Brewer instruments from most sites. Forecasts are run twice daily, roughly at 1800 GMT (afternoon run) and 0200 GMT (evening run). Both runs produce forecasts for the next day; the evening run also produces a two-day forecast which is currently being tested.

The second stage of the forecast is the determination of the UV Index field from the ozone field. This is done from the statistical relationship of the dependence of UV flux values on total ozone and solar

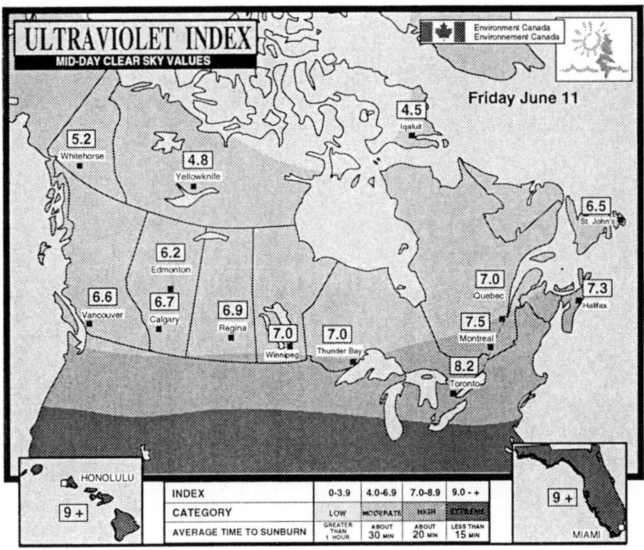


Figure 12. Map of Canada showing Ozone Watch Values issued June 1, 1993.

zenith angle for clear sky situations. This dependence was described earlier and is summarized in Figure 7. The forecasts are moderated according to forecast cloud conditions. Further details regarding the forecast of the ozone field and the determination of the UV Index field from the ozone field are given elsewhere [7,11].

## SUMMARY

The Canadian ozone monitoring network, as well as contributing reliable data for both national and international studies on the ozone layer, provides the basis for extensive public information on ozone and ultraviolet radiation.

## REFERENCES

1. Carswell, A., A. Ulitsky, and D.I. Wardle. Lidar measurements of the Arctic stratosphere. Atmosphere-Ocean, in press.

2. Evans, W.F.J., J.B. Kerr, D.I. Wardle, and C.T.McElroy. 1989. Modernization of the Canadian ozone monitoring network. Ozone in the Atmosphere, Proceedings of the Quadrennial Ozone Symposium, pp.780-783.

3. Farman, J.C., B.G. Gardiner, and J.D. Shankin. 1985. Large losses of total ozone in Antarctica reveal seasonal $ClO_x/NO_x$ interaction. Nature 315:207-210.

4. Kerr, J.B., I.A. Ashbridge, and W.F.J. Evans. 1988.

Intercomparison of total ozone measured by the Brewer and Dobson spectrophotometers at Toronto. J. Geophys. Res. 193:11129-11140.

5. Kerr, J.B., I.A. Ashbridge, and W.F.J. Evans. 1989. Long-term intercomparison between the Brewer and the Dobson spectrophotometers at Edmonton. Ozone in the Atmosphere, Proceedings of the Quadrennial Ozone Symposium, pp.105-108.

6. Kerr, J.B., C.T. McElroy, D.W. Tarasick, and D.I. Wardle. 1992. The Canadian Ozone Watch and UV-B Advisory programs. Proceedings of the Quadrennial Ozone Symposium, Charlottesville, Virginia.

7. Kerr, J.B. 1991. Trends in total ozone at Toronto between 1960 and 1991. J. Geophys. Res. 96:20703-20709.

8. McKinley, A., and B.L. Diffey. 1987. A reference action spectrum for ultraviolet induced erythema in human skin. *In* W.F. Passchier and B.F.M. Bosnajakovic (eds.), Human Exposure to Ultraviolet Radiation: Risks and Regulations, Elsevier, Amsterdam, pp.83-87.

9. Stolarski, R.S., P. Bloomfield, R.D. McPeters, and J.R. Herman. 1991. Total ozone trends deduced from Nimbus 7 TOMS data. Geophys. Res. Lett., 18:1015-1018.

10. Stolarski, R.S., R. Bojkov, L. Bishop, C. Zerofos, J. Staehelin, and J. Zawodny. 1992. Measured trends in stratospheric ozone. Science 256:342-349.

11. Wilson, L.J., M. Vallee, D.W. Tarasick, J.B. Kerr, and D.I. Wardle. 1992. Operational forecasting of daily total ozone and UV-B radiation levels. AES Research Internal Report No. MSRB/ARQX 92-004.

12. World Meteorological Organization. 1988. *Report of the International Ozone Trends Panel 1988,* WMO Global Ozone Research and Monitoring Project Report No.18, Geneva.

13. World Meteorological Organization. 1991. *Scientific Assessment of Ozone Depletion: 1991,* WMO Global Ozone Research and Monitoring Project Report No. 25, Geneva.

# THE CANADIAN OZONE WATCH AND UV INDEX

Anne O'Toole

Environment Canada, 4905 Dufferin Street
Downsview, Ontario M3H 5T4 Canada

Key words: Brewer spectrophotometer, Canadian Green Plan, cataracts, clouds, erythema, ozone watch, UV, UV Index, skin cancer, sunburn

**ABSTRACT**

The Green Plan, developed by the Canadian Government in 1990, provides Canadians with information on the environment including information on UV levels, called the UV Index, so they may guard themselves from excess exposure to the sun. Information from ozone measurements by Brewer spectrophotometers across Canada along with weather predictions and meteeorological information to predict the depth of ozone, thus the probable UV and health risks.

## INTRODUCTION

In 1990 the Canadian government developed the Green Plan which defines Canada's goals and priorities for environmental protection and education. One of the main goals of the Green Plan is to provide Canadians with information on the environment and information on actions they can take to improve environmental quality. With this knowledge, individuals can choose actions that will benefit the environment. They can also use the information to make decisions on how to protect themselves from environmental hazards.

Under the Green Plan, Environment Canada stated that, by 1993, it would provide information on ultraviolet levels so that the public could guard itself against excessive exposure to direct sun. During the spring of 1992, however, Canadians became very concerned over the possibility of increased ozone depletion and the resulting potential increases in UV levels. There was a demand for the most current scientific information available. The challenge of delivering this information was met by using the dissemination systems and scientific support of Environment Canada's Weather Service.

The general public in Canada and elsewhere expects the public weather programs to provide information required for their safety, security, and well-being. Traditional weather information considers safety from the perspective of protection from immediate physical harm. With a forecast of a severe weather event, such as a hurricane or tornado, individuals and communities can take specific steps to seek protection from the storm. In a more general way, the environment and changes to it can affect our heath or well-being over the longer term. Information on the environment can be used to help individuals choose lifestyles that can, overall, reduce risks to their health.

## OZONE WATCH

On February 3, 1992, NASA reported the potential for severe ozone depletion over the Arctic, possibly spreading into mid-latitudes for the coming Spring.

NATO ASI Series, Vol. I 18
Stratospheric Ozone Depletion/
UV-B Radiation in the Biosphere
Edited by R. H. Biggs and M. E. B. Joyner

Overwhelming media attention and public fear was generated by this announcement. People were concerned that the "ozone hole" was going to cover Canada and that the increases in UV would be immense. The difference between ozone depletion, a small but significant general thinning in the amount of ozone, and the concept of an "ozone hole," in other words, little or no ozone over the area was difficult to communicate. Another complication in explaining the issue was that stratospheric ozone depth varies dramatically from day to day in concert with changing weather patterns.

In response to the public's need for information, Environment Canada began to provide Ozone Watch as a vehicle to increase public awareness and reduce the fears that are often generated by misunderstanding. Stratospheric ozone is measured with a Brewer Spectrophotometer at about twelve locations across Canada. For Ozone Watch, at the end of each week a two-week average of current ozone values is compared to the pre-1980 averages for the same time period. Pre-1980 averages were chosen to represent the natural historic level of the ozone layer.

Ozone Watch, issued once a week, provides ongoing information on the normal short term variations in the ozone layer and also provides an indication of the degree of ozone depletion. (See pages 280 and 281 for examples of the graphic used as part of the Ozone Watch.) Environment Canada supplies the Ozone Watch to the media in the same manner as its regular weather reports. Ozone is naturally variable—the Ozone Watch shows these increases and decreases each week.

Over Canada in 1992 the average ozone values reported were lower overall than averages for any preceding year, although the values were not dramatically lower and on any individual day there were no low-value records broken. Despite this, it appears that there were more days with lower ozone values and on the days when stratospheric ozone depth was naturally high the magnitude of the values was not quite as high as observed in similar situations in past years. During the first months of 1993, ozone values across Canada were consistently low. Record values, about 25% below pre-1980 normals, were observed in Edmonton, Toronto, and in the high Arctic as far north as Resolute Bay. These very low values occurred during the late winter and spring when UV levels are naturally not as high as in the summer. However, in late March, UV Index levels similar to those expected in mid-May were observed.

The concern over ozone depletion is high mainly because of the potential for increased amounts of damaging ultraviolet radiation from the sun. The most recent scientific information indicates that for every 1% decrease in ozone one would expect a 1.1% to 1.4% increase in UV. UV radiation occur naturally and exposure to UV has always been a health concern. Excessive exposure to the sun's ultraviolet radiation leads to sunburn and, in some cases, can lead to skin cancer and cataracts. Many people in Canada have adopted lifestyles that promote excessive sun exposure. The incidence of skin cancer, which is related to exposure to the sun's ultraviolet radiation, is increasing dramatically. It is felt that if people continue to expose themselves excessively to the sun and if ozone depletion results in increased ultraviolet radiation, then the health implications could be very significant.

## THE UV INDEX

As part of the efforts to educate Canadians on the effects of the sun, the UV Index was developed and introduced in the spring of 1992. The UV Index is a measure of the erythemal (sunburning) UV radiation. The UV Index provides a forecast of the expected intensity of UV for the coming day. The UV Index is defined on a scale related to the total intensity of the portion of the UV spectrum that affects humans. The Canadian scale generally runs from 0 to 10. A rating of 10 to 12 would be typical of a mid-summer, sunny-sky mid-day in the tropics.

The goal of the UV Index program is to inform Canadians of the effects of ultraviolet radiation and increase their awareness of how to protect themselves from overexposure to the sun. The strategy for delivering the product and the public education that went with its introduction was developed in consultation with a number of health organizations and the medical community. For example, the Canadian Cancer Society, Canadian Association of Optometrists, Canadian Dermatology Association, Occupational Health and Safety groups, and government health agencies, dermatologists and opthamologists all contributed.

The UV Index serves as a guide to help people make decisions on how and when to guard against excessive exposure to the burning radiation of the sun.

TABLE 1. UV Index categories and sunburn context.

| UV INDEX | 0 - 3.9 | 4.0 - 6.9 | 7.0 - 8.9 | > 9.0 |
|---|---|---|---|---|
| CATEGORY | LOW | MODERATE | HIGH | EXTREME |
| SUNBURN CONTEXT | More than 1 Hr | About 30 Minutes | About 20 Minutes | Less than 15 Minutes |

The UV Index is provided daily for locations across Canada combined with sun-awareness health information. The UV Index forecasts and real-time measurements are available from Environment Canada weather offices across the country. In addition, a private-sector weather channel, The Weather Network, provide both the forecasts of the UV Index from Environment Canada and hourly reports the UV Index measurements obtained from their own network of UV monitors.

Forecasts of the UV Index for any day and the next day are determined using the following:

1. Ozone measurements from Brewer spectrophotometers across Canada;

2. Weather prediction models and meteorological relationships to predict the depth of ozone; and

3. Sun angles and the stratospheric ozone depth predictions are used to calculate the UV Index. A relationship between ozone depth and UV was determined from physical principles and the algorithms were fitted to past data. The UV intensity is weighted with respect to the effects on human erythema (sunburn). The weighting curve is known as the McKinley-Diffey Action Spectrum.

4. The UV Index forecasts are adjusted for the forecast sky condition. Cloudy and rainy conditions lead to lower UV Index values.

The higher the value of the Index, the more intense the UV. Four categories or ranges are identified with the UV Index. These categories, in turn, have a time-to-sunburn context to help relate the UV Index to health issues. The sunburn relationship, however, does not reflect other health concerns. As well, even these general sunburn times are only valid for some fair-skinned individuals and could be misleading. However, the sunburn context has helped the public understand the UV Index values. As the public becomes more accustomed to the UV Index numbers, an attempt may be made to move completely away from sunburn times to more general health linkages.

The UV Index is delivered along with the weather forecast from Weather Offices across the country. Newspapers, radio, and television deliver the UV Index on a daily basis. The sky condition in the weather forecast was an important companion to the UV Index and both types of information are used together. The forecasts of the UV Index include only the effects of thick clouds but not elevation, air pollution, or type of surface. The cloud conditions have a great influence on the UV reaching the surface. When conditions are mostly sunny, the forecast clear-sky UV Index value is quite representative; when skies are overcast, however, the situation is more complex. For a bright day, the UV Index will be similar to the sunny, clear-sky values, while for overcast days, the forecast UV Index is reduced to account for the blocking by clouds. Observations can be used to determine the UV Index under actual cloud conditions.

The UV Index can be determined for any location on earth and forecasts could be done in a similar manner if the appropriate data sets (stratospheric ozone measurements and numerical model output) are available. Although a reference of 10 is used to provide a context for Canadians, the UV Index is open-ended. Practically speaking, a value of 12, equivalent to 300 milliwatts/meter/meter, is the highest likely value anywhere on the earth at sea level.

## HEALTH EFFECTS OF ULTRAVIOLET

UV exposure is linked to sunburn, skin ageing, several types of skin cancer, certain types of cataracts, and, in some people, allergic reactions to the sun. Avoiding excessive exposure to the sun is particularly important for children and teenagers. It is therefore critical that parents and care-givers of children be aware of sun-smart behavior.

It is also important to keep the concerns in context. A healthy lifestyle promotes outdoor activities. The message is simply that it is easy to act in sun-smart ways and it is worthwhile to be aware of the effects of the sun's ultraviolet radiation and act accordingly. The basic sun-smart messages are:

Avoid the midday sun;
Cover-up—wear a hat and clothes that cover the arms and legs;
Wear UV rated sunglasses;
Use a UV protecting sunscreen.

## CONCLUSION

Public feedback clearly shows the acceptance of the UV Index as a "weather" parameter. In recent surveys, people indicated that they wanted the UV Index reports in time to plan their activities for the day. A UV Index included in the weather report provides a prompt for people to remember the simple actions that can be taken in planning their outdoor activities to protect both themselves and their children from the sun.

# UV MONITORING PROGRAM AT THE U.S. EPA

William F. Barnard and Larry T. Cupitt

Atmospheric Research and Exposure Assessment Laboratory
U.S. EPA, Research Triangle Park, NC

Key words: AREAL, Brewer spectrophotometer, cancer, cloud cover, EPA, NIST, NOAA, ozone, USDA, UV exposure index, WMO

**ABSTRACT**

The U.S. Environmental Protection Agency (EPA) has initiated a program to develop and operate a monitoring network to measure ultraviolet (UV) radiation flux at the earth's surface and to provide for public awareness of exposure to UV radiation through adaptation and publication of a predictive UV "exposure Index." This Index, similar to the one developed and publicized by Canada in 1992, is being developed in conjunction with the National Weather Service for possible public distribution this year. Current plans call for a fifteen-site monitoring network to measure UV intensities on a spectrally-resolved basis, together with other parameters (e.g., total column ozone, cloud cover, interfering tropospheric pollutants, etc.) that affect UV flux. Eleven urban sites will provide data to validate the Index predictions in proximity to ~25% of the U.S. population. The combination of urban (polluted) and four rural (less polluted) sites will enable EPA to improve the algorithms upon which the Index predictions are based. A series of instrumental intercomparisons with other agencies are also planned to ensure comparability of data and good quality assurance/quality control. EPA's intent is to coordinate its efforts with other federal agencies and with the international community.

## INTRODUCTION

In the fall of 1987, EPA's Atmospheric Research and Exposure Assessment Laboratory (AREAL) undertook a program to measure UV radiation at the earth's surface. At that time, EPA efforts were focused on developing a standard reference instrument and a quality assurance/quality control (QA/QC) program for the monitoring community. In coordination with experts from the National Oceanic and Atmospheric Administration (NOAA) and the National Institute of Standards and Technology (NIST), specifications for a reference instrument were developed. Procurement efforts were initiated to obtain four such instruments from different vendors. EPA's program called for a process of characterization and evaluation in coordination with NOAA and NIST. The intent was to designate the most stable instrument as a reference standard and to use the three others as transfer standards for comparison with UV monitoring equipment used by other agencies or researchers. Subsequently, AREAL purchased one instrument through a competitive procurement contract under an Interagency Agreement (IAG) with NIST. NIST was to obtain and characterize the spectroradiometer. Since Research Support Instruments (RSI) had the most experience at building similar instruments, they were awarded the contract for its construction in the spring of 1990. The instrument was completed by RSI in June, 1992, and it is presently being characterized by NIST.

AREAL obtained funding to purchase a Brewer spectrophotometer in the spring of 1992. This instrument was bought to establish a stratospheric ozone monitoring site at Research Triangle Park, NC, (RTP) and to enable AREAL to join the other Brewer sites operated world-wide and reporting their data to the World Meteorological Organization (WMO) through

NATO ASI Series, Vol. I 18
Stratospheric Ozone Depletion/
UV-B Radiation in the Biosphere
Edited by R. H. Biggs and M. E. B. Joyner

the Atmospheric Environment Service (AES) in Toronto, Canada. In the early 1980s, this Brewer network began to replace the Dobson spectrophotometers which were then used for making stratospheric ozone measurements. Of the more than ninety Brewer instruments currently operating around the world, approximately sixty are currently equipped to make measurements of both UV-B and stratospheric ozone. The Brewer network is coordinated by AES of Canada which currently operates a network of twelve Brewer spectrophotometers in their monitoring programs to measure total column ozone concentrations, UV-B flux irradiance, and to verify their UV Index. AES maintains a triad of calibrated Brewers, as well as a traveling standard, to provide a measure of system accuracy and precision. Periodically, locations around the world are checked to maintain the accuracy of the Brewers in the field. The traveling standard instrument is certified against the triad before and after each trip to ensures the accuracy, precision, and homogeneity of the Brewer data reported to the WMO Global Ozone Observing System ($GO_3OS$). Instrument comparisons are done annually, and conferences of Brewer operators are held every other year. AREAL is presently integrating the RTP site into the existing Brewer network and has begun sending column ozone data to the $GO_3OS$ in Toronto. This provides AREAL scientists not only a major data base management system already in place, but a QA/QC system for making measurements of known quality.

## PRESENT STATUS

Early in the summer of 1992, in response to a request from EPA's Administrators, the Office of Research and Development (ORD)/AREAL began developing plans to establish a UV Exposure Index and Monitoring Network. The focus of the program is to inform the public of the potential for UV exposure on the basis of daily predictions of an "exposure Index." The monitoring network is intended, primarily, to verify the predictions of potential human UV exposure (the Index); to assess and improve the algorithms for predicting the UV flux, accounting for pollutants and other factors in populated areas; and to provide high-quality, spectrally-resolved UV radiation measurements that can be convolved with a variety of human health or biological action curves, either known now or to be developed in the future. Planning is currently underway to implement the Index and Network, including establishing agreements with appropriate federal agencies and other researchers, selecting and procuring the monitoring instrumentation, designing the monitoring network and the field sites, establishing quality assurance/quality control methodology, and so on.

At present an agreement is being negotiated between EPA and the NWS at NOAA's Climate Analysis Center (CAC) in Suitland, MD to begin work on the Index. The Index, which is presently being derived from the one developed by Canada's AES, with additional input from John Frederick at the University of Chicago, is intended to inform the U.S. population of predicted UV radiation levels. The AES has been publishing a UV Index since June, 1992, and has developed extensive experience that the NWS has drawn on in developing the U.S. Index. The NWS will produce a daily gridded analysis of the available NOAA satellite column ozone data utilizing standard gridding techniques. The gridded total column ozone data will then be matched and correlated against the appropriate meteorological parameters of the analysis/forecast system, thus providing a parameterization for the forecast fields. The values of total column ozone and cloud cover will be input to an algorithm which will determine the UV radiation at the ground for that time-step. Initially, only noonday, clear-sky values will be computed and the 500 millibar geopotential height will be used as the fundamental ozone-forecast parameter, similar to the Canadian model. In ensuing months, the inclusion of cloud parameters and other predictive parameters such as tropopause height and stratospheric temperature will be used to improve the predictive model.

The evolving U.S. Index is being checked against last year's ground-based UV-B data from Canada. The NWS-calculated Index values using satellite data from 1992 is being correlated against the measured ground-based UV data from each of the Canadian sites. This process is leading to the development of a quality assurance/quality control data base, providing the Index with a solid scientific footing. This provides the NWS a means of testing and verifying the Index algorithms.

Predictions for 1993 are being correlated with data from AREAL's Brewer monitoring site in Research Triangle Park, NC, allowing for a lower latitude test of the AES Index's algorithms. The output from the Brewer's UV-B data system can be

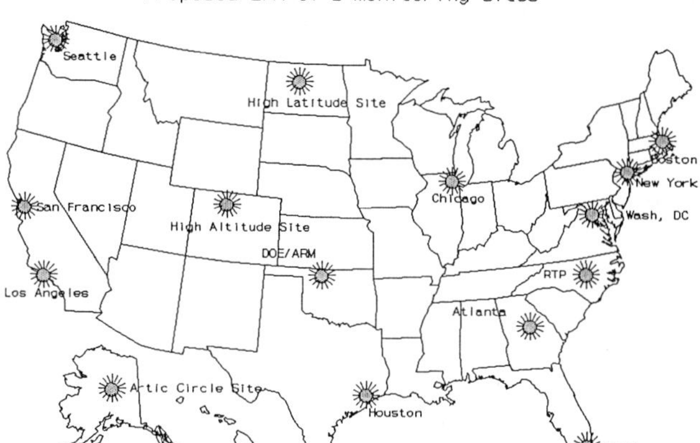


Figure 1.

expressed in many fashions, as has been discovered in attempting to correlate AREAL's data with the NWS Index predictions. Data processing within the Brewer allows for convoluting the raw data with at least three different action curves: the Diffey curve, the erythemal curve, and another curve with a UV-A corrections factor. To date, correlations with the Diffey convolved UV-B data and the predicted values of UV-B from the RTP site have shown very good agreement. Once the data demonstrate that the Index, as predicted by the NWS, correlates well with actual ground-based values, EPA will negotiate with NOAA to publish and disseminate the Index.

The means by which the Index may be disseminated would be decided through the cooperative efforts of AREAL and NWS. At present, the Index would likely be published as the value for clear sky conditions at solar noon.

The preliminary monitoring network design plans call for an EPA network with fifteen research-grade monitoring sites, augmented by additional broadband measurements. Eleven of the research-grade sites will be located in urban areas that cover more than 25% of the U.S. population. Some of these sites will be located in cities where cancer registries are presently being maintained either by the National Cancer Institutes or the International Agency for Research on Cancer (IARC). Four sites will be located in background (low pollution) areas, as shown in Figure 1. The background sites will represent a high altitude site, a high latitude site in the 48 contiguous states, and if feasible, a polar circle site to detect possible Arctic ozone holes should they be forecast.

EPA's needs for background sites are consistent with the requirements specified by a recent U.S Department of Agriculture (USDA) site selection committee meeting. AREAL intends to coordinate site selection with USDA, NOAA, NIST, and any other government agencies making similar measurements, to ensure UV-B measurements can complement each other and to avoid duplication of effort. Technical coordination between EPA and other U.S. agencies has been underway since 1987. Experts from NOAA and NIST were critical to EPA's early efforts. An EPA technical staffer now serves on the USDA Site Selection Committee. EPA staff have frequent discussions with federal UV monitoring experts through its role on the Observations and Data Working Group (WG-1) subcommittee of the Committee on Earth and Environmental Sciences and through the recent formation of a Technical Advisory Panel (TAP), consisting of experts from federal agencies, academia and industry. The TAP reviews the proposed EPA program plan and recommends any necessary changes.

AREAL has already entered into Interagency Agreements (IAG) with NOAA and NIST relevant to this program, and intends to expand the scope of those agreements. EPA is establishing IAGs or Memoranda of Understanding (MOU) with the NWS (NOAA), USDA, and the Department of Energy (DOE) pursuant to the development of the UV Index and the operation of the background monitoring sites. EPA is currently developing a Cooperative Agreement with the University of Georgia to develop a diode array spectrophotometer, to establish and operate a network of monitoring sites, and to develop QA/QC documentation for the site/instrument operation.

EPA anticipates that the network, when fully established, will consist of a framework of research-grade instruments, augmented by broad-band instruments or devices that represent the erythemal response used in the Index calculations. The network will collect data on UV intensity and on a variety of meteorological and environmental parameters. To accomplish the design goals, the intensity of the UV-B flux must be measured with high accuracy and precision across the spectral region of concern. Measurements must be made with sufficient spectral resolution to allow matching the observed intensities with current or future "action" curves for a diversity of effects (sunburn, skin cancer, cataracts, terrestrial and aquatic effects, materials damage, etc.). The research-grade instruments will provide spectrally resolved intensity measurements across the UV-B and a portion of the UV-A range. Data from the network will be compiled and analyzed to estimate the impact of various factors on the UV flux at the earth's surface. Parameters to be investigated include: solar angle, altitude, cloud coverage and cloud density, physical parameters of the atmosphere, total column ozone, nitrogen dioxide, and sulfur dioxide, aerosol scattering, and tropospheric concentrations of ozone, nitrogen dioxide, and sulfur dioxide. Network data will be used to improve the algorithms for prediction of the Index, for extrapolation to other human health end points (other than erythema or sunburn), and for comparison/standardization of EPA's research efforts on effects of UV radiation.

The research-grade instruments will also be used for long-term trend analysis. Rigorous QA/QC procedures and comparisons with other federal agencies (e.g., the National Science Foundation, USDA, NIST, NOAA, and DOE) and with the international community will ensure data that are compatible with the USDA network and the WMO/Canadian global network.

## FUTURE DIRECTIONS

EPA's UV program will focus on improving the Agency's ability to assess human exposures and effects, including: (1) enhancing the accuracy, reliability and applicability of the Index and the predictive algorithms upon which it is based; (2) operating EPA's UV monitoring network in order to verify and improve the Index calculations and predictions and to acquire reliable data for convolving with human effects "action curves"; and (3) exchanging data and QA/QC procedures with other UV researchers within EPA, in other U.S. agencies and institutions, and in the international community. The focus of EPA's efforts will be human exposure and effects, combined with a concerted effort at QA/QC.

AREAL is planning a number of projects to enhance the development of the Index, to ensure the quality of the data within the UV-B monitoring community, and to obtain devices which will provide human health researchers with a potentially inexpensive means to determine the exposure of individuals to UV-B radiation. AREAL is considering two cooperative agreements, one to compare the spectral response of DNA dosimeters to the data from AREAL's spectrophotometer, and the other to develop a personal dosimeter to detect ultraviolet radiation exposure.

A project will begin this fall to determine a relationship between cloud cover and UV-B transmission. These data will then be utilized in the algorithms for predicting the Index, which will again be tested against other ground-based monitoring sites. A similar project to determine the transmission properties of various types of aerosols is in the planning stages and should begin in the spring of 1994. This project involves utilizing two spectrophotometers, one located on a mountain peak and the other in a nearby valley. Data from aerosol samplers would be correlated with the difference in the UV-B flux measurements between the two instruments to evaluate the effects of aerosols on the transmission of UV-B radiation. This experiment would be conducted once in the eastern part of the US and again in the western states to account for the differing characteristics of aerosols across the continent.

Facilities are also available at RTP for the RSI instrument when it is delivered by NIST in 1993. Intercomparison of the RSI-NIST instrument and the

AREAL Brewer spectroradiometer are planned for later this fiscal year. In addition, NIST is planning an instrument intercomparison in the fall of 1993 at NIST. The instruments would first be characterized in the laboratory and then operate for two weeks to ensure sufficient data to determine their respective responses to the solar spectrum.

A group from NOAA, NIST, USDA, and EPA met recently to discuss the possibility of jointly funding the NOAA laboratory in Boulder, Colorado, to become a central calibration and instrument auditing facility for all U.S. agencies making UV measurements. This facility would be utilized to perform the routine characterization and calibration of the instruments deployed in the various monitoring networks around the country. The facility would be supported by the agencies mentioned above and through a set of fees established for instrument characterizations, audits, and calibrations. While the facility would be under NOAA's direction, it would have close technical ties to NIST for radiometric standards and technical guidance.

EPA's intent is to establish a network whose sites are in the most advantageous locations possible to satisfy the needs of the Agency while still serving the needs of others in government and academia with the UV-B information necessary for their respective research programs.

**Acknowledgments**
The information in this document has been funded wholly by the US Environmental Protection Agency. It has been subjected to the agency's peer and administrative reviews, and it has been approved for publication as an EPA document. Mention of trade names does not constitute endorsement or recommendation for use.

# NOAA/EPA SURFACE ULTRA-VIOLET FLUX INDEX

Craig S. Long
NOAA/NWS/NMC/Climate Analysis Center
5200 Auth Road, Washington, DC 20233

Key words: EPA, NOAA, skin cancer, UV flux index

**ABSTRACT**

The National Oceanic and Atmospheric Administration (NOAA) and the Environmental Protection Agency (EPA) have joined together to research, validate and present to the general public an Ultraviolet Index (UV Index). The UV Index is modeled after a similar index created by the Canadian Atmospheric Environment Service (AES), which has been in effect since 1992. Differences between the NOAA/EPA and the AES UV Indexes as well as current and future improvements will be discussed.

## INTRODUCTION

To varying degrees, the public is aware of the decay of the ozone layer by man-made chlorofluorocarbons. The public is also aware of the resulting risks of melanoma, cataracts, and early aging of the skin due to increased ultraviolet (UV) radiation at the surface. The Environmental Protection Agency (EPA) and the National Oceanic and Atmospheric Administration (NOAA) are collaborating in a venture to increase the public's awareness by creation of a Ultraviolet Index (UV Index). The UV Index is intended to be part of the daily weather forecast, like the air quality index. By providing the public with a UV forecast in conjunction with a public awareness campaign, it is hoped that the increasing number of skin cancer cases will decrease.

Several countries already have either a strong public awareness campaign in place, like Australia, or a UV Index of their own. The Canadian Atmospheric Environment Service (AES) has been providing the Canadian public with a UV Index since early 1992. The NOAA/EPA UV Index will be similar to that produced by the AES in that the range of the scales will coincide with one another. However, the methodologies by which the Indexes are arrived at differ.

## METHODOLOGIES

The AES makes use of a network of eight Brewer instrument sites throughout Canada. These Brewer instruments detect the amount of incoming solar radiation between the wavelengths of 290 and 325nm (the UV-B range). These instruments are also capable of measuring the amount of total ozone in the vertical column above the site. The total UV flux is created by applying an erythemal weighting function (4) to the radiances and integrating over the 290 to 400nm spectral range (the range of both UV-B and UV-A). Since the Brewer instrument only observes the UV-B wavelengths between 290 and 325nm, an adjustment must be added to the flux integration to include the UV-A wavelengths in order to arrive at the

NATO ASI Series, Vol. I 18
Stratospheric Ozone Depletion/
UV-B Radiation in the Biosphere
Edited by R. H. Biggs and M. E. B. Joyner

full spectral flux. A regression between the "clear sky/noon" UV flux, the total ozone overhead and the solar zenith angle has been created from an existing climatology at each site. The total ozone observations are assimilated into a numerical meteorological model that expands the observations to a grid covering all of Canada. Correlations are made between the total ozone amounts and several meteorological parameters. The numerical model is run for the next day and forecasted ozone amounts are determined. Using the regression mentioned above, UV fluxes are determined using their relationship with the total ozone forecasted. The UV flux is multiplied by a scale factor, resulting in an index number between 0 and 10 (personal communication, J. Kerr, 1993), where 0 indicates no UV radiation reaching the surface and 10 implies a relatively large amount of UV radiation reaching the surface.

The method by which the NOAA/EPA UV Index will be determined differs in many ways. Instead of measuring total ozone from a ground-based instrument, the total ozone as determined from the solar backscattering ultraviolet II instrument (SBUV/2) on board a NOAA polar orbiting environmental satellite will be used. The quality of the SBUV/2 data is known to be reliable (1) and its relationship with collocated measurements on the ground has been studied (5). The use of a satellite-based means of observing the total ozone allows full global coverage (the SBUV/2 instrument, however is restrained to sunlit portions of the earth). The SBUV/2 is a "nadir only" looking instrument. Hence, there are appreciable gaps between successive orbits (up to 25° of longitude at the equator). However, if the observations for an entire day are analyzed using a modified Cressman scheme (2), a global analysis can be derived. Knowing each grid point's total ozone, the grid point's latitude, and the day of the year a three dimensional look-up table derived from a radiative transfer model (3) can be used to determine that grid

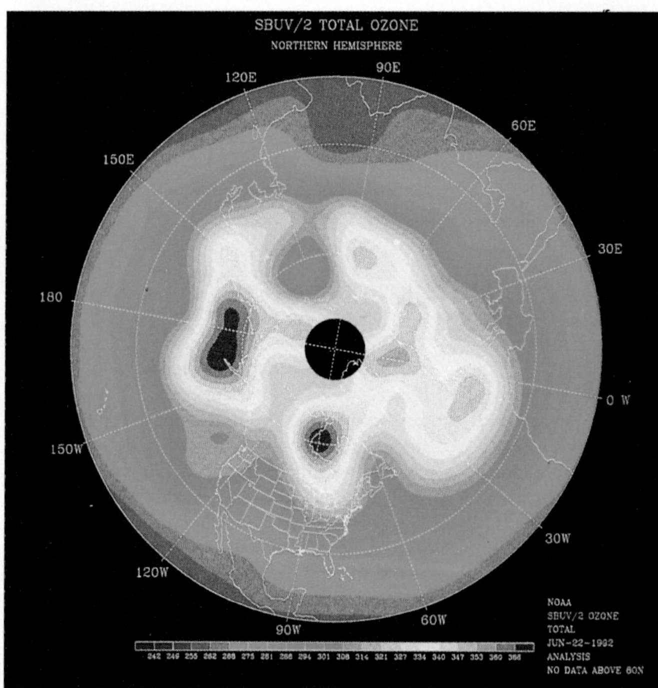


Figure 1. Northern Hemispheric analysis of total ozone as detected by the Solar Backscattering Ultra-Violet II instrument on board the NOAA 11 polar orbiting satellite for June 22, 1992.

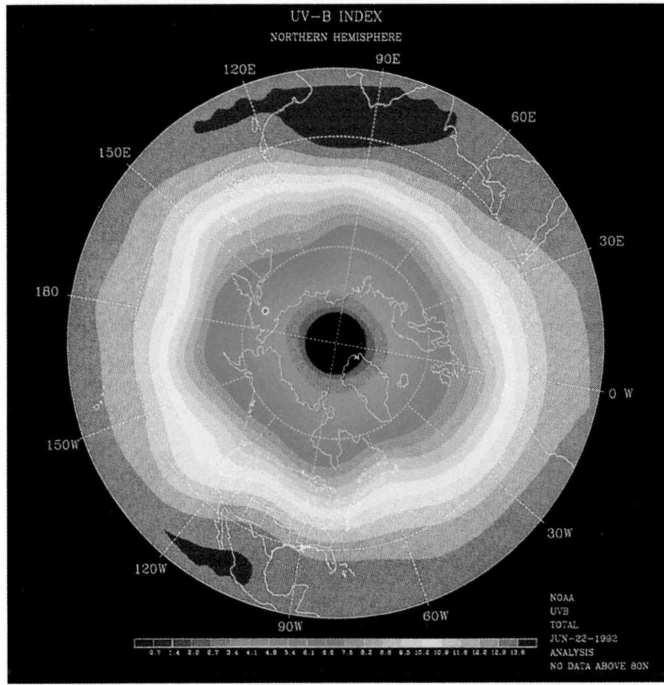


Figure 2. Northern Hemispheric analysis of UV Index for June 22, 1992, resulting from a radiative transfer model (3) using the SBUV/2 data in Figure 1 as input.

point's UV flux. The same scaling factor as used by the AES is applied to the flux to arrive at an index value. The NOAA/EPA UV Index initially will be a clear sky-noon value.

## TRIAL CASE

Figure 1 shows the northern hemisphere's analysis of SBUV/2 derived total ozone for June 22, 1992. The observed total ozone values range from 235 to 373 Dobson Units (DU). The highest concentrations are near the pole. The black area around the North Pole indicates the portion of the earth not properly illuminated for the SBUV/2 algorithm to make a valid observation. The concentrations of total ozone have troughs and ridges just as temperature and height fields do. In fact the 50mb temperature field correlates quite well with the SBUV/2 total ozone, especially during the winter. In the tropics the total ozone concentrations become very zonal in orientation, with the minima towards the equator.

Applying a trial radiative transfer model to the above ozone field produces the analysis of the UV Index in Figure 2. The deep troughs and ridges are severely smoothed out to the point of almost having constant zonal values. This indicates that the UV flux and thus the UV Index is highly dependent upon the solar zenith angle and the total ozone concentration around a latitude circle plays a minor role in the UV flux determination. The range of the UV Index extends from 0.0 to 14.3. This is well beyond the peak value believed to exist by AES.

UV Index values determined by the trial radiative transfer model have been found to produce higher UV flux values than that observed by the Brewer network in Canada. The ratio of UV fluxes derived from the radiative transfer model to Brewer observations ranges from 1.1 to 1.2. Even by applying a liberal correction to the model-derived fluxes, the resulting UV index in the tropics would still be above 10.0. NOAA, the EPA, and AES are collaborating to improve the radiative transfer model to correlate to a greater degree with the Brewer observations.

## IMPROVEMENTS

Creation of a UV flux or Index is useless unless it is properly validated. The EPA has a Brewer instrument currently taking observations in Raleigh, North Carolina. The EPA plans to increase the number of sites in the United States to provide a network of UV observations. The radiative transfer model has been improved upon to agree more with the Canadian Brewer observations. The current UV

## References:

1. Gleason, J.F., P.K. Bhartia, J.R. Herman, R. McPeters, P. Newman, R.S. Stolarski, L. Flynn, G. Labow, D. Larko, C. Seftor, C. Wellemeyer, W.D. Komhyr, A.J. Miller, and W. Planet. 1993. Record low global ozone in 1992. Science 260:523-526.

2. Finger, F.G., H.M. Woolf, and C.E. Anderson. 1965. A method for objective analysis of stratospheric constant-pressure charts. Mon. Wea. Rev. 87:367-374.

3. Frederick, J.E., and D. Lubin. 1988. The budget of biologically active ultraviolet radiation in the Earth-atmosphere system. J. Geophys. Res. 93:3825-3832.

4. McKinlay, A.F., and B.L. Diffey. 1987. A reference action spectrum for ultraviolet-induced erythema in human skin. In: W.R. Passchler and B.F.M. Bosnajakovic (eds.), Human Exposure to Ultraviolet Radiation: Risks and Regulations Elsevier, Amsterdam. pp.83-87.

5. Planet, W. 1991. NESDIS operational ozone data products. In: E. Barron, S. Cloetingh, A. Henderson-Sellers, D. Liverman, M. Meybeck, B. Moore, and P. Pirazzoli (eds.), Palaeo-geography, Palaeoclimatology, Palaeoecology (Global and Planetary Change Section), Elsevier, Amsterdam. pp.55-60.

# THE USDA GLOBAL CHANGE RESEARCH PROGRAM

Boyd W. Post

USDA/CSRS 329, 901 D Street SW, Washington, DC 20250-2200

## INTRODUCTION

Global Change research within the Department of Agriculture is coordinated in the office of the Assistant Secretary for Science and Education. Five agencies are primarily responsible for this work, but several others are represented on the USDA Global Task Force. These agencies each have their separate budgets for Global Change activities and conform to the budgetary policies of the agency and USDA.

USDA's objective is to assure the security of food, fiber, and forest products through sustainable management systems. This objective is being achieved through a long-term strategy for research, education, technology transfer, extension activities, and policy analysis to assist in sustaining U.S. agriculture and forestry in a manner compatible with future global environmental changes.

In the broadest sense, this research deals with assessing how global change affects U.S. and international agriculture and forestry, and how agriculture and forestry contribute to global change. Included are basic research on biological responses to increasing greenhouse gases and tropospheric ozone, improvement of plant and animal germplasm to adapt to global change, and development and implementation of plans for systems to mitigate increases in greenhouse gases. These include research on applications of agricultural meteorology to improve agricultural and management decisions. New knowledge and technologies development at federal and state research institutions will be transferred to the production sector through the extension system.

USDA is part of the Subcommittee on Global Change Research under the Committee on Earth and Environmental Sciences of the Federal Coordinating Council for Science, Engineering, and Technology. As an agency, the USDA also participates in international global change research. Policies that guide the overall food and fiber system are derived from this vast federal-state-private agricultural and forestry research system.

## ISSUES

Agricultural research has always been focused on change. Weather and the resulting climate have been the subject of agricultural research and development from the beginning. Plant and animal scientists have long studied weather variables such as temperature and precipitation as they affect the physical and biological mechanisms of ecosystems. Our productive agriculture and forestry have benefitted greatly from this research and, with the potential changes the earth faces, research efforts become even more critical to maintaining supplies of food and fiber.

Carbon dioxide levels have risen markedly in recent years. This rise may result in substantial increases in yield of crops and forest growth as well as improved water use efficiency in plants. Simultaneously, air pollution, atmospheric deposition, tropo-

NATO ASI Series, Vol. I 18
Stratospheric Ozone Depletion/
UV-B Radiation in the Biosphere
Edited by R. H. Biggs and M. E. B. Joyner

spheric ozone, methane, and nitrous oxide have become increasing concerns as their levels of concentration have increased. Ultraviolet radiation levels may be affected by the destruction of stratospheric ozone and may increase in the future.

The effects of these changing variables on the earth's climate is not certain. However, in view of the current world food storage capacity of three to six months, increases in annual climate fluctuation are potential problems. Global atmospheric circulation patterns and regional precipitation distribution as a result of these changes are also uncertain; regional variability is particularly uncertain.

## THE FUTURE

Until the present, research had enabled agriculture and forestry to adapt to climate regimes in various parts of the country and to increase production capacity. Because great differences are expected compared to past experiences, extrapolation from the past is not feasible. USDA cannot wait until change occurs to develop response strategies. As mentioned earlier, USDA's global change research programs are composed of two parts:

1) the effects of global change factors on forest, pastoral, and agricultural ecosystems, and

2) the effects of the forest, pastoral, and agricultural ecosystems on factors that influence global change. The research elements of the U.S. Global Change Research Plan cover a very broad scope of research including "Ecological Systems and Population Dynamics" which is the highest priority for USDA.

USDA strategies will augment existing research in data and information management and support new programs where needed. Research programs such as development of plant varieties and animal breeds, genome mapping, and genetic engineering will continue to contribute to the adaptive ability of agriculture and forestry. New commodity storage and food processing technologies being jointly developed by state, federal, and private research efforts will help U.S. agriculture and forestry adapt to long-term change and will help to minimize agricultural contributions to such change. In addition, alternative strategies must be developed to sustain supplies of food and fiber.

Models will be used to examine future systems and estimate changes in crop yields that changes in these variables may produce. Data management in the era of global change will challenge the most advanced levels of computer technology.

These comments and observations are from the introduction section of the USDA Global Change Strategies Plan (USDA, 1990), and represent the combined inputs of the USDA Global Change Task Force.

# MODELING

# WORKSHOP ON MODELING: SUMMARY

Max L. Bothwell

National Hydrology Research Centre, Environment Canada
Saskatoon, Saskatchewan S7N 3H

At the outset, it was noted that the absence of aquatic ecosystem modeling in the morning session likely reflected the view that such an undertaking with our present state of knowledge was either not possible or was very risky.

Nevertheless, such an endeavor is recommended in the future because modeling can help reveal unknown potential interactions within systems. These interactions might compound the effects of UV-B damage or possibly compensate for increased UV-B damage to primary production. In addition, this approach will highlight important areas for future research. Because of limited resources, it would be necessary to restrict such a modeling effort to ecosystems that are either globally most important such as marine pelagic ecosystems or those thought to be most susceptible to future increases in UV-B such as bodies of shallow water, wetlands, rivers and streams, and the littoral zones of lakes.

Ecosystem models to examine:

1. Marine pelagic systems
2. Estuaries
3. Lakes
4. Wetlands
5. Rivers and streams

Approach:

Because each has unique characteristics, each type of aquatic ecosystem would need to be modeled separately. Models of each of these systems either are available already or are thought to be under development. Some of these models are more sophisticated and better-tested than others.

Submodels or processes to examine:

1. Underwater UV-B photon budget
   a. Existing information is inadequate to describe the underwater irradiance field in the UV region. There is a need to quantify the role(s) of different water mass properties that influence it.

2. Carbon cycling
   a. Inhibition of algal photosynthesis and potential for species succession
   b. Effect of UV-B on grazing by zooplankton
   c. Vertical flux of carbon as influenced by the above-two processes
   d. Secondary production
   e. Tertiary production (birds, mammals, fish)

3. Nitrogen cycling
   a. Nitrogen fixation
   b. Nitrate uptake and assimilation
   c. Nitrogen transformations including nitrate to nitrite, denitrification, and so on
      i. biological
      ii. photochemical

4. Microbial loop
   a. How UV-B impacts bacterial metabolism, growth, and recycling of nutrients within the mixed layer and within "marine snow" particulates

NATO ASI Series, Vol. I 18
Stratospheric Ozone Depletion/
UV-B Radiation in the Biosphere
Edited by R. H. Biggs and M. E. B. Joyner

b. How UV-B impacts microzooplankton and processes within the microbial loop.

5. Photochemical processes

a. Kinetics of free-radical production and their role in various biological processes

6. Photobiogeochemical processes

a. Need to examine the role of UV-B in the cycling of iron in the upper layers of the ocean

b. Need to address the role of UV-B in photo-oxidation of vitamins such as Vitamin $B_{12}$

c. Need to quantify photodegradation of DOC and the impacts of bacterial metabolism and the nitrogen cycle

The purpose of this exercise is to identify the degree of importance that increases in UV-B will have on ecosystem processes, with the understanding that some of the effects will be indirect and may only reveal their significance during manipulation or sensitivity tests on an operational model.

# PLANT EFFECTS AND MODELING RESPONSES TO UV-B RADIATION

L.H. Allen, Jr.

U.S. Department of Agriculture, Agricultural Research Service
University of Florida, Gainesville, FL 32611 USA

Key words: agriculture, crop responses, DNA damage, DNA repair, modeling, plant responses, UV-B radiation

## INTRODUCTION

Modeling plant responses to the environment began with pioneering efforts of de Wit (1965), which followed the development of computers suitable for the tasks and the numerical techniques required. In the USA, the earliest efforts with agronomic crops began with Duncan et al. (1967). These efforts have continued with a range of models of ecological system and agricultural crop models.

Most of the early efforts in modeling were directed toward predicting ecosystem or crop responses to a range of aerial and soil environments found in nature. Some efforts have been directed toward modeling the effects of air pollutants on vegetation (Adams et al., 1988; Moseholm, 1988). Probably a lot more effort has been expended in predicting the effects of rising carbon dioxide effects on crops (Duncan et al., 1967; Waggoner, 1969; and Allen et al., 1971). These efforts have grown beyond those early predictions to recent studies on the combination of effects of $CO_2$ and climate change factors (mainly temperature and precipitation) on crop production (Peart et al., 1989; Curry et al., 1990a, 1990b; Rosenzweig, 1989; Ritchie et al., 1989). Little modeling work, however, has been conducted on the effects of UV-B radiation on terrestrial plants (Olszyk, this volume), probably because of the lack of clear-cut, repeatable response functions of this potentially damaging radiation across a range of exposures (Fiscus, this volume). Damage to a tissue or organ may be cumulative with time, and may cross a couple of thresholds as postulated by Caldwell and Nachtwey (1975). The direct effects of light and $CO_2$ on photosynthesis, of temperature on cell division, and on tissue, organ, and whole plant development and growth, and of soil water availability on overall plant size and biomass accumulation, are probably more amenable for modeling plant growth. Certainly, the processes and response functions of plant growth and development have to be worked out well on these factors before other factors such as UV-B radiation can be successfully introduced into plant models.

## LEVEL OF ORGANIZATION/ AGGREGATION

Where do you start in plant growth modeling? Probably the pivotal point is the individual whole plant when terrestrial systems are to be simulated. However, the individual plant is certainly impacted by other plants growing in its immediate vicinity. Furthermore, the individual plant develops from the balanced integration of its parts at ever decreasing levels of organization.

One scheme for hierarchial levels of organization might be as follows:

1. Global
2. Ecosystem
3. Diverse Community
4. Monoculture
5. Individual Plant
6. Organs/Tissues
7. Cellular
8. Organelle
9. Molecular/Biochemical

Most modeling efforts would attempt to interface

NATO ASI Series, Vol. I 18
Stratospheric Ozone Depletion/
UV-B Radiation in the Biosphere
Edited by R. H. Biggs and M. E. B. Joyner

quantitative interactions across only about three levels of organization. For example, in predicting crop yields, models typically focus on organ and tissue responses that provide for whole plant growth in a monoculture of individual plants. These models may apparently utilize information down to the molecular level in developing a leaf photosynthesis model, but in reality, all of these photosynthesis models may be viewed as whole-leaf constructs with theoretical underpinnings.

The levels of organization can be viewed from the standpoint of the model rather than the plant. Such a scheme might be one as follows:

1. Statistical models. For example, regional crop yields vs. weather.
2. Plant ecological community models.
3. Plant monoculture models
   a. Descriptive constructs
   b. Quantitative process constructs

The quantitative process-level models aggregate from lower levels to higher levels. Usually crop models aggregate from leaf-level gas exchange ($CO_2$ uptake) to higher levels.

Most of the plant growth and yield models require a common (or at least similar) set of dynamic input conditions. Many require a similar set of conceptual, and usually computational, modules for implementation of the model. The following outline covers the way that many plant growth and yield models function.

Typical structure of current models of plant growth and yield outline.

1. Weather data. Light, temperature, rainfall.
2. Soils data. Soil type, physical properties, depth, and so on.
3. Phenology component. A schedule of stages of development of the plant driven primarily by photoperiod and temperature.
4. Photosynthetic component
   a. Leaf photosynthetic rates as a function of several parameters, especially light, temperature, plant water status, leaf nitrogen content, and others.
   b. Light absorption model for the leaves of the canopy.
5. Photoassimilate allocation (and biomass accumulation)
   a. Shoots: Mainstems, branches or tillers, leaves
   b. Root system
   c. Reproductive organs
6. Soil and soil water
   a. Water availability/distribution/uptake by roots
   b. Nutrient availability/distribution/uptake by roots
   c. Root length distribution
7. Flowering, seed growth to maturity, and final yield

More information can be found on model structure, development, and predictive capabilities in publications by Boote and Loomis (1991), Goudriaan (1988), Reynolds and Acock (1985), and Reynolds et al. (1992).

## MODELING PROGRAMS RELATED TO CLIMATE CHANGE

Some air pollutant assessments and global climate change assessments have been taken beyond the impacts on yields at local experimental sites onward to regional and national economic analyses of the impacts of climate change (e.g., Adams et al., 1988; 1990). Assessments of global climate change and elevated $CO_2$ effects have received increasing attention. The U.S. Department of Energy has a long term program on Global Climate Change that began in 1979 (Dahlman, 1993). A large part of this DOE program is devoted to vegetation effects. The US EPA financed a series of studies that utilized and adapted pre-existing models to assess the impacts of climate change on crop productivity in the US. An overall review was provided by Smith and Tirpak (1989). Crop productivity impacts of doubled $CO_2$ and concomitant climate changes were modeled for several key crops in the southeast (soybean and maize (Peart et al., 1989 and Curry et al., 1990a, 1990b), the Great Lakes area (soybean, maize, Ritchie et al., 1989), the Great Plains (wheat, Rosenzweig et al., 1989), and California (Dudek, 1989). The economic impacts of changes in soybean and maize production were further analyzed by Adams et al. (1990). A report of global assessments sponsored by USAID and EPA is currently in press (Rosenzweig et al., 1994). These programs have taken information on crop growth and productivity for various weather station sites and various soils, and aggregated information to regional, national, and global scales by subjective summation over blocks of common land-use areas.

## MODELING UV-B EFFECTS: WHERE DO WE BEGIN?

Modeling the effects of UV-B radiation in a process-level crop growth model could be approached from either descriptive parameterizations or from functional relationships. The whole process is more difficult than direct cause and effect relationship such as the response of leaf photosynthetic rates to light or to $CO_2$. The first types of information needed are: Where does the damage occur? How does the damage occur? How does the damage impact photosynthesis and growth? The answers to these questions may vary from species to species. Damage may occur to developing organs such as leaves which could reduce the overall leaf area of a plant canopy and cause secondary effects of reduced plant size and biomass accumulation. Growth habit such as internode length responses to UV-B radiation may also provide secondary effects on the final harvest level of plant response. The damage may be cumulative as visualized by Caldwell (1975), and lead to loss of photosynthetic light-harvesting leaf area throughout the progression of the life cycle of the plant. Plants may synthesize protective pigments (e.g., Beggs et al., 1986), or repair the damage caused by UV-B radiation so that visible damage is minimized, but at some undetermined metabolic cost to the plant. High total solar irradiance with respect to UV-B irradiance is known to result in much less damage to crop plants than to plants that have been exposed to artificially-elevated UV-B radiation under low photosynthetically-active radiation, or PAR. As the crop canopy develops, the lower leaves become shielded from UV-B radiation, but they also receive less PAR. Finally, some plants have meristematic tissues exposed to solar radiation, whereas other species have the meristematic tissues imbedded within other tissues. Clearly, the targets of exposure of terrestrial plants to UV-B radiation are much more complex than for single cell organisms or small undeveloped organisms such as larval stage marine organisms.

So again, where do we begin? We can take some clues from previous modeling efforts.

### LEVEL OF ORGANIZATION

*Descriptive parameterization*. An example of this process is the parameterization of leaf photosynthetic rates of $C_3$ plants and $C_4$ plants to a doubling of atmospheric $CO_2$ concentration. Curry et al. (1990a; 1990b) gave $C_3$ plants a photosynthetic amplification factor of 1.35 and $C_4$ plants a photosynthetic amplification factor of 1.10. Such a process could be used in modeling UV-B effects on photosynthesis. However, better experiments on photosynthetic rates with duration of exposure are needed to apply this type of descriptive parameterization to one (albeit important) factor of photosynthesis, since the UV-B effect attenuation factor could change with progression of time.

*Functional relationships*. One example of this process is the Farquhar model of the effect of RUBISCO enzyme and RUBP substrate on controlling leaf photosynthetic rates. To apply this level of process control of photosynthetic rates would require a great deal more gas exchange and enzyme effect studies on leaves.

### RADIATION IN PLANT CANOPIES

Light penetration models can be used to predict the distribution and interception of solar UV-B radiation in plant canopies (Allen et al., 1975). The parameters that are needed for use in these existing models are leaf reflectance and transmittance. Because of UV-B absorbing pigments, leaf epidermal transmittance can be near zero, and leaf reflectance is usually low, about 4 to 6% of incident radiation across the 280 to 360 nm range (Allen et al., 1975). Citrus and agave have low epidermal transmission for UV-B radiation, whereas kalanchoë have values of epidermal transmission that range from about 15% at 280 nm up to 40% at 330 nm.

The penetration of UV-B radiation throughout a plant canopy was predicted by Allen et al. (1975) using 5% reflectance and 1% transmittance. The model output gave the distribution with respect to height of downwelling direct beam, diffuse skylight source, and total direct beam plus diffuse skylight source UV-

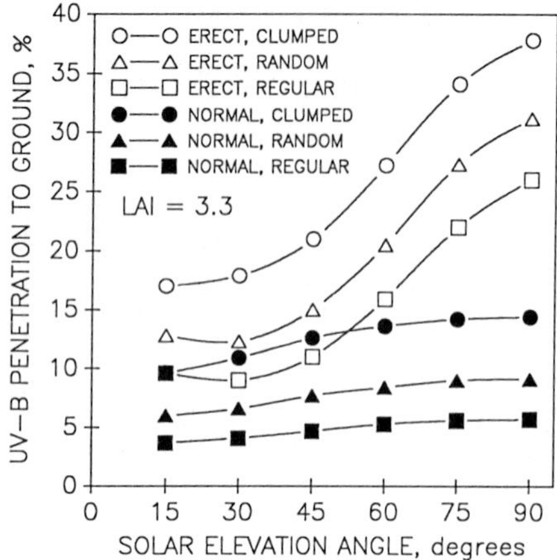


Figure 1. Penetration of solar UV-B radiation to ground level through two types of leaf angles (erectophile and normal) and two distributions of leaf material (clumped and random) at a range of solar elevation angles (adapted from Allen et al. 1975).

B radiation. In addition, it gave the components transmitted downward by the leaves that intercepted UV-B radiation, and the total amount of UV-B reflected upward.

Figure 1 shows the penetration of UV-B radiation to ground level computed for 6 canopy types with a leaf area index (LAI) of 3.3. The computations were made with solar elevation angles from 15 to 90°. This figure shows that erectophile leaves allowed much more solar UV-B to penetrate to ground level than "normally" displayed leaves. A planophile leaf angle distribution absorbs only slightly more solar UV-B radiation than the "normal" leaves (Allen et al., 1975).

The model gave predicted downward and upwelling fluxes of UV-B radiation in the plant canopy that could be used to predict radiation on various target organisms in a plant canopy. Because of the low transmissivity of UV-B radiation by plant leaves, the predicted areas of lowest exposure were the undersides of leaves near the top of the canopy.

## SPECULATIONS ON RESPONSE OF TERRESTRIAL PLANTS TO UV-B RADIATION

Figure 2 presents a typical set of curves of an action spectrum, the solar irradiance, and the biologi-

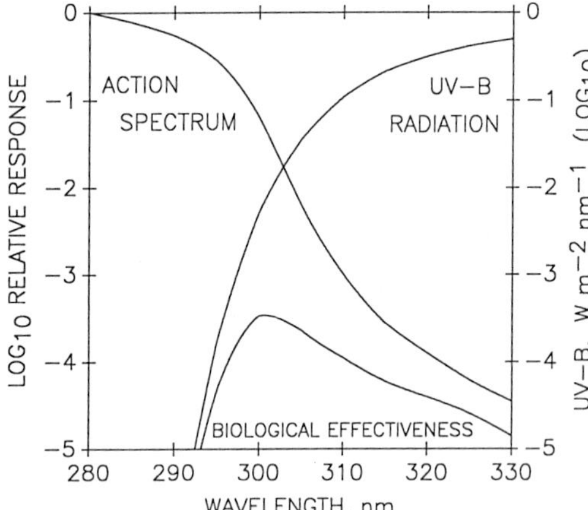


Figure 2. Qualitative illustration of the resultant Biological Effectiveness of UV-B radiation when crossed with an action spectrum.

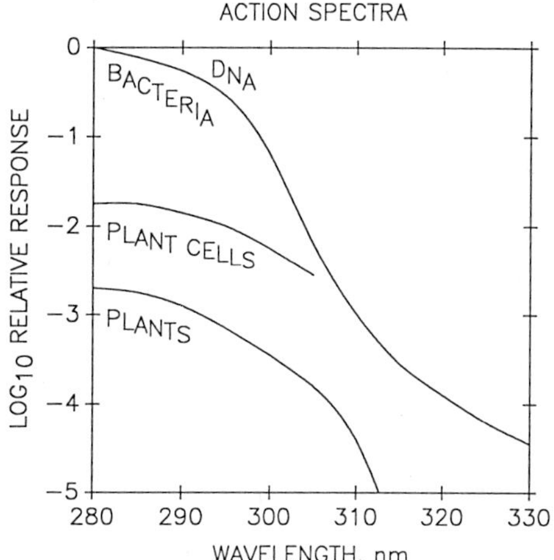


Figure 3. Qualitative action spectra illustrating the lower damage response of higher plant cells and the lower damage that may be expected of intact plants growing under natural conditions.

cal effectiveness of UV-B radiation. However, as pointed out several times during this NATO Advanced Research Workshop, higher plant cells may not demonstrate the same type of response to UV-B radiation as DNA or naked bacterial cells (Figure 3). The relatively lower response of plant cells to UV-B radiation is almost certainly due to more protective pigments, as pointed out in several of the papers in this volume. One could speculate further that responses of cells in intact plants may show even less damaging response to UV-B radiation. (See also action spectra presented by Caldwell et al., 1986.)

The action spectrum of response to UV-B radiation has been shown to depend on whether monochromatic UV-B irradiance was used alone or whether "white light" was added to monochromatic UV-B radiation (Figure 4). Other participants in this workshop discussed this phenomenon. In white light, the UV-B action spectrum decreases much more rapidly with increasing wavelength, implying (or demonstrating) that photorepair was assisted greatly by other wavelengths in the white light. The absolute action spectrum for UV-B damage may not be as great even at low wavelengths for biological materials exposed to white light, as implied by the lower curves in Figure 4.

Many of the experiments conducted on agricultural crops have shown little response to exposure to enhanced UV-B radiation (e.g., Fiscus, this Workshop). In order to model the effects of UV-B, experi-

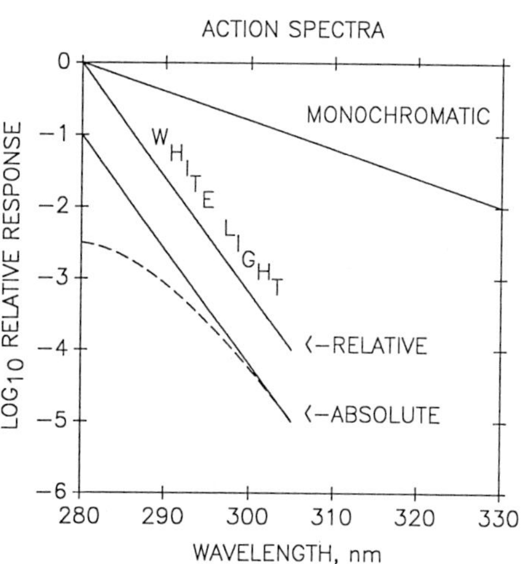


Figure 4. Qualitative action spectra from monochromatic UV-B radiation only, white light plus monochromatic UV-B radiation, and scenarios for further reduction in the action spectrum due to protection mechanisms that may be present in field plants.

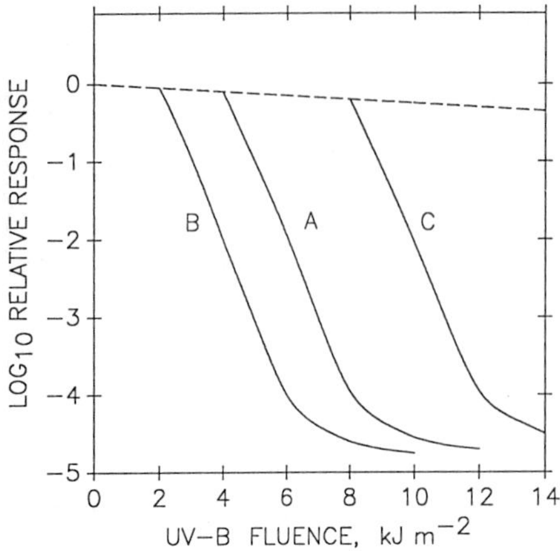


Figure 5. Qualitative fluence responses that may help explain damage responses of plants to UV-B radiation. *Curve A* illustrates a condition with one plant filtering and tolerance level to UV-B radiation. With less amount of stimulus or genetic capacity to repair damage or to synthesize protective pigments, the response may be represented by *Curve B*. With more stimulus or greater genetic capacity to repair damage or to synthesize protective pigments, the response may be represented by *Curve C*. These types of gross response curves are need by modelers to begin to be able to predict the plant to field scale responses to UV-B radiation. (Adapted from T. Coohill, Workshop participant.)

ments should be conducted at higher fluences in order to determine the nature of the damage relationships across an interval of time. Such an approach may require new thought on the best way to conduct studies in order to avoid the trap of using experimental conditions that predestine the outcome due to other unplanned, unnatural limits on plant growth and development.

Figure 5 shows an idealized type response to increasing UV-B fluence. *Curve A* represents a response that could occur from 4 to 8 kJ $m^{-2}$. If solar PAR is increased, position of the response could move to high UV-B fluences (*Curve B*). On the other hand, if solar PAR is decreased, the response could shift to lower UV-B fluence levels (*Curve C*). This same type of sensitivity to UV-B radiation could change with other plant growth limitations which could impair the ability to synthesize and maintain protective pig-

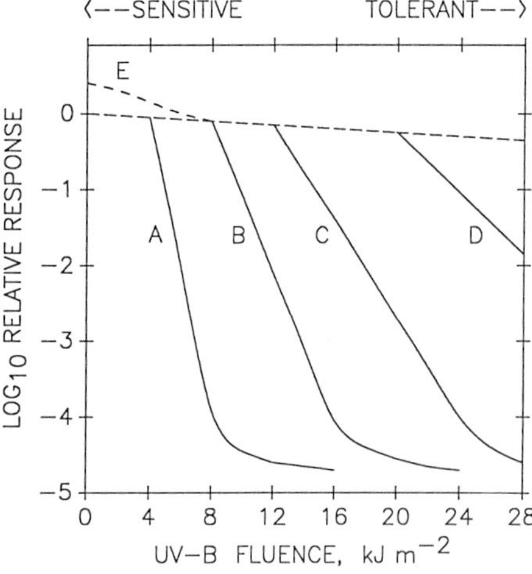


Figure 6. Qualitative fluence response curve to UV-B radiation that may be more realistic under outdoor conditions for plants that thrive under conditions of high level of exposure. The slope of the damage response curve is postulated to decrease under high levels of UV-B that are accompanied with an increase in the capability of the plant to generate filtering capacity or to otherwise tolerate high levels of UV-B radiation. Such conditions might develop as plants thrive across a season when solar irradiance increases, and thus produces more photo-assimilate, part of which can be partitioned into metabolic processes for overall protection against (repair), or tolerance of (pigmentation), UV-B radiation as illustrated by transition from *Curve A* to *B* and *C* and *D*. This type of fluence response may also represent the range expected by inherently sensitive and tolerant plants. The short dashed curve at the top of the figure near at very low fluences, *Curve E*, represents enhanced growth responses that are often seen when UV-B irradiance is kept very low. The shape of this curve is arbitrary and may not be realistic.

ments, repair DNA damage, maintain photoxidative quenching molecules or free radical scavenging enzymes, or synthesize new protein.

A more likely scenario for the action spectrum is illustrated in Figure 6. Here, the figure illustrates that the threshold for damage could be shifted to higher UV-B fluences, and then the shape of the damage response function could become flatter as we go from *Curve A* to *Curve B* and *Curve C* and *Curve D*.

For whole plant responses, it is very possible that each increment of UV-B fluence could induce synthesis of more protective pigments, encourage photorepair opportunities, or stimulate synthesis of more proteins to replace damaged protein.

The curves in Figure 6 may also represent the dynamic response to greater amounts of solar photosynthetically active radiation and generation of photoassimilates, and perhaps other releases from limitations to growth. Of course, at some unknown high levels of UV-B, the ability of plants to shield from UV-B, tolerate damage, or repair damage, should be exceeded beyond recovery.

Experiments should be designed to determine where the real thresholds to UV-B exposure lie for a wide range of plants. With the recent loss of ozone presumably caused by the eruption of Mount Pinatubo in the Philippines, it behooves us to be able to define real limitations of UV-B radiation to plant growth, development, and yield under real world environmental conditions, particularly with high solar photosynthetically active radiation levels.

As we move the future, elevated $CO_2$ concentrations may also provide a greater resource for plants in terms of photoassimilate accumulation rate and storage. Part of the extra fixed energy of the plants may then be available for metabolism responses that would limit or delay adverse effects of greater exposure to UV-B radiation.

Plant modelers should be engaged in designing experiments that will provide information both for inputs into plant models, and for verification of model predictions.

### Acknowledgments

I wish to thank all the workshop participants, particularly Manfred Tevini, Thomas Coohill, Janet Bornman, Betsy Sutherland, and John Hays, for providing stimulating ideas about UV-B radiation processes and effects that can be adapted for development of approaches for modeling plant responses to UV-B stresses. Any misdirected interpretations are due to my own imaginations.

## REFERENCES

Adams, R.M., J.D. Glyer, and B.A. McCarl. 1984. The NCLAN economic assessment: Approaches, findings, and implications. p. 473-504. *In* W.W. Heck, O.C. Taylor, and D.T. Tingey (eds.), Assessment of crop loss from air pollutants. Proceedings of the international conference, Raleigh, NC, USA. Elsevier Applied Science, London.

Adams, R.M., C. Rosenzweig, R.M. Peart, J.T. Ritchie, B.A. McCarl, J.D. Glyer, R.B. Curry, J.W. Jones, K.J. Boote, and L.H. Allen, Jr. 1990. Global climate change and U.S. agriculture. Nature 345:219-224.

Allen, L.H., Jr., S.E. Jensen, and E.R. Lemon. 1971. Plant response to carbon dioxide enrichment under field conditions: A simulation. Science, 173:256-258.

Allen, L.H., Jr., H.W. Gausman, and W.A. Allen. 1975. Solar ultraviolet radiation in terrestrial plant communities. J. Environ. Qual. 4:285-294.

Beggs, C.J., U. Schneider-Ziebert, and E. Wellmann. 1986. UV-B radiation and adaptive mechanisms in plants. p. 235-259. *In* R.C. Worrest and M.M. Caldwell (eds.), Stratospheric ozone reduction, solar ultraviolet radiation, and plant life. NATO ASI Series G: Ecological Sciences, Vol. 8. Springer-Verlag, Berlin.

Boote, K.J., and R.S. Loomis. 1991. The prediction of canopy assimilation. Chapter 7. *In* K.J. Boote and R.S. Loomis (eds.), Modeling Crop Photosynthesis--From biochemistry to Canopy. CSSA Special Publication No. 19. American Society of Agronomy, Madison, WI.

Caldwell, M.M., and D.S. Nachtwey. 1975. Introduction and Overview. p. 1-1 to 1-30. *In* D.S. Nachtwey, M.M. Caldwell, R.H. Biggs, P. Cutchis, J.J. Gwiazdowski, and A.J. Grobecker (eds.), Impacts of Climatic Change on the Biosphere. CIAP Monograph 5, Part 1: Ultraviolet Radiation Effects. Final Report DOT-TST-75-55, Department of Transportation, Washington, DC.

Caldwell, M.M., L.B. Camp, C.W. Warner, and S.D. Flint. 1986. Action spectra and their key role in assessing biological consequences of solar UV-B radiation change. p. 87-111. *In* D.S. Nachtwey, M.M. Caldwell, R.H. Biggs, P. Cutchis, J.J. Gwiazdowski, and A.J. Grobecker (eds.), Impacts of Climatic Change on the Biosphere. CIAP Monograph 5, Part 1: Ultraviolet Radiation Effects. Final Report DOT-TST-75-55, Department of Transportation, Washington, DC.

Curry, R.B., R.M. Peart, J.W. Jones, K.J. Boote, and L.H. Allen, Jr. 1990a. Simulation as a tool for analyzing crop response to climate change. Trans. ASAE 33:981-990.

Curry R.B., R.M. Peart, J.W. Jones, K.J. Boote, and L.H. Allen, Jr. 1990b. Response of crop yield to predicted changes in climate and atmospheric $CO_2$ using simulation. Trans. ASAE 33:1381-1390.

Dahlman, R.C. 1993. $CO_2$ and plants: revisited. Vegetatio 104/105:339-335.

de Wit, C.T. 1965. Photosynthesis of leaf canopies. Versl. Landbouwk. Onderz. (Agr. Res. Rep.) 663, Pudoc, Wageningen, 57 pp.

Dudek, D.J. 1989. Climate change impacts upon agriculture and resources: A case study of California. p. 5-1 to 5-38. *In* J.B. Smith and D.A. Tirpak (eds.), The Potential Effects of Global Climate Change on the United States, Appendix C (Agriculture), Vol. 1. Report EPA-230-05-89-053, U.S. Environmental Protection Agency, Washington, DC.

Duncan, W.G., R.S. Loomis, W.A. Williams, and R. Hanau. 1967. A model for simulating photosynthesis in plant communities. Hilgardia 38:181-205.

Goudriaan, J. 1988. The bare bones of leaf-angle distribution in radiation models for canopy photosynthesis and energy exchange. Agric. For. Meteorol. 43:155-169.

Moseholm, L. 1988. Analysis of air pollution plant exposure data: The soft independent modelling of class analogy (SIMCA) and partial least squares modelling with latent variable (PLS) approaches. Environ. Pollution 53:313-331.

Peart, R.M., J.W. Jones, R.B. Curry, K.J. Boote, and L.H. Allen, Jr. 1989. Impact of climate change on crop yield in the Southeastern USA: A simulation study. p. 2-1 to 2-54. *In* J.B. Smith and D.A. Tirpak (eds.), The Potential Effects of Global Climate Change on the United States, Appendix C (Agriculture), Vol. 1. Report EPA-230-05-89-053, U.S. Environmental Protection Agency, Washington, DC.

Reynolds, J.F., and B. Acock. 1985. Modeling approaches for evaluating vegetation responses to carbon dioxide concentration. p. 33-51. *In* B.R. Strain and J.D. Cure (eds.), Direct Effects of Increasing Carbon Dioxide on Vegetation. U.S. Department of Energy, Carbon Dioxide Research Division, DOE/ER-0283, Washington, DC.

Reynolds, J.F., J.Chen, P.C. Harley, D.W. Hilbert, R.L. Dougherty, and J.D. Tenhunen. 1992. Modeling the effects of elevated $CO_2$ on plants: extrapolating leaf response to a canopy. Agric. For. Meteorol. 61:69-94.

Ritchie, J.T., B.D. Baer, and T.Y. Chou. 1989. Effect of global climate change on agriculture: Great Lakes region. p. 1-1 to 1-42. *In* J.B. Smith and D.A. Tirpak (eds.), The Potential Effects of Global Climate Change on the United States, Appendix C (Agriculture), Vol. 1. Report EPA-230-05-89-053, U.S. Environmental Protection Agency, Washington, DC.

Rosenzweig, C. 1989. Potential effects of climate change on agricultural production in the Great Plains: A simulation study. p. 3-1 to 3-43. *In* J.B. Smith and D.A. Tirpak (eds.), The Potential Effects of Global Climate Change on the United States, Appendix C (Agriculture), Vol. 1. Report EPA-230-05-89-053, U.S. Environmental Protection Agency, Washington, DC.

Rosenzweig, C., L.H. Allen, Jr., L.A. Harper, S.E. Hollinger, and J.W. Jones. 1994. Climate change and international impacts. ASA Special Publication No. ASA-CSSA-SSSA, Madison, Wisconsin. (In press)

Smith, J.B., and D.A. Tirpak (eds.). 1989. The potential effects of global climate change on the United States. Report EPA-230-05-89-050, U.S. Environmental Protection Agency, Washington, DC.

Waggoner, P.E. 1969. Environmental manipulation for higher yields. p. 343-373. *In* J.D. Eastin, F.A. Haskins, C.Y. Sullivan, and C.H.M. van Bavel (eds.) Physiological aspects of crop yield. ASA-CSSA, Madison, Wisconsin.

# THE USE OF CROP SIMULATION MODELS IN THE ECONOMIC ASSESSMENT OF GLOBAL ENVIRONMENTAL CHANGE

Wayne I. Park

U.S. Department of Agriculture, Agricultural Research Service
University of Florida, Gainesville, Florida 32611

Key words: crop simulation, global climate change, modeling, ozone depletion, UV-B

## INTRODUCTION

Changes in our environment raise concerns about the potential impacts on the quality of human life. One of the primary concerns of agricultural research into the impact of global environmental changes, such as ozone depletion, is the possible impairment of our ability to feed ourselves. Thus, research is directed towards understanding the responses of crop plants to environmental changes. However, a full impact assessment includes an evaluation of the economic implications of potential changes. In order to conduct such an assessment, methods are needed to translate the results of basic research into relevant economic terms. One tool which has proven to be valuable in this process is a crop simulation model. As a user of crop models, I would like to make a few general comments on the economic assessment of the impact of environmental change and the role that crop simulation models play.

## USE OF CROP SIMULATION MODELS

Changes in our environment have the potential to affect crop production directly and indirectly. Obviously, the first concern of scientific research is to determine the direct effects of changes on the plant processes which affect economic yields. However, as the direct effects are perceived by crop producers, they will adjust their production practices in an attempt to reduce any negative impact on their well-being. Adjustments may be made in input use, cultural practices, or even changes in crops grown at a particular location [Kaiser *et al.*, 1993]. In any event, adaptations on the part of producers may partially mitigate the undesirable direct effects. Crop simulation models have the potential to provide information regarding the likely outcomes of alternative decisions available to the producer. The information provided by the models may be used to assess which alternatives are more likely to be adopted by crop producers. Thus, the simulation models can be used in economic analysis to anticipate producer behavior under alternative environmental conditions.

An economic evaluation of the impact of UV-B radiation on agriculture can follow procedures employed to examine the effects of increases in carbon dioxide and global climate change. Lovell and Smith [1985] identified three characteristics of global climate change which affect the way economic analysis should be conducted. These characteristics are also relevant to the economic assessment of increases in UV-B radiation. The first characteristic is that agricultural production is stochastic. Crop producers must make production decisions prior to the realization of weather variables. Their decisions depend on the expected value of the outcome, and the perceived risk involved. Therefore, the fact that UV-B effects may be manifested only under well-watered conditions could be a significant economic factor. However, the negligible effects which have been identified in field experiments suggest that UV-B may have little

NATO ASI Series, Vol. I 18
Stratospheric Ozone Depletion/
UV-B Radiation in the Biosphere
Edited by R. H. Biggs and M. E. B. Joyner

economic significance to the producer when compared to other factors.

The second characteristic is that the changes will occur gradually with sufficient time for adaptations to occur. Producers will have time to adapt to changes if alternatives are available. Moreover, crop plants will have time to adapt. Selection of high yielding varieties by breeders may naturally include varieties with a tolerance to significant UV-B radiation. Again, the small measured effect and the long-time horizon suggest that other economic factors are likely to overshadow any detrimental effects of UV-B.

The third characteristic identified by Lovell and Smith is that the change is a global phenomenon. While the impact on yields will be manifested at the farm level, it is the aggregate effect across all farms which will determine if there are significant price movements. The need to perform economic analysis on an aggregate level requires crop models which can provide representative yields for a large region. This becomes a challenge to adequately represent annual variations in regional crop yields with models which are dependent on site specific information. For this type of analysis, simple models which require less detailed information may be aggregated more readily.

Let me briefly mention a couple of studies which have used output from crop models to examine the economic implications of climate change. Adams *et al.* [1988] examined the economic implications for the U.S. agricultural sector of changes in average yields due to climatic changes. One of their findings was that the estimate of economic impact was greatly reduced when the direct effects of carbon dioxide on crop yields was included in the analysis. In a more recent study, McCarl *et al.* [1993] show that estimates of changes in producer and consumer welfare are also sensitive to changes in the variability of crop yields. Based on these two results, research into the effects of UV-B radiation should consider the interactions with other anticipated environmental changes, and be concerned with the effect on yields under a variety of growing conditions to be able to provide information on the temporal and spatial distribution of crop yields.

Given an appropriate experimental base, crop simulation models can be used to estimate the distribution of crop yields. For example, Park and Sinclair [forthcoming] use a simple, mechanistic model to show that a warmer drier climate in Urbana, Illinois would tend to be associated with lower average corn yields and a flatter, more symmetric, probability distribution function. Furthermore, higher levels of carbon dioxide were shown by Park and Moss [in review] to increase the mean, variance, and skewness of corn and soybean yields directly opposite to the effects of warmer temperatures and reduced rainfall. In the case of soybeans, the increased carbon dioxide dramatically increased the variability of yields. The differential impact of carbon dioxide on $C_3$ and $C_4$ crops could significantly alter land allocation decisions and optimal input use under uncertainty.

## CONCLUSIONS

In conclusion, the tools are available to consider many of the economic implications of changes in the distribution of crop yields. Simulation models can provide a useful link between basic research and economic assessment by estimating yields of major crops grown under alternative environmental conditions. The type of information desired includes changes in mean yields and variability of crop yields, as well as responses to alternative inputs. An analysis of global change must consider the stochastic nature of agricultural production, the fact that the change will be gradual, and the need for regional estimates of crop yields. Finally, if the effect of increases of UV-B radiation on crop yields is as small as has been indicated at this conference, UV-B may not have a significant economic impact on the agricultural sector.

## REFERENCES

Adams, R. M., B. A. McCarl, D. J. Dudek, and J. D. Glyer. 1988. Implications of Global Climate Change for Western Agriculture. West. J. of Ag. Econ. 13:348-356.

Kaiser, H. M., S. J. Riha, D. S. Wilks, D. G. Rossiter, and R. Sampath. 1993. A Farm-Level Analysis of Economic and Agronomic Impacts of Gradual Climate Warming. Am. J. of Ag. Econ. 75:387-398.

Lovell, C. A. Knox, and V. Kerry Smith. 1985. Microeconomic Analysis. Climate Impact Assessment, ed. R. W. Kates, J. H. Ausubel, and M. Berberian. London: John Wiley and Sons.

McCarl, B. A., q. He, D. K. Lambert, M. S. Kaylen, and C.-C. Chang. 1993. The Stochastic Agricultural Sector Model: Applications to Global Climate Change and Farm Program Revision. Quantifying Long Run Agricultural Risks and Evaluating Farmer Responses to Risk. Southern Regional Project S-232. Jekyll Island, Georgia.

Park, W. I., and C. B. Moss. The Impact of Carbon Dioxide on the Temporal Distribution of Corn and Soybean Yields. In review.

Park, W. I., and T. R. Sinclair. Consequences of Climate and Crop Yield Limits on the Distribution of Corn Yields. Rev. of Ag. Econ. (forthcoming).

# MODELING THE IMPACTS OF UV-B RADIATION ON ECOLOGICAL INTERACTIONS IN FRESHWATER AND MARINE ECOSYSTEMS

Horacio E. Zagarese and Craig E. Williamson

Department of Earth and Environmental Sciences
Lehigh University,
Bethlehem, PA 18015-3188

KEY WORDS: UV-B radiation, zooplankton, vertical migration, pigmentation, environmental gradients, light, community structure, aquatic ecosystem.

**ABSTRACT**

Recent data demonstrating global increases in biologically damaging UV-B radiation raise the need for knowledge of how natural communities and ecosystems will respond to these environmental changes. These responses are likely to differ in terrestrial versus aquatic environments, and in freshwater versus marine environments due to fundamental differences in the structure and function of these ecosystems.

Here we develop a simple conceptual model of how zooplankton are likely to respond to changes in UV-B radiation. Recognizing the selective pressures of direct damage from UV-B radiation and the variety of responses that aquatic organisms exhibit in response to high levels of UV-B, we focus on the implications of these responses for ecological interactions at the community and ecosystem levels in freshwater and marine environments.

## INTRODUCTION

Depletion of stratospheric ozone has led to increases in ambient levels of biologically damaging UV-B radiation on a global scale. These increases have been most pronounced in the spring Antarctic ozone hole and least pronounced in tropical regions with statistically significant trends showing an increase in biologically effective UV-B of 10% or more per decade over wide regions in the north and south temperate regions [36,61]. In 1992 global levels of total ozone were 2-3% lower than in any of the previous 13 years [18].

Living organisms can and do respond to the selective pressures placed on them by high levels of damaging solar radiation [20,52]. This ability of living organisms to respond to UV-B radiation will reduce the direct effects of tissue damage on the one hand, but on the other hand will lead to fundamental changes in the ecological interactions between an organism and its environment (indirect effects). For example, avoidance of high levels of UV-B may require that an organism sacrifice access to optimal food habitats or suffer increases in predation risk. The persistence of species in communities under conditions of elevated UV-B will depend on their ability to balance the direct and indirect effects of UV-B radiation to minimize negative impacts on their fitness.

The ecological consequences of the response of

NATO ASI Series, Vol. I 18
Stratospheric Ozone Depletion/
UV-B Radiation in the Biosphere
Edited by R. H. Biggs and M. E. B. Joyner

zooplankton to UV-B radiation are the central focus for this paper. A simple model is developed to describe the sequence of potential responses of zooplankton to ambient levels of UV-B radiation. The environmental constraints and ecological implications of these responses to UV-B are considered, and examples are used to illustrate how these responses may lead to parallel, but often fundamentally different, indirect effects in freshwater versus marine ecosystems. The potential interaction between the selective pressures from increased UV-B radiation and other anticipated agents of environmental change such as global warming and acid precipitation are also discussed.

## THE MODEL

Natural levels of UV-B radiation are known to damage a wide variety of biochemicals in living organisms [19], but the primary target molecule for UV-B damage in living organisms is generally considered to be DNA [10,49,50,52,54]. The model presented here focuses on DNA as a target molecule for clarity of presentation and to provide a common currency for comparing organisms in different taxa. This model could be extended to apply to many other UV-B sensitive biochemicals as well.

Zooplankton have a variety of response mechanisms that can either minimize the dose of UV to which they are exposed or reduce the negative consequences of exposure. These responses can occur at three levels: behavioral avoidance, production of protective compounds, and photorepair [66]. These levels of response can be arranged according to the sequence in which they can occur (Figure 1). The first two levels of response attempt to minimize the dose of UV radiation that reaches the DNA with internal or external filters. The third level reduces the negative consequences of DNA damage through molecular repair processes.

When there are high ambient levels of UV-B, the first response option for zooplankton is to use the external filter (water and the substances in it, Figure 2) available to them to reduce their exposure to UV-B. They can do this by either permanently migrating

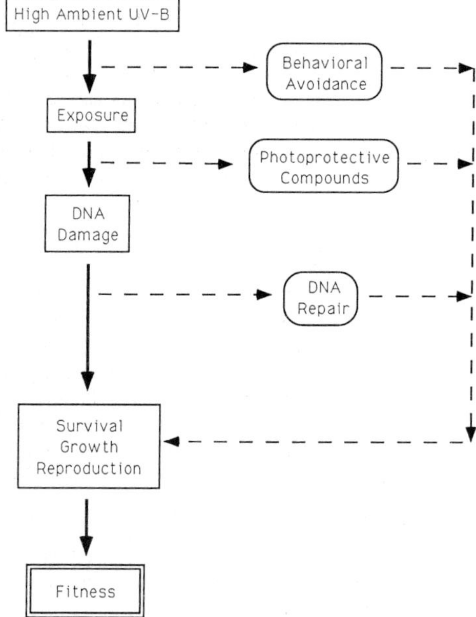


Fig. 1. Flow diagram of direct (solid lines) and indirect (dashed lines) effects of UV-B radiation on zooplankton.

out of the surface layers of the water column or by undergoing a diel vertical migration wherein they avoid high UV-B levels in the surface waters during the day. Although it has fallen out of favor recently, damage from short wavelength solar radiation was one of the earliest hypotheses proposed to explain the diel vertical migration of zooplankton [28].

The second level of response available to zooplankton is to manufacture or sequester UV-B absorbing compounds in their bodies to protect vital tissues from damage. Zooplankton may reduce photodamage with a variety of photoprotective compounds including compounds that absorb primarily blue light, such as astaxanthin [4,21], and compounds that block out a broader range of wave-lengths, such as melanin [25]. Some zooplankton also possess mycosporine-like amino acids that absorb in the UV-B range[8,30,57]. We will refer to this level of response as an internal filter (Figure 3).

The third level of response available to zoo-plankton is repair of DNA after UV-B damage occurs. Repair mechanisms have been described (photorepair, excision repair, postreplication repair, etc.) in a wide variety of plants and animals, including zooplankton. Some (if not all) of these mechanisms appear to be unspecific, in the sense that they can repair the damaged DNA regardless of the etiology of the dam-

External Filter - $K_w Z$
(Water)

1. Filter Density (water extinction coefficient - $K_W$)

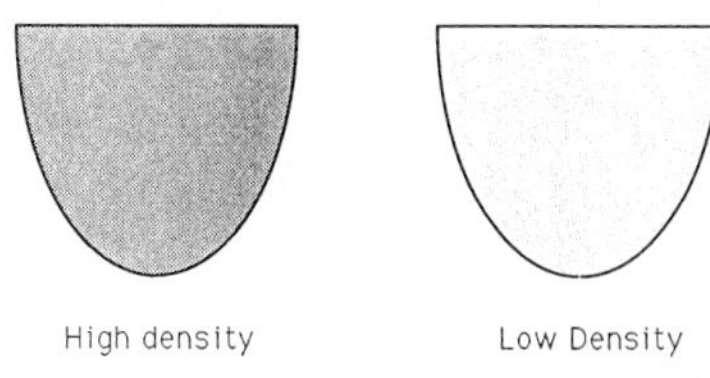


2. Filter Thickness (vertical position - Z)

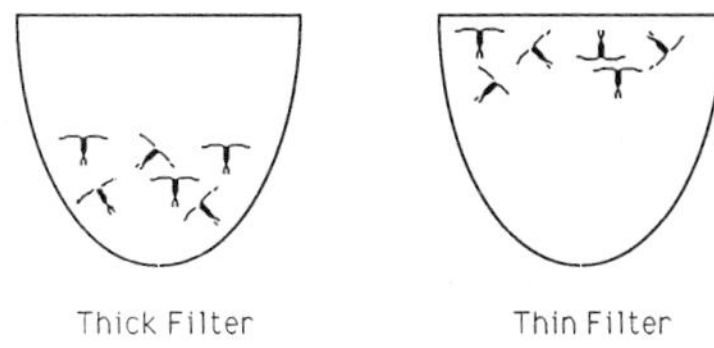


Fig. 2. Diagrammatic representation of external filter response option for zooplankton. Filter density is dependent upon the extinction coefficient of the water ($K_w$), while filter thickness is a function of the distance of the organism from the water surface (Z).

INTERNAL FILTER - $K_p$ P h
(protective compounds)

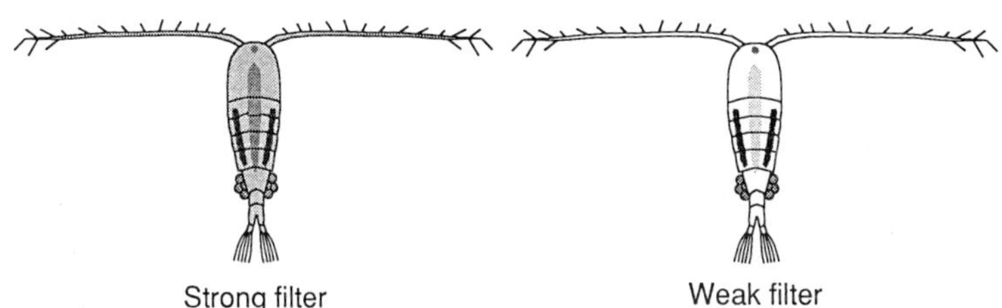


Fig. 3. Diagrammatic representation of internal filter response option for zooplankton. Filter density is dependent upon the molar specific extinction coefficient of the photoprotective compounds ($K_p$), while filter thickness is a function of the thickness of the body tissue containing the protective compounds (h) and the concentration of these compounds (P).

Table 1. Environmental constraints and ecological costs of zooplankton responses to UV-B radiation.

| RESPONSE TO HIGH UV-B | ENVIRONMENTAL CONSTRAINTS | ECOLOGICAL COSTS |
|---|---|---|
| I. EXTERNAL FILTER | | |
| migrate to stay deep | basin depth<br>low oxygen | demographic<br>- temperature?<br>- predation?<br>- food? |
| diel vertical migration | basin depth<br>low oxygen<br>swimming speed | demographic<br>- temperature?<br>- predation?<br>- food?<br>energetic |
| II. INTERNAL FILTER | | |
| colored pigments (melanin, carotenoids) | visual predation | energetic |
| colorless compounds | | energetic |

age. In zooplankton, repair is usually, but not always, enhanced by exposure to longer wavelengths of light [14,51,55]. As far as we know, there is no information on how the effectiveness of repair mechanisms relates to previously-experienced UV exposure, on how it varies across environmental gradients or with variation among taxa. Future research should address these topics.

Responses to UV-B radiation at each of these three levels mitigate the direct effects of exposure to UV-B radiation. However, these responses also have a variety of environmental constraints which are likely to impose serious ecological costs on the organism (Table 1). Because of these costs, the indirect effects of UV-B radiation on zooplankton (dashed arrows in Figure 1) may be at least as important as the direct effects of tissue damage (solid arrows in Figure 1) in maximizing fitness.

Assuming that UV-B attenuation obeys the Lambert-Beer law, the amount of UV-B radiation that reaches the DNA can be described as a function of the attenuation coefficient of the ambient water ($K_w$), the depth of an organism in the water column (Z), the concentration of UV-B protective compounds in the organism (P), the molar-specific extinction coefficient of these protective compounds ($k_p$), and the thickness of the tissue (or cytoplasm) containing the protective compounds (h):

$$I_{DNA} = \frac{I_0}{e^{(K_w Z + k_p Ph)}} \qquad (1)$$

The mortality increase due to $I_{DNA}$ is given by:

$$d_{UV\text{-}B} = G \cdot I_{DNA} \qquad \text{(after Diffey, 1991)} \qquad (2)$$

where G is a wavelength-specific measure of DNA sensitivity (weighting function), and t is the exposure time. The above equation does not take into consideration the effect of repair mechanisms. To account for repair mechanisms, we propose:

$$d_{UV\text{-}B} = (G \cdot I_{DNA}) - q \qquad (3)$$

where q is the rate of DNA repair. The value of q varies between zero and $q_{max}$ (the maximum rate of repair that a cell can achieve). The repair rate (q) is probably a species-specific function of several factors including the amount of damaged DNA, ambient temperature, and the intensity of photoactive radiation (320-450 nm). Several different functional relationships between mortality rate and total dose have been proposed [19].

The mortality rate due to UV-B can be incorporated into a general model of population growth. Consider, for instance, in the standard equation for exponential population growth $N_0$ and $N_t$ are population size at times zero and t, respectively, and r is the exponential growth rate of the population, then:

$$N_t = N_0\, e^{rt} \qquad (4)$$

where r is a function of the instantaneous birth (b) and death (d) rates:

$$r = b - d \qquad (5)$$

If mortality due to UV-B radiation ($d_{UV\text{-}B}$) is the only source of mortality in a population, then,

$$d = d_{UV\text{-}B} \qquad (6)$$

Obviously UV-B is not the only source of mortality for zooplankton, and other sources of mortality can be added to this relationship. For example, mortality due to predation ($d_{PR}$) can be included such that:

$$d = d_{UV\text{-}B} + d_{PR} \qquad (7)$$

This approach is useful because r is perhaps the most useful quantity to use if we want to estimate the effects of ecological variables on the evolutionary fitness of an organism. Thus, changes in population size can be modeled as a function of UV-B intensity and several other possible sources of mortality:

$$N_t = N_0\, e^{(b - d_{UV\text{-}B} - d_{PR} - d_{...})t} \qquad (8)$$

The persistence of a species of zooplankton (or any other living organism) through time is dependent upon its ability to maximize its fitness (estimated by r). If organisms do not actively respond to elevated incident UV-B radiation ($I_0$), then their DNA will be exposed to correspondingly higher levels of UV-B ($I_{DNA}$). This will result in an increase in UV-B related death rates ($d_{DNA}$), and ultimately lower fitness (solid lines in Figure 1).

If organisms do respond actively to elevated UV-B, then these responses are likely to have an associated cost. These indirect pathways include internal filters, external filters, or repair mechanisms (dashed lines in Figure 1). The relative con-tribution of direct versus indirect effects of UV-B on fitness will be determined largely by the ability of a given species to actively respond to elevated UV-B or an environmental correlate (such as high visible light) before extensive DNA damage to the organism occurs.

Little information is available on the ability of invertebrates to actively respond to UV-B radiation. Some species respond to UV light by avoidance (water flea *Daphnia*, UV <334 nm [40]; pluteus larvae of sand dollar *Dendraster*, UV-B <315 nm, [47], while others exhibit no response to UV-B <315 nm (euphausiid shrimp *Thysanoessa* [14]; copepod *Epilabidocera*, zoea larvae of crab *Hemigrapsus*, and larvae of shrimp *Pandalus* [13]). In these latter cases, not only did the invertebrates not respond actively to UV-B, but they were continuously attracted to bright levels of visible light even after they had received lethal doses of UV-B radiation [13]. This suggests that organisms which are exposed to, but which cannot actively respond to, high levels of UV-B radiation, will suffer fitness losses primarily through direct DNA damage (Figure 1).

## APPLICATION OF THE MODEL

The internal and external UV-B filters of zooplankton can be measured and, to some extent, experimentally manipulated. We recently performed some experiments in which we manipulated external filter thickness and density by incubating natural zooplankton communities with and without Mylar (half-transmittance point 323 nm) at several depths in the water column of an oligotrophic and a eutrophic north temperate lake (Williamson et al., in review). Based

Table 2. Mortality rates due to UV-B in a natural zooplankton community incubated for three days at four different depths in Lake Giles, Pike County, PA, July 13-16, 1992 with (experimentals, = e) and without (controls, = c) Mylar filters to exclude UV-B radiation. Mortality rates are calculated as $r_c - r_e$", $day^{-1}$, using initial and final densities, and assuming differences among treatments are due to UV-B manipulation alone (equation 4). "Total" means all individuals died in the UV-B exposed treatments so mortality rates could no be estimated.

| Depth (m) | *Daphnia* | *Diaphanosoma* | *Diaptomus* | *Keratella* |
|---|---|---|---|---|
| 0.5m | total | total | - 0.14 | - 0.09 |
| 1.1m | - 0.40 | - 0.20 | - 0.10 | + 0.04 |
| 2.6m | - 0.14 | - 0.26 | - 0.04 | + 0.05 |
| 6.0m | - 0.08 | 0.0 | - 0.01 | - 0.06 |

on this work we can use the final densities in the control and experimental treatments of the oligotrophic lake and equations 4-6 to calculate mortality rates due to UV-B radiation ($d_{UV-B}$, Table 2).

These mortality rates demonstrate that different zooplankton species vary in their sensitivity to short-term exposures to UV-B radiation and that this sensitivity to UV-B can be experimentally estimated and used in the model presented here. It must be emphasized, however, that these mortality rates are short-term sensitivities that represent the selective pressures on zooplankton to respond to UV-B, and that in nature zooplankton are likely to respond with either external or internal filter responses that modify UV-B exposure. Longer-term responses of aquatic communities may differ from short-term responses because of either complex trophic interactions or differences in the inherent resistance or ability of the various species to adapt to potentially damaging UV-B [5].

## MINIMIZING DNA EXPOSURE TO UV-B

Zooplankton that are able to actively respond to high levels of UV-B radiation may reduce the exposure of their DNA to UV-B ($I_{DNA}$) at either of two levels: by using the external filter consisting of the surrounding medium ($K_wZ$ in equation 1), or by using an internal filter consisting of UV-B absorbing compounds ($k_pPh$). Both of these filters have a density component ($K_w$ and $k_pP$ respectively) and a depth or thickness component (Z and h). Zooplankton are not able to influence $I_0$, the only other component contributing to $I_{DNA}$ (equation 1).

### EXTERNAL FILTER RESPONSE

The external filter response option consists of increasing the amount of water between an organism and the UV-B source (increasing Z in equation 1). This is accomplished by either long-term vertical habitat shifts or diel migrations into and out of the surface waters as UV-B levels fluctuate over a 24h cycle. These will be collectively referred to here as vertical distribution responses. The environmental constraints and ecological costs involved in these vertical distribution responses will depend on basin morphometry and existing vertical gradients in temperature, oxygen, food quantity, food quality, and predators (visual and tactile) in the water column. Many of these gradients are driven largely by longer-wavelength solar radiation.

*Habitat Boundaries*. One obvious constraint on vertical distribution responses to UV-B is the maximum depth of the basin. This is likely to be an important constraint in shallow, clear lakes where zooplankton cannot migrate deep enough to avoid damaging levels of UV-B. This constraint will be unimportant in open oceans or in more turbid lakes where UV-B is attenuated rapidly. The thermocline may also provide a distinct lower boundary for vertical distribution, particularly for warm stenotherms. In more productive aquatic systems that are thermally stratified, respiration and decomposition may deplete oxygen in deeper strata, thus placing further constraints on the depths to which zooplankton can migrate to obtain a refuge from UV-B. These habitat boundary constraints may force species to use either an internal filter or repair mech-anisms to prevent or recover from UV-B damage, or otherwise risk being eliminated from certain systems. Further study is needed to estimate the attenu-ation depth of biologically damaging levels of UV-B in order to determine the importance of habitat boundaries. For example, the more rapid attenuation of UV-B that is expected in productive aquatic systems may make a deep anoxic boundary irrelevant to vertical distribution responses to UV-B.

*Thermal Stratification.* Several types of vertical gradients are likely to influence the costs of vertical distribution responses to UV-B. For example, in thermally stratified systems the warmer temperatures of surface waters offer distinct demographic advantages such as shorter generation times and higher reproductive rates in many zooplankton species [39,44,60,64].

Lakes generally have higher concentrations of dissolved and particulate matter than marine systems. Thus both UV-B and longer wavelengths of light will attenuate more rapidly in lakes than in oceans. Combined with the much greater fetch of oceans, this will lead to steeper thermal stratification and a much more dense external UV-B filter (higher $k_w$) in lakes than in oceans. This means that freshwater zooplankton that use diel vertical migration to minimize their exposure to UV-B will not need to migrate as far as organisms in marine environments to attain the same level of protection from UV-B, and thus will likely incur a lower energetic cost. On the other hand, the steeper thermal gradients in freshwater (often 5-10°C over a few meters depth) versus marine environments suggest that the demographic costs resulting from lower temperatures may be more severe in freshwater than in marine systems.

*Vertical Gradients in Food Quality and Quantity.* Zooplankton feed primarily on phytoplankton and components of the microbial loop (bacteria, protozoans). In thermally stratified systems, vertical gradients in factors such as light and nutrient availability are likely to influence both the quantity and the quality of these food resources. Examples of gradients in food availability include subsurface chlorophyll maxima [10,26], metalimnetic peaks in microbial biomass [3,46], and changes in the nutritional quality of algae as a result of differing light and nutrient ratios available within the water column [1,53,58]. UV-B photoinhibition of primary productivity by algae [57] and secondary productivity by bacteria in the surface waters of aquatic systems may contribute to these gradients in food availability. Zooplankton that undergo vertical migration or habitat shifts may thus experience changes in the availability of food resources with subsequent impacts on their survival, growth, reproduction, and ultimately fitness.

Although not an adaptive response, an interesting negative feedback loop exists between zoo-plankton and the density of the external water filter ($k_w$ in equation 1) through their grazing activities. Increased densities of zooplankton will increase grazing rates which will decrease phytoplankton and other particulates in the water column. The result will be a decrease in the density and protective properties of the external filter, resulting in increased UV-B irradiance and potential decreases in zooplankton abundance in exposed surface waters. Interestingly, large zooplankton that are strong vertical migrators could reduce or destroy the external filter for small zooplankton species (protozoans, rotifers, nauplii) that are not strong swimmers and therefore tend to remain in the surface waters.

The spring clear-water phase that is often observed in lakes is of particular interest in this respect. When zooplankton communities are dominated by large cladocerans such as *Daphnia*, zooplankton grazing can decrease algal biomass by fourfold or more, resulting in a doubling of water clarity (Secchi depth [33,59]). These clear-water events develop over periods of only one or two weeks and may last for a month. Although UV-B radiation has not been measured during clear-water phases, it is likely that the timing of these events, which follow the spring algal bloom and just precede the summer solstice, will lead to substantial short-term increases in UV-B radiation during periods when reproductive rates of zooplankton and many fish are high.

*Predation.* Vertical gradients in predation risk from both tactile and visual predators are common in aquatic systems. They are likely to be ecologically important for zooplankton responding to vertical gradients in UV-B radiation because predators often play a central role in determining the vertical distribution and size structure of zooplankton communities [31,32,43].

Predation rates of visually-feeding fish are an increasing function of visible light intensity [9,62]. Thus we would expect predation risk from visual predators to decrease with depth as well as vary over diel cycles in a manner similar to mortality risk from UV-B radiation. However, because UV-B is attenuated more rapidly in the water column than visible light, there are likely to be differences in the range and slope of the vertical gradients in UV-B radiation and predation risk from visual predators. Tactile planktonic predators, on the other hand, are not dependent upon light to feed, tend to be vulnerable to visual predators, and consequently often seek refuge in the deeper, darker layers of the water column during the day. Both types of predators may themselves be vulnerable to UV-B.

Predation risk can be expressed as a mortality rate coefficient and integrated into the standard population growth rate equation along with UV-B mortality rates (equations 4-8). Prey can respond to predation risk either by reducing their exposure to high predator densities through vertical migrations that minimize their spatial and temporal overlap with predators, or by reducing their vulnerability to predators with defense mechanisms such as spines, armor, or behavioral escape responses [63,65]. Zooplankton may exhibit analogous migration or vulnerability responses to damaging levels of UV-B radiation: they can reduce their exposure to high UV-B levels in surface strata through vertical habitat shifts or diel vertical migrations (increase Z), or by reducing their vulnerability to UV-B exposure with protective compounds (increase P or $k_p$).

INTERNAL FILTER RESPONSE.

If zooplankton are unable to minimize their exposure to UV-B radiation through vertical distribution responses, an alternative option is to modify their internal UV-B filter by producing or sequestering protective compounds, although the production of photoprotective compounds is costly to zooplankton [25]. In addition, for large zooplankton in environments with visual predators such as fish, there will be costly ecological constraints: visual predators prefer pigmented copepods and cladocerans over their pale conspecifics [6,22,35]. The distribution of melanin-pigmented *Daphnia* [27] is, in fact, restricted to environments with sparse or no fish populations, suggesting that the threat of visual predation makes pigmentation infeasible for large, non-evasive, species. Under most natural conditions the use of pigmentation by large zooplankton may be constrained by the metabolic costs of pigment production as well as the risk of visual predation.

EXTERNAL OR INTERNAL FILTER RESPONSES?

What happens when the environmental conditions are such that zooplankton can employ more than one protective mechanism? The observation that protective compounds are lost when they are not exposed to radiation [27], but can be induced *de novo* after continuous irradiation [8] suggests that irradiation is required for the development of photoprotective compounds, and that production of these compounds comes at some cost to the organism. Therefore, organisms that remain deep enough during the day will not be exposed to solar radiation and will not develop photoprotective compounds. In fact, the vertical position of zooplankton has been shown to be inversely related to the concentration of protective compounds (i.e. the more pigmented animals occur at shallower depths and vice versa) in copepods [24] and *Daphnia* (in lakes with no visual predators [25]).

The next question, then, is whether one mechanism is used preferentially over the other. From equation 1 we would expect that the concentration of photoprotective compounds will be inversely proportional to the water extinction coefficient. It follows that the safe depth to which zooplankton must migrate to avoid UV-B damage should decrease with decreasing water transparency. Byron (1982) found in a series of Colorado lakes that carotenoid concentration in calanoid copepods increased with decreasing depth of the lake basin. However, when he compared body pigmentation to chlorophyll concentrations, he found that highly pigmented copepods were found only in lakes that were either extremely clear, or extremely shallow. This suggests that copepods use internal filter responses that increase $k_p$ only when there is a lower habitat boundary that prevents migration to a safe depth, or when $K_w$ is extremely small and external filter responses become very costly.

The presence of photoprotective compounds or symbiotic algae are common in ciliates (e.g. *Stentorina* in planktonic species of *Stentor*, symbiotic *Chlorella* in *Ophrydium naumani*). Due to the rather limited ability of these slow-swimming organisms to adjust their depth in the water column (Z), we predict an inverse relationship between the concentration of chlorophyll in the ambient water and the concentration of UV-B protective compounds (P). Unfortunately, we have been unable to find quantitative data on pigment concentrations with simultaneous records of water extinction coefficient estimates to test this hypothesis.

## MINIMIZING MORTALITY FROM UV-B AT FIXED EXPOSURE LEVELS

### REPAIR MECHANISMS

Repair mechanisms have been described (photorepair, excision repair, post-replication repair, and so on) in a wide variety of plants and animals. Some (if not all) of these mechanisms appear to be nonspecific in the sense that they can repair the damaged DNA regardless of the etiology of the damage. Little or no information is available on the environmental constraints or ecological costs of repair mechanisms.

### LIFE HISTORIES

The portion of DNA involved in reproduction (germinal DNA) is usually separated from the remaining (somatic) DNA in metazoans. Such separation often occurs at a very early stage in the ontogeny of the individuals. If only the somatic DNA were damaged, organisms with short life spans would be less affected (at relatively low UV intensities) since they would probably die anyway before accumulating a lethal UV dose. Their offspring could rebuild the somatic DNA from the undamaged germinal DNA. Since this strategy is useless unless the germinal DNA remains undamaged, short-lived metazoans would focus their efforts on protecting germinal and embryonic DNA (e.g., ovaries and eggs).

Moreover, the timing of hatching should be important in these species. For species with a generation time of about 24 hours, such as several rotifers at summer temperatures, the offspring that hatch in the afternoon or evening will be unexposed to UV radiation during the best part of their lives. That might give them time to grow and produce eggs before suffering UV damage. Synchronous egg hatching has been observed in *Keratella americana*, *Brachyonus calyciflorus* and *B. havanaensis* only during the summer. Interestingly, the maximum rates of hatching occur in the afternoon or early evening [37], immediately preceding prolonged periods of low UV-B.

On a slightly different note, bacteria are more sensitive to UV-B damage in exponential growth phase than when in a stationary growth phase [17]. This implies that UV-B sensitivity of microbes may vary with resource supply in natural ecosystems.

## UV-B AND THE SIZE STRUCTURE OF ZOOPLANKTON COMMUNITIES

Body size is an important factor in zooplankton communities due to the importance of physiological relationships between metabolic rates and body size [48], the importance of size in determining vulnerability to tactile versus visual predators [67], and the importance of size in determining the impact of zooplankton grazing on the chloro-phyll:phosphorus relationship [7,34,45].

Body size is also likely to be important in determining the ecological effects of UV-B radiation in zooplankton communities. For example, small slow-swimming zooplankton species will be able to utilize the external filter response more effectively in strongly stratified lakes than in the generally more weakly stratified oceans due to the steeper attenuation gradient of UV-B in lakes. The intensity of the stratification will be less of an obstacle for larger, faster swimming zooplankton.

Visually-feeding fish tend to prefer larger, more conspicuous prey, while tactile planktonic predators tend to prefer smaller, less conspicuous prey [42,67]. Because conspicuousness is influenced by pigmentation, the potential for zooplankton to employ photoprotective compounds may also be a function of body size. Large pigmented zooplankton would be easy prey for visual predators, while pigmentation in small zooplankton might not be so effective at increasing the vulnerability of these prey to visual predators.

## COMPLEX ECOLOGICAL INTERACTIONS AND ENVIRONMENTAL CHANGE

One might ask what may be the potential impact of other environmental changes that are co-occurring with the global changes in UV-B radiation? One such environmental change that is likely to interact strongly with changes in UV-B is acid precipitation. Lakes with low pH may be acidic due to either natural organic acids (humic and tannic acids), or to anthropogenic acids [2]. The high concentrations of organics in naturally acidified waters are likely to cause rapid attenuation of UV-B radiation in the water column and suggest that these lakes will be less vulnerable to future changes in UV-B radiation. Anthropogenically acidified lakes, on the other hand, tend to have very low concentrations of organic substances, low productivity, and very clear waters, and are likely to be much more vulnerable to future changes in UV-B radiation. Oceans are well buffered and are unlikely to be influenced by changes in acid deposition.

Some interesting experiments were performed earlier in the century that suggest interactive effects of pH and UV radiation on the phototactic response of some zooplankton. Moore (1912) found that *Daphnia* showed a negative phototaxis when exposed to UV radiation (<334 nm) from a mercury lamp. This negative phototaxis was reversed when either carbonated water or mineral acids (HCl) were added to the medium.

The importance of temperature in determining the ecological costs of UV-B radiation impacts suggests that global climate changes may also have interactive effects with future changes in UV-B. Modeling efforts to assess changes in thermal habitats for fish and zooplankton are currently underway to assess the impact of global warming on them [15,38]. The relationship between body size and zooplankton responses to temperature suggest that global warming may also influence the size structure of zooplankton communities in lakes [41]. Similar modeling efforts should be carried out to assess how the vertical habitat boundaries of zooplankton and fish will be modified under different scenarios of increased UV-B radiation.

Burchard and Dworkin (1966, as cited in Hairston 1979b) suggested that photodamage is a physical process with a rate independent of temperature. More recently, Quintern et al. (1991) showed that the response of biofilms based on defective strains of *Bacillus subtilis* to UV radiation was independent of temperature between -20 and 70°C. Unlike DNA damage, enzymatic repair mechanisms are expected to be temperature dependent. Photoprotection by carotenoids in bacteria has been shown to be less efficient at low temperatures [23] and spatial [6] patterns of carotenoid concentration in copepods show an inverse relationship with temperature. This raises the possibility that lower concentrations of protective compounds are required during warmer periods because of the increased efficiency of repair mechanisms at higher temperatures.

## FUTURE DIRECTIONS

The above model provides a framework for addressing several questions that need to be resolved if we are to understand the mechanisms, constraints, costs, and ultimately the impacts of UV-B radiation on zooplankton communities. These questions include:

1) What is the relative importance (costs and consequences) of direct versus indirect effects of UV-B on zooplankton?

2) How important are internal versus external filter responses in determining impacts of elevated UV-B on zooplankton communities?

3) What is the relative importance of UV-B radiation and visual predation in determining the spatial and temporal distribution of zooplankton communities? These potential sources of mortality are likely to be correlated in nature.

4) What are the molecular repair mechanisms available to zooplankton and how do they vary across taxa and with variations in light, temperature, and other environmental variables?

To answer these questions, experimental and comparative studies will need to be accompanied by good optical data on the attenuation of UV-B in the water column and corresponding data on the sensitivity of plankton to UV-B levels at various depths. Hydrologic mixing models may also be useful in determining the timing of zooplankton exposures to UV-B. Future UV research will benefit from the collaboration of marine and freshwater scientists. While many of the more compelling global scale questions dealing with the impacts of UV-B on aquatic ecosystems must be answered in the oceans, lakes can serve as effective model systems for understanding larger scale marine communities and ecosystems for several reasons:

1) The mechanisms underlying many physical, chemical, and biological processes in aquatic communities and ecosystems are similar in lakes and oceans.

2) Lakes have discrete boundaries, are easily monitored, can be experimentally manipulated, and are convenient models for educating students.

3) Lakes occur over a variety of elevations, and exhibit diverse water chemistry and optical characteristics over small geographic areas as well as over short distances vertically within the water column.

## REFERENCES


1. Ambler, J.W. 1986. Effect of food quantity and food quality on egg production of *Acartia tonsa* Data from East Lagoon, Galveston, Texas. Estuarine Coastal Shelf Sci. 23:183-196.

2. Baker, L. A., A.T. Herlihy, P.R. Kaufmann, and J.M. Eilers. 1991. Acidic lakes and streams in the United States: The role of acidic deposition. Science 252:1151-1154.

3. Bennett, S.J., R.W. Sanders, and K.G. Porter.1990. Heterotrophic, autotrophic, and mixotrophicnanoflagellates: Seasonal abundances and bacterivory in a eutrophic lake. Limnol. Oceanogr. 35:1821-1832.

4. Bidigare, R.R. 1989. Potential effects of UV-Bradiation on marine organisms of the southern ocean: Distributions of phytoplankton and krill during austral spring. Photochem. and Photobiol. 50:469-477.

5. Bothwell, M.L., D. Sherbot, A.C. Roberge, and R.J. Daley. 1993. Influence of natural ultraviolet radiation on lotic periphytic diatom community growth, biomass accrual, and species composition: Short-term versus long-term effects. J. Phycology. 29:24-35.

6. Byron, E.R. 1982. The adaptive significance of calanoid copepod pigmentation: A comparative and experimental analysis. Ecology 63:1871-1886.

7. Carpenter, S., et al. 1991. *In* Cole, J., G. Lovett, and S. Findlay (eds.) Comparative Analysis of Ecosystems: Patterns, Mechanisms, and Theories. New York: Springer-Verlag. 375 pp.

8. Carreto, J.I., M.O. Carignan, G. Daleo, and S.G. De Marco. 1990. Occurrence of mycosporine-like amino acids in the red-tide dinoflagellate *Alexandrium excavatum*: UV-photoprotective compounds? J. Plankton Res. 12:909-921.

9. Confer, J.L., G.L. Howick, M.H. Corzete, S.L. Kramer, S. Fitzgibon, and R. Landesberg. 1978. Visual predation by planktivores. Oikos 31:27-37.

10. Cullen, J.J. 1982. The deep chlorophyll maximum: Comparing vertical profiles of chlorophyll a. Can. J. Fish. Aquat. Sci. 39:791-803

11. Cullen, J.J., P.J. Neale, and M.P. Lesser. 1992. Biological weighting function for the inhibition of phytoplankton photosynthesis by ultraviolet radiation. Science. 258:646-650.

12. Damkaer, D.M., D.B. Dey, G.A. Heron, and E.F. Prentice. 1980. Effects of UV-B radiation on near-surface zooplankton of Puget Sound. Oecologia (Berlin) 44:149-158.

13. Damkaer, D.M., and D.B. Dey. 1982. Short-term responses of some planktonic crustacea exposed to enhanced UV-B radiation. *In* J. Calkins (ed.), The Role of Solar Ultraviolet Radiation in Marine Ecosystems. Series IV: Marine Sciences, New York: Plenum. pp.417-427.

14. _____. 1983. UV damage and photoreactivation potentials of larval shrimp, *Pandalus platyceros*, and adult euphausiids, *Thysanoesa raschii*. Oecologia (Berlin) 60:169-175.

15. De Stasio, B.T., N. Nibbelink, and P. Olsen. 1993. Diel vertical migration and global climate change: Adynamic modeling approach to zooplankton behavior. Verh. Int. Verein. Limnol. in press.

16. Diffey, B.L. 1991. Solar ultraviolet radiation effects on biological systems. Med. Biol. 36:299-328.

17. Eisenstark, A. 1982. Action spectra and their role in solar UV-B studies. *In*:J. Calkins (ed.), The Role of Solar Ultraviolet Radiation in Marine Ecosystems. Series IV: Marine Sciences, New York, Plenum. pp.157-159.

18. Gleason, J.F., P.K. Bhartia, J.R. Herman, R. McPeters, P. Newman, R.S. Stolarski, L. Flynn, G. Labow, D. Larko, C. Seftor, C. Wellemeyer, W.D. Komhyr, A.J. Miller, and W. Planet. 1993. Record low global ozone in 1992. Science 260:523-526.

19. Häder, D., and M. Tevini. 1987. General photobiology. Oxford: Pergamon Press. 323 pp.

20. Häder, D., and R.C. Worrest. 1991. Effects of enhanced solar ultraviolet radiation on aquatic ecosystems. Photochem. Photobiol. 53:717-725.

21. Hairston, N.G. 1976. Photoprotection by carotenoid pigments in the copepod *Diaptomus nevadensis*. Proc. Nat. Acad. Sci. USA 73:971-974.

22. _____. 1979a. The adaptive significance of color polymorphism in two species of *Diaptomus* (Copepoda). Limnol. Oceanogr. 24:15-37.

23. _____. 1979b. The effect of temperature on carotenoid photoprotection in the copepod *Diaptomus nevadensis*. Comp. Biochem. Physiol. 62A:445-448.

24. Hairston, N.G., Jr. 1980. The vertical distribution of diaptomid copepods in relation to body pigmentation. *In* W.C. Kerfoot (ed.), Evolution and Ecology of Zooplankton Communities, University Press of New England, Hanover, N.H. pp.98-110.

25. Hebert, P.D.N., and C.J. Emery. 1990. The adaptative significance of cuticular pigmentation in *Daphnia*. Funct. Ecol. 4:703-710.

26. Herman, A.W. 1989. Vertical relationships between chlorophyll, production and copepods in the eastern tropical Pacific. J. Plankton Res. 11:243-261.

27. Hobaek, A., and H. Wolf. 1991. Ecological genetics of Norwegian *Daphnia*. II. Distribution of *Daphnia longispina* genotypes in relation to short-wave radiation and water colour. Hydrobiologia 225:229-243.

28. Huntsman, A.G. 1924. Limiting factors for marine animals. I. The lethal effect of sunlight. Canadian Biology 2:83-88.

29. Karanas, J.J., R.C. Worrest, and H. Van Dyke. 1981. Impact of UV radiation on the fecundity of the copepod *Acartia clausii*. Mar. Biol. (Berlin) 65:125-133.

30. Karentz, D., F.S. McEuen, and W.C. Dunlap. 1991. Survey of mycosporine-like amino acid compounds in Antarctic marine organisms: Potential protection from ultraviolet exposure. Mar. Biol. (Berlin) 108:157-166.

31. Kerfoot, W.C. (ed.). 1980. Evolution and Ecology of Zooplankton Communities. Univ. Press of New England, Hanover, NH. 800 pp.

32. Kerfoot, W. C., and A. Sih, eds. 1987. Predation: Direct and indirect impacts on aquatic communities. University Press of New England, Hanover, NH. 386 pp.

33. Lampert, W., W. Fleckner, H. Rai, and B.E. Taylor. 1986. Phytoplankton control by grazing zooplankton: A study on the spring clear-water phase. Limnol. Oceanogr. 31:478-490.

34. Leibold, M.A. 1989. Resource edibility and the effects of predators and productivity on the outcome of trophic interactions. Am. Nat. 134:922-949.

35. Luecke, C., and J.O. O'Brien. 1981. Phototoxicity and fish predation: Selective factors in color morphs in *Heterocope*. Limnol. Oceanogr. 26:454-460.

36. Madronich, S. 1992. Implications of recent total atmospheric ozone measurements for biologically active ultraviolet radiation reaching the earth's surface. Geophys. Res. Let. 19:37-40.

37. Magnien, R.E., and J.J. Gilbert. 1983. Diel cycles of reproduction and vertical migration in the rotifer *Keratella crassa* and their influence on the estimation of population dynamics. Limnol. Oceanogr. 28:957-969.

38. Magnuson, J.J., J.D. Maisner, and D.K. Hill. 1990. Potential changes in the thermal habitat of Great Lakes fish after global climate warming. Trans. Amer. Fish. Soc. 119:254-264.

39. McLaren, I.A. 1963. Effects of temperature on growth of zooplankton, and the adaptive value of vertical migration. J. Fish. Res. Board. Can. 20:685-727.

40. Moore, A.R. 1912. Concerning negative phototro-

pism in *Daphnia pulex*. Journal of Experimental Zoology 13:573-575.

41. Moore, M., and C. Folt. 1993. Zooplankton body size and community structure: Effects of thermal and toxicant stress. Trends in Ecology and Evolution 8:178-183.

42. O'Brien, W.J., and D. Kettle. 1979. Helmets and invisible armor: Structures reducing predation from tactile and visual planktivores. Ecology 60:287-294.

43. Ohman, M.D. 1990. The demographic benefits of diel vertical migration by zooplankton. Ecology 60:257-281.

44. Orcutt, J.D., and K.G. Porter. 1983. Diel vertical migration by zooplankton: Constant and fluctuating temperature effects on life history parameters of *Daphnia*. Limnol. Ocean. 28:720-730.

45. Pace, M.L. 1984. Zooplankton community structure, but not biomass, influences the phosphorous-chlorophyll a relationship. Can. J. Fish. Aquat. Sci. 41:1089-1096.

46. Pace, M.L., and J.D. Orcutt, Jr. 1981. The relative importance of protozoans, rotifers, and crustaceans in a freshwater zooplankton community. Limnol. Oceanogr. 26:822-830.

47. Pennington, J., and R.B. Emlet. 1986. Ontogenetic and diel vertical migration of a planktonic echinoid larva, *Dendraster excentricus* (Eschscholtz): Occurence, causes, and probable consequences. J. Exp. Mar. Biol. Ecol. 104:69-95.

48. Peters, R.H. 1983. The ecological implications of body size. 329 pp. Cambridge Studies in Ecology, E. Beck, H.J.B. Birk, and E.F. Connor (eds.), 2. Cambridge University Press.

49. Quintern, L.E., G. Horneck, U. Eschweiler, and H. Bucker. 1992. A biofilm used as ultraviolet-dosimeter. Photochem. Photobiol. 55:389-395.

50. Regan, J.D., W.L. Carrier, H. Gucinski, B.L. Olla, H. Yoshida, R.K. Fujimura, and R.I. Wicklund. 1992. DNA as a solar dosimeter in the ocean. Photochem. Photobiol. 56:35-42.

51. Ringelberg, J., A.L. Keyser, and B.J.G. Flik. 1984. The mortality effect of ultraviolet radiation in a red morph of *Acanthodiaptomus denticornis* (Crustacea: Copepoda) and its possible ecological relevance. Hydrobiologia 112:217-222.

52. Ronto, G., S. Gaspar, and A. Berces. 1992. Phages T7 in biological UV dose measurement. J. Photochem. Photobiol. B. Biol. 12:285-294.

53. Rothhaupt, K.O. 1991. Variations on the zooplankton menu: A reply to the comment by Smith. Limnol. Oceanogr. 36:824-827.

54. Setlow, R.B. 1974. The wavelengths in sunlight effective in producing skin cancer: A theoretical analysis. Proc. Nat. Acad. Sci. 71:3363-3366.

55. Siebeck, O. 1978. Ultraviolet tolerance of planktonic crustaceans. Verh. Internat. Verein. Limnol. 20:2469-2473.

56. Smith, R.C. 1989. Ozone, middle ultraviolet radiation and the aquatic environment. Photochem. Photobiol. 50:459-468.

57. Smith, R.C., B.B. Prézelin, K.S. Baker, R.R. Bidigare, N.P. Boucher, T. Coley, D. Karentz, S. MacIntyre, H.A. Matlick, D. Menzies, M. Ondrusek, Z. Wan, and K.J. Waters. 1992. Ozone depletion: Ultraviolet radiation and phytoplankton biology in Antarctic waters. Science 255:952-959.

58. Smith, V.H. 1991. Competition between consumers. Limnol. Oceanogr. 36:820-823.

59. Sommer, U., Z.M. Gliwicz, W. Lampert, and A. Duncan. 1986. The PEG*- model of seasonal succession of planktonic events in fresh waters. Arch. Hydrobiol. 106:433-471.

60. Stich, H.B., and W. Lampert. 1984. Growth and reproduction of migrating and nonmigrating *Daphnia* species under simulated food and temperature conditions of diurnal vertical migration. Oecologia. 61:192-196.

61. Stolarski, R., R. Bojkov, L. Bishop, S. Zerefos, J. Staehelin, and J. Zawodny. 1992. Measured trends in stratospheric ozone. Science 256:342-349.

62. Vinyard, G.L., and W.J. O'Brien. 1976. Effects of light and turbidity on the reactive distance of bluegill (*Lepomis macrochirus*). J. Fish. Res. Board Can. 33:2845-2849.

63. Williamson, C.E. 1993. Linking predation risk model with behavioral mechanisms: Identifying population bottlenecks. Ecology 74:320- 331.

64. Williamson, C.E., and N.M. Butler. 1987. Temperature, food, and mate limitation of copepod reproductive rates: Separating the effects of multiple hypotheses. J. Plankton Res. 9:821-836.

65. Williamson, C.E., M.E. Stoeckel, and L.J. Schoeneck. 1989. Predation risk and the structure of freshwater zooplankton communities. Oecologia 79:76-82.

66. Worrest, R.C. 1982. Review of literature concerning the impact of UV-B radiation upon marine organisms. *In* J. Calkins (ed.), The role of solar ultraviolet radiation in marine ecosystems. Series IV: Marine Sciences, New York: Plenum. pp.429-457.

67. Zaret, T.M. 1980. Predation and freshwater comunities. Ann Harbor, Michigan. 187 pp.

# MODELING THE LATITUDE-DEPENDENT INCREASE IN NON-MELANOMA SKIN CANCER INCIDENCE AS A CONSEQUENCE OF STRATOSPHERIC OZONE DEPLETION

Steven A. Lloyd and Edward S. Im

Chemistry Department, Harvard University
Cambridge, MA 02138

Donald E. Anderson, Jr.

The Johns Hopkins University Applied Physics Laboratory
Laurel, MD 20723

Key Words: action spectrum, atmosphere, biologically accumulated dosage, biologically effective dosage, cataracts, DNA, modeling, multiple scattering, non-melanoma skin cancers, ozone depletion, ozone layer, UV-B

**ABSTRACT**

The principal medical concern of stratospheric ozone depletion is that ozone loss will lead to the enhancement of ground-level UV-B radiation. The incidence of non-melanoma (basal and squamous cell) skin cancers is correlated with cumulative lifetime UV exposure. Global ozone climatology (40°S to 50°N latitude) was incorporated into a radiation field model to calculate the biologically accumulated dosage (BAD) of UV-B radiation integrated over days, months, and years. The slope of the annual BAD as a function of latitude was found to correspond to epidemiological data for non-melanoma skin cancers for 30°N to 50°N. Various ozone loss scenarios were investigated. It was found that that a small ozone loss in the tropics can provide as much additional biologically effective UV-B as a much larger ozone loss at higher latitudes. Also, for ozone depletions of >5%, the BAD of UV-B increases exponentially with decreasing ozone levels.

## INTRODUCTION

The principal medical concern of stratospheric ozone depletion is that ozone loss will lead to enhanced levels of UV-B radiation which, in turn, may damage biological systems, many of which are very sensitive to UV-B flux. One of the marvels of nature is that the absorption spectrum of DNA roughly parallels that of ozone, so DNA absorbs strongly in the UV-B region; it is as if ozone provides a perfect umbrella to shield biological matter, wavelength-for-wavelength, from the sun's damaging UV-B. Figure 1 shows the absorption spectra for DNA [24] and ozone [16] as well as the solar flux at the "top of the atmosphere" [5] and at the earth's surface [2] for a "normal" atmosphere and one in which the ozone

NATO ASI Series, Vol. I 18
Stratospheric Ozone Depletion/
UV-B Radiation in the Biosphere
Edited by R. H. Biggs and M. E. B. Joyner

layer has been depleted by 30%. The earth's "ozone layer," spread out principally in the lower stratosphere between 15 and 25 km altitude, is all that separates life on earth from the sun's damaging ultraviolet rays.

Our atmosphere is transparent to the least damaging of these rays, the UV-A region (wavelengths from 320 to 400 nm), since ozone does not absorb at those wavelengths. UV-C (<280 nm) is absorbed completely by molecular oxygen as well as by ozone, and hence none reaches the earth's surface. UV-B (280-320 nm), however, is only partially absorbed by the ozone layer; therefore, the ground-level flux of UV-B depends critically on the amount of ozone overhead. It is the overlap of the ground-level flux and the relevant action spectrum (the wavelength-dependent measure of biological effects) that is important in predicting the consequences of enhanced UV-B. The biologically effective dosage (BED) is simply the overlap integral of these two terms: a sharply decreasing function times a sharply increasing function. Note that if one were to shift the ground-level flux curve towards the ultraviolet several nanometers by removing a portion of the ozone column, the overlap integral would increase dramatically. Note also that shorter wavelength photons are relatively much more damaging to the DNA molecule than longer wavelength photons.

Potential biological consequences of stratospheric ozone depletion include the increase in the incidence of non-melanoma skin cancers, hastening of cataract formation, disruption of the aquatic food chain, and crop damage [29]. Previous studies have dealt mostly with the biological consequences of relatively small ozone losses [8,19]. In 1985, Farman et al. announced their findings of seasonal, dramatic ozone depletions in the lower stratosphere over Antarctica [9] subsequently confirmed by archived NASA satellite data [25]. The 1987 Antarctic Atmospheric Ozone Experiment sampled the polar ozone layer during the formation of the ozone hole and confirmed that chlorine- and bromine-catalyzed reactions are responsible for the >50% loss of ozone over the South Pole each austral spring [3,4]. Further concern was raised by the findings of the Ozone Trends Panel that ozone is also thinning at mid-latitudes, as measured by the NASA Total Ozone Mapping Spectrometer (TOMS) satellite instrument [30]. The panel concluded not only that ozone is thinning over heavily populated mid-latitudes on the order of 1-6% per decade, but also the trend is increasing. The observed ozone reductions in the southern hemisphere raise the possibility of continued significant seasonal ozone perturbations over the next few decades. We have therefore carried out calculations to investigate the biological effects of comparatively large ozone losses.

Of all the health-related consequences of enhanced UV insolation, non-melanoma (basal and squamous cell) skin cancers provide the most firmly established link between UV exposure and biological consequences. Epidemiological data indicate that the risk of non-melanoma cancers is correlated with cumulative lifetime UV exposure [22]. These carcinomas are thought to be the result of UV-induced pyrimidine dimers or other covalent modification of the DNA strand [20].

Since DNA appears to be the absorbing chromophore responsible for carcinogenesis, its absorption spectrum can be used as the appropriate action spectrum. A numerical model was developed to correlate epidemiological data of non-melanoma skin cancer incidence with annual biologically accumulated dosage (BAD) of UV-B, which is a strong function of latitude [23].

## MODEL ASSUMPTIONS

An atmospheric radiation field model was developed at the Naval Research Laboratory [1,15] to specifically address questions of chemical photolysis rates, ground level fluxes and nadir intensities at twilight [2]. The photons are initially deposited in a spherical atmosphere (including refraction effects), while isotropic multiple scattering is calculated using a plane-parallel approximation. Comparison with a full spherical model indicates that this approach to multiple scattering is accurate for solar zenith angles ($\chi$) up to 95 degrees.

For a given model atmosphere and ground albedo (surface reflectivity), the model generates a series of solar zenith angle-, altitude- and wavelength-dependent amplification factors. The amplification factor is defined as a multiplicative factor such that, when multiplied by the solar flux incident upon the top of the atmosphere, it yields the total flux into a unit volume element at a given altitude. The amplification factor is broken into a direct attenuated flux term and a diffuse or multiple scattering flux term. In the UV-B region, the radiation field is dominated by multiply scattered light. This component of the source function can be integrated and added to the product of the direct flux and the cosine of the solar zenith angle to yield the ground level flux across the earth's surface.

Two model atmospheres were adopted in this study; one is the 1976 U.S. Standard Atmosphere for $N_2$, $O_2$, and $O_3$. The second reference ozone model used in the latitude-dependent studies are taken from a compilation of five satellite data sets (Nimbus 7 LIMS, Nimbus 7 SBUV, AE-2 SAGE, SME UVS and SME IR) spanning the period November 1978 to December 1983 [13]. This model incorporates monthly zonal mean ozone mixing ratios for pressure altitudes above the 20 mbar level as a function of presssure and latitude (40°S to 50°N).

The standard deviation for these data is typically less than 10%, which is remarkable, considering that several entirely different techniques are used to measure the vertical structure of ozone. Nimbus 7 SBUV data also indicate that the interannual variability of ozone vertical structure is typically less than 10%. Lower altitude ozone levels were reconstructed using the difference between the above zonal monthly values and Nimbus 7 TOMS zonal monthly total column ozone for the period November 1978 to September 1982.

Simulated ozone depletions were produced by scaling back the whole ozone profile by a multiplicative factor; for smaller ozone losses (<30%) and solar zenith angles <80°, the flux calculations are not very sensitive to the morphology of stratospheric ozone distribution. An ultraviolet ground albedo of 0.04, suitable for fields, pasture and grasslands, was adopted [6]. Integrated solar fluxes are taken from the World Meteorological Organization's 1985 recommended values [5]. Cloud cover and continental haze are not included. Clear sky conditions are sufficient to qualitatively examine the relationship between increasing levels of ozone depletion and the biological consequences thereof; while cloud cover may affect the absolute value of the ground-level flux, it would have little impact on the ratio of the fluxes for different column ozone abundances [14].

## BIOLOGICAL ASSUMPTIONS

The biological assumptions adopted in this study are as follows:

First, it is assumed that UV-B light can damage the DNA molecule, via pyrimidine dimer formation or other covalent modification of the DNA helix [10];

Second, photochemical modification of DNA can result in deleterious biological consequences, such as non-melanoma skin cancers [21] and damage to crops [27] and aquatic life [28].

Setlow first reported that the absorption spectrum of DNA parallels the action spectrum for skin erythema in the UV-B region [24]. Peak et al. have more recently reported that the action spectra for lethality and mutagenesis of *E. coli* exposed to UV light also correspond to the DNA absorption spectrum [17], leading to the conclusion that these effects proceed via the photochemical alteration of the DNA molecule. Third, the biological effects of stratospheric ozone depletion scale linearly with changes in exposure to ground-level biologically effective dosages (BED's) of UV-B light. Several studies have shown that the incidence rates for both non-melanoma and melanoma skin cancers correlate inversely with latitude. Both Rundel [18] and Scotto et al. [22] took this a step further and compared data on the incidence of basal and squamous cell skin cancers in the United States to establish the linearity of the UV-B dose-response relationship [18].

This study also arrives at the same conclusion by comparing modelled BED's of UV-B light at 30°–50° north latitude with epidemiological data from the National Cancer Institute's study of non-melanoma skin cancer incidence rates from June 1977 to May 1978 (see Figure 9). In assessing the health effects of stratospheric ozone loss, numerical models are useful in predicting changes in ambient UV insolation; however, other factors such as individual susceptibility, genetic predisposition, and personal behavior defy simple mathematical analysis.

## BIOLOGICAL DOSAGES (BED and BAD)

The biologically effective dosage (BED), which corresponds to the "DNA effective irradiance" in the calculations of Frederick and coworkers [14], is a function of both solar zenith angle ($\chi$, 0° is overhead, 90° is the horizon) and column ozone abundance, and is defined as:

$$\mathbf{BED}(\chi, O_3) = \underset{\lambda}{\mathbf{S}}\ \mathbf{F}(\lambda) \propto \mathbf{A}(\chi, \lambda, O_3) \propto \mathbf{R}(\lambda)$$

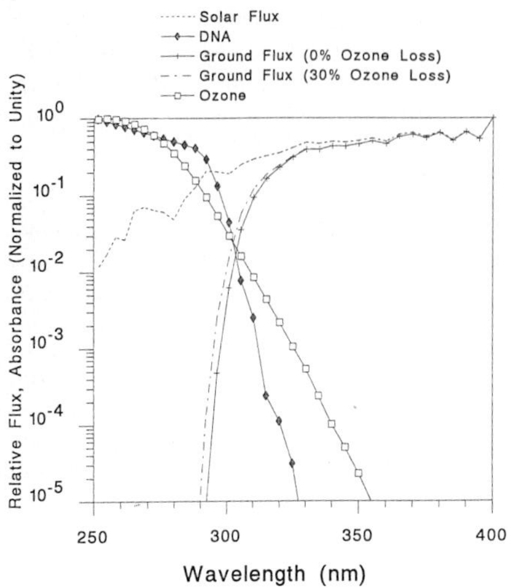


Figure 1. Relative Solar Flux, Ground-Level Flux, and Absorption Spectra of DNA and Ozone.

where F is the wavelength-dependent solar flux per unit time per wavelength bin (photons $cm^{-2}$ $s^{-1}$ $bin^{-1}$) at the top of the atmosphere, A is the amplification factor for the integrated solar flux across a unit surface area ($cm^2$) at sea level, R is the wavelength-dependent risk factor or action spectrum, and the sum is over UV wavelengths (250-400 nm). A related quantity, the biologically accumulated dosage (BAD), is a measure of the extent to which bio-organisms and ecosystems will respond to varying levels of cumulative ground-level UV-B flux over some period of time. The BAD is defined as:

$$\mathbf{BAD}(O_3) = \sum_{\chi}\sum_{\lambda} \mathbf{F}(\lambda) \infty \mathbf{A}(\chi,\lambda,O_3) \infty \mathbf{R}(\lambda)$$

$$= \sum_{\chi} \mathbf{BED}(\chi, O_3)$$

which is effectively the BED summed over some period of time (hours, days, months, seasons or years). The BAD is a function only of overhead ozone abundances. The summation is accomplished by adding up the BED each few minutes as the sun rises and sets at a given location, using a solar ephemeris model to calculate the solar zenith angles at each time interval. While many studies have shown comparisons of the BED or its equivalent at fixed times (usually local noon) to analyze the relative risk of various biological responses, the integrated BAD is a more appropriate model for phenomena which depend on the total accumulated UV-B flux. Such biological phenomena include basal and squamous cell carcinomas as well as various types of cataracts (cortical [26], posterior subcapsular [7], and severe [11] or brunescent [31] nuclear cataracts).

## RESULTS

Figure 2 indicates the relative contribution of the diffuse (multiply scattered) portion of the UV radiation field for the US Standard Atmosphere (solid line) and an atmosphere with a 30% ozone loss (dashed line). The ratio of multiple scattering to direct attenuated solar flux is not a strong function of ozone loss.

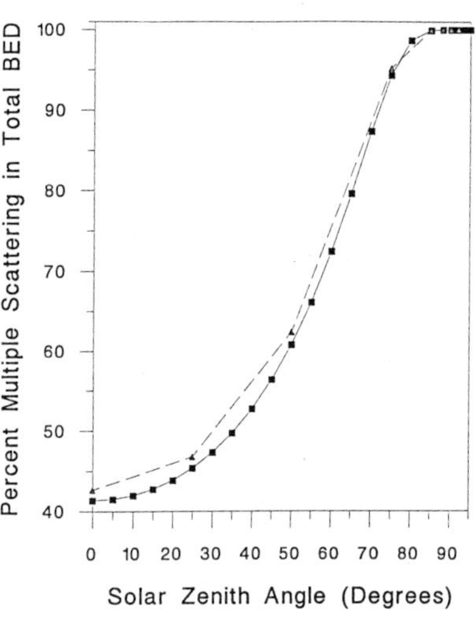


Figure 2. Multiple Scattering Component of Biologically Effective Dosage (BED).

Over 41% of the BED is multiply scattered for $\chi$=0° (overhead sun), and more than 50% for $\chi > 35°$. The UV-B region responsible for most biological responses is largely multiply scattered. If one takes a telescopic spectrometer set at 300 nm and points it towards the sun, one finds that the direct solar flux is not significantly greater than if one were to point the spectrometer at any other point in the sky; therefore UV-B photons are throroughly randomized and isotropically distributed by the time they reach the earth's surface. In practical terms, even on a cloudy day with little or no direct solar flux, the diffuse or multiply scattered component of the UV radiation field may still be significant; therefore, one can sometimes get a severe sunburn even on a cloudy day.

Figures 3 and 4 show the solar zenith angle dependence of the BED for the present atmosphere and for a 30% ozone loss. When the sun is reasonably high in the sky ($\chi$<60°), the BED scales roughly linearly with solar zenith angle, and then falls off rapidly as the sun sets. From Fig. 4 one can extract equivalent dosages; for example, a 30% ozone depletion at $\chi$=50° will give the same BED per unit time as the present atmosphere at $\chi$=38°, or the present atmosphere at $\chi$=0° is equivalent to an 30% ozone loss at $\chi$=34°. Because the BED drops off logarithmically with increasing solar zenith angles, a small percentage ozone loss at a low solar zenith angle can result in a much more significant absolute increase in the BED than a large ozone loss at a higher zenith angle.

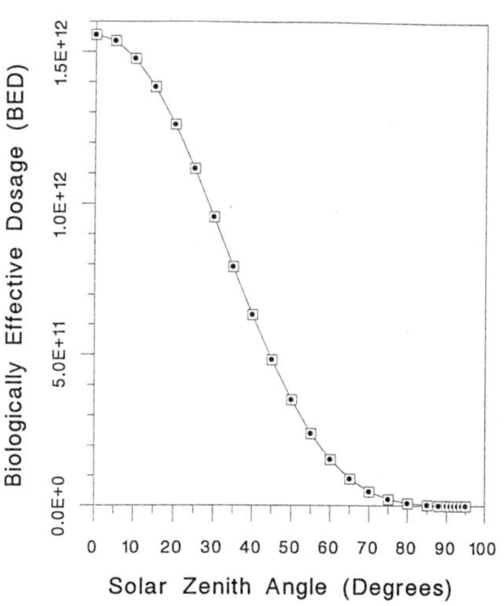


Figure 3. Biologically Effective Dosage (BED) of UV-B as a function of Solar Zenith Angle.

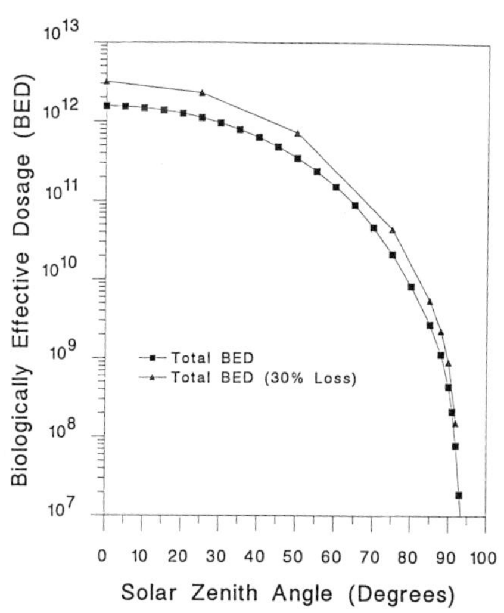


Figure 4. Biologically Effective Dosage (BED) of UV-B as a function of Solar Zeneth Angle.

There is much confusion in the technical as well as popular literature concerning the numerical relationship between a given ozone loss and the resulting consequences. Figures 5-7 show the modelled relationship between ozone losses and the resulting increases in total ground-level UV flux (all wavelengths below 400 nm, Fig. 5), UV-B flux (280-320 nm, Fig. 6), and the biologically effective dosage (BED, Fig. 7).

For small ozone losses (<5%), each 1% decrease in ozone abundance results in a 0.04% increase in the total ground-level UV flux, a 0.53% increase in the ground-level UV-B flux, and a 2.1% increase in the BED. Note that while ozone loss has a significant

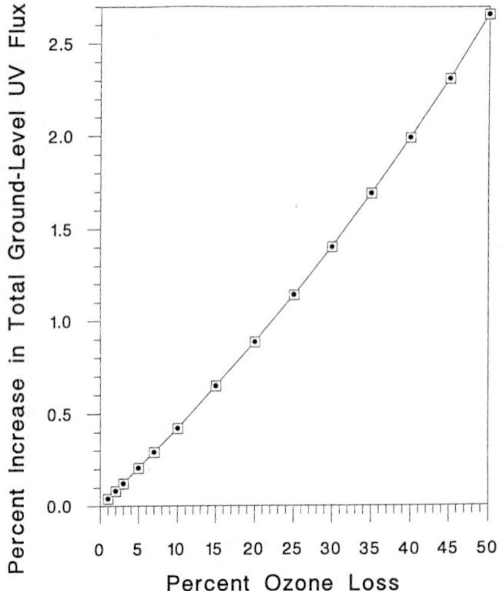


Figure 5. Increase in Total Ground-Level UV-B Flux (<400 nm) as a function of Ozone Loss.

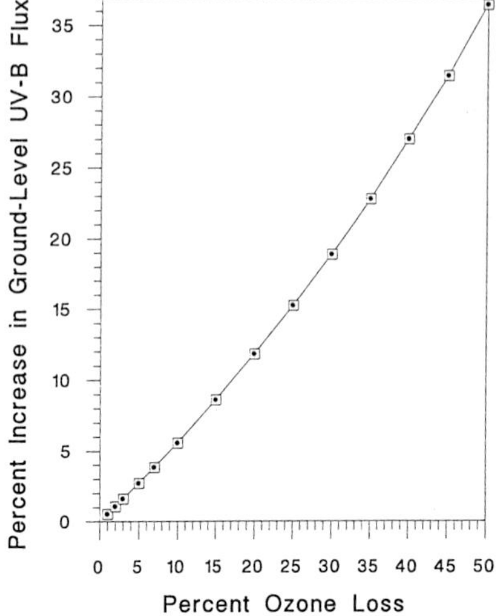


Figure 6. Increase in Ground-Level UV-B Flux (280-320 nm) as a function of Ozone Loss.

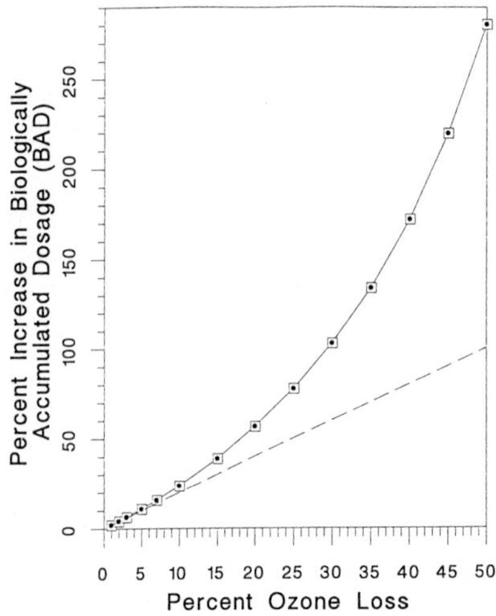


Figure 7. Increase in Biologically Accumulated Dosage (BAD) of UV-B as a function of Ozone Loss.

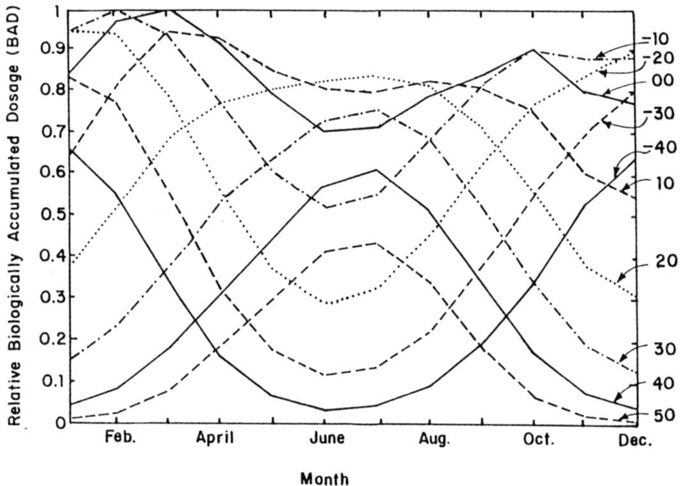


Figure 8. Relative Biologically Accumulated Dosage (BAD) of UV-B per Month at 10 Latitudes.

effect on the UV-B flux and the resulting BED, it has comparatively little impact on the total UV flux, since the majority of the UV flux at ground level is in the UV-A region (320-400 nm), which is not significantly enhanced by ozone depletion.

Setlow (1974) was the first to calculate the relationship between ozone loss and biological responses; he concluded that a 1% decrease in atmospheric ozone results in a 2% increase in the BED [24]. However, since that time both the popular press and considerable academic literature have (naively) generalized this result, concluding that the percentage increase in the BED for any amount of ozone depletion is simply twice the percentage depletion. While this 2:1 "rule of thumb" ratio (represented by the dashed line in Fig. 7), is reasonably accurate for relatively small ozone losses, it is seriously in error for larger ozone losses; this was also the conclusion of the 1975 Federal Task Force on the Inadvertent Modification of the Stratosphere (IMOS). For ozone losses of up to 5%, the increase in the BED with decreasing ozone is roughly linear. At larger ozone losses the overlap integral between the action spectrum and the enhanced solar flux begins to increase exponentially; thus a 10% ozone reduction results in a 24% increase in the consequences, a 30% loss will result in a doubling, and a 50% loss yields a quadrupling. In the light of significant polar ozone losses observed over the past decade (the so-called "Antarctic ozone hole"), this non-linearity of the biological consequences of ozone depletion should not be overlooked.

The seasonality of the BAD is quite pronounced. Figure 8 indicates the relative BAD per month for ten latitudes, normalized to unity at the equator in March. The summer/winter ratio is small near the equator, but can be as large as a factor of 50 at 50°N latitude. This figure also allows one to compare monthly exposure risks; for example, during the summer months one receives an equivalent dosage of biologically effective UV-B at c=30°N (the latitude of New Orleans) as one would at the equator, and July at c=50°N (just south of London) gives an equivalent exposure to c=30°N in April.

The BAD for a given latitude, integrated over a year, represents the annual effective dosage of UV-B; this corresponds to the relative risk of non-melanoma skin cancer incidence (see Fig. 9). Also plotted in Fig. 9 are the annual age-adjusted incidence rates for non-

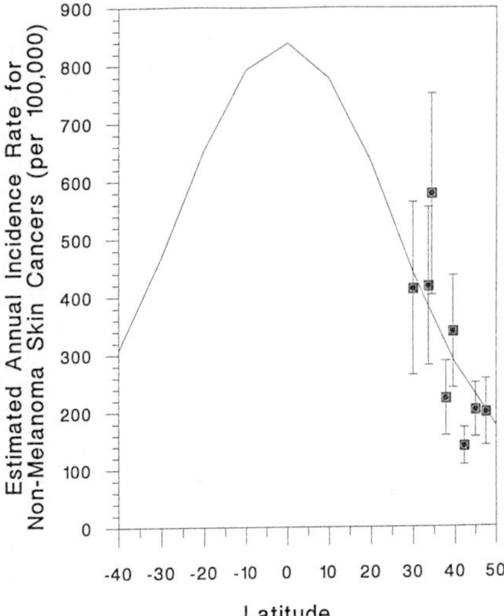


Figure 9. Latitude Dependence of the Estimated Annual Incidence Rate for Non-Melanoma Skin Cancers.

melanoma skin cancer (both basal and squamous cell) for the white population of North America (30°-50°N latitude) during the National Cancer Institute's 1977-78 survey of nonrecurring skin cancers. Incidence rates are per 100,000 population. The incidence for males is plotted as the upper error bar for each location, the female incidence rate as the lower limit. Data from the 1971-72 Third National Cancer Survey, although slightly lower, also fit the modelled BAD well. For temperate latitudes, a 10% increase in non-melanoma skin cancer incidence corresponds to a 1.8° equator-ward shift in latitude, or moving approximately 250 miles towards the equator.

One subtle conclusion to be drawn from this modeling is that a small ozone loss at the equator provides as much additional biologically-effective UV-B as a considerably larger ozone loss at higher latitudes (Figure 10). For example, a 10% ozone loss at the equator will produce the same absolute annual increase in the BAD as a 30% loss at c=50°N latitude. NASA has recently reported ozone losses of 2-3% in the tropics during 1991-92, possibly as a consequence of the eruption of Mt. Pinatubo [12]. A 2% ozone loss in the Tropics is commensurate with a 6% ozone loss at c=50°N latitude; therefore, we can say that most of the Northern Hemisphere and Tropics experienced a roughly equivalent absolute increase in the biologically accumulated dosage of UV-B during this period.

While the chlorine burden of the atmosphere is expected to begin to decrease after the turn of the millennium (assuming uniform compliance with international accords on the manufacture of fully halogenated hydrocarbons such as CFCs), the long atmospheric lifetime of the CFCs means that the gradual recovery of the earth's ozone layer will span several generations. Since many of the health consequences of enhanced UV-B are related to cumulative exposure over several decades, this means that the medical community must prepare to deal with these issues well into the next century.

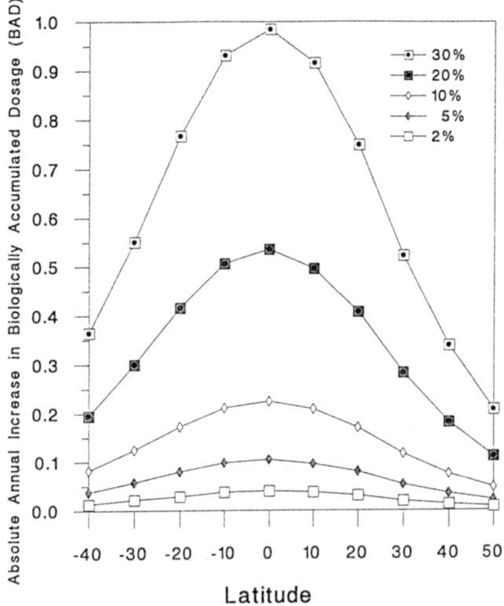


Figure 10. Latitude Dependence of the Absolute Annual Increase in the Biologically Accumulated Dosage (BAD) of UV-B.

**Acknowledgments**
This work was funded by NASA grant NGT-30031.

## REFERENCES

1. Anderson, D.E. 1993. Planet. Space Sci. 31:1517-1523.

2. Anderson, D.E., and S.A. Lloyd. 1990. J. Geophys. Res. 95:7429-34.

3. Anderson, J.G., et al. 1989. J. Geophys. Res. 94:11480-520.

4. Anderson, J.G., D.W. Toohey, and W.H. Brune. 1991. Science 251:39-52.

5. Atmospheric Ozone 1985. 1986. World Meteorological Organization Global Ozone and Monitoring Network Report No. 16, Washington, D.C.

6. Blumthaler, M., and W. Ambach. 1988. Photochem. Photobiol. 48:85-88.

7. Bochow, T.W., et al. 1989. Arch. Ophth. 107:369-372.

8. Dave, J.E., and P. Halpern. 1976. Atmos. Envir. 10:547-555.

9. Farman, J.C., B.G. Gardiner, and J.D. Shanklin. Nature 315:207-210.

10. Francis, A.A., W.L. Carrier, and J.D. Regan. 1988. Photochem. Photobiol. 48:67-71.

11. Hiller, R., R.D. Sperduto, and F. Ederer. 1986. Am. J. Epid. 124:916-25.

12. Gleason, J.F., et al. 1993. Science 260:523-526.

13. Keating, G.M., and D. F. Young. 1985. *In* K. Labitzke, J.J. Barnett, and B. Edwards (eds.), Handbook for the Middle Atmosphere Program, Vol. 16 (ICSU Sci. Comm. on Solar-Terrestrial Physics), pp.205-229.

14. Lubin, D., J.E. Frederick, and A.J. Krueger. 1989. J. Geophys. Res. 94:8491-8496.

15. Meier, R.R., D.E. Anderson, and M. Nicolet. 1982. Planet. Space Sci. 30:923-933.

16. Molina, L.T., and M.J. Molina. 1986. J. Geophys. Res. 91:14501-14508.

17. Peak, M.J., et. al. 1984. Photochem. Photobiol. 40:613-620.

18. Rundel, R.D. 1983. Photochem. Photobiol. 38:569-575.

19. Rundel, R.D., and D.S. Nachtwey. 1983. Photochem. Photobiol. 38:577-591.

20. Saito I., H. Sugiyama, and T. Matsura. 1983. Photochem. Photobiol. 38:735.

21. Scotto, J. 1986. *In* J.G. Titus (ed.), Effects of Changes in Stratospheric Ozone and Global Climate, Vol. 2, Washington, D.C.: US EPA, pp.33-61.

22. Scotto, J., T.R. Fears, and J.F. Fraumeni. 1982. *In* D. Schottenfeld and J.F. Fraumeni (eds.), Cancer Epidemiology and Prevention, Philadelphia: Saunders, pp.254-276.

23. Scotto, J., and J.F. Fraumeni. 1982. *In* D. Schottenfeld and J.F. Fraumeni (eds.), Cancer Epidemiology and Prevention, Philadelphia: Saunders, pp.996-1011.

24. Setlow, R.B. 1974. Proc. Nat. Acad. Sci. USA 71:3363-3366.

25. Stolarski, R.S., et al. 1986. Nature 322:808-811.

26. Taylor, H.R., et al. 1988. N.E.J. Med. 319:1429-1433.

27. Teramura, A.J. 1986. *In* J.G. Titus (ed.), Effects of Changes in Stratospheric Ozone and Global Climate, Vol. 2, Washington, D.C.: US Environmental Protection Agency, pp.255-262.

28. Thomson, B.E. 1986. *In* J.G. Titus (ed.), Effects of Changes in Stratospheric Ozone and Global Climate, Vol. 2, Washington, D.C.: US Environmental Protection Agency, pp.203-209.

29. Titus, J.G., (ed.). 1986. Effects of Changes in Stratospheric Ozone and Global Climate, Vols. 1 & 2, Washington, DC: US Environmental Protection Agency.

30. WMO. 1990. World Meteorological Organization Report of the International Ozone Trends Panel, 1988, World Meteorological Organization Global Ozone and Monitoring Network Report No. 18, Washington, DC: WMO.

31. Zigman, S., M. Datiles, and E. Torczynski. 1979. Invest. Ophth. 18:462-467.

# MODELING OF EFFECTS OF UV-B RADIATION ON ANIMALS

Edward C. DeFabo

Department of Dermatology
The George Washington University Medical Center
Washington, DC 20037

**ABSTRACT**

Animal models employed in the study of photoimmunology have centered mainly on the mouse although some recent studies have used the American opossum (*Monodelphis domestica*), a non-placental mammal. Evidence now exists showing that humans also exhibit UV-induced immune suppression. Indeed, the photoreceptor responsible for initiating UV immune suppression, urocanic acid (UCA), is found in the stratum corneum of both mouse and man, it undergoes similar trans-to-cis photoisomerization in both species when exposed to sunlight, and it appears to show similar dose-responsive kinetics in both species. Consequentally, the mouse seems to be an excellent model for studying UV immune suppression.

Several mouse strains are now available that have provided us with important information about how UV immune suppression works. First, there are important genetic differences in susceptibility to immune suppression between strains. For example, the C57BL/6 strain is much more susceptible to UV immune suppression than the Balb/c, requiring approximately one-sixth the dose of UV-B for 50 percent suppression. A dominant autosomal gene plays a role in determining high susceptibility. Second, a mutant mouse strain, deficient in the skin photoreceptor UCA, shows significantly impaired UV immune suppression, strongly indicating the role of UCA as the initial light absorber. Third, feeding Balb/c mice 10 percent dietary histidine increased the trans isomer of UCA in skin and significantly enhanced sensitivity to UV immune suppression. This again indicates a central role of UCA in regulating immune suppression. Finally, backcrossing mice that are high in susceptibility to UV suppression with mice that show low susceptibility results in a new mouse strain that is congenic for high susceptibility and can be used for investigating the mechanism of UV-B photocarcinogenesis. If UV immune suppression plays a critical role in skin cancer formation, as is indicated, this strain should play an important role in helping to understand this mechanism. These animal models will be described and their role in understanding the implication of increased UV-B on human health as a result of stratospheric ozone depletion will be discussed.

NATO ASI Series, Vol. I 18
Stratospheric Ozone Depletion/
UV-B Radiation in the Biosphere
Edited by R. H. Biggs and M. E. B. Joyner

# OZONE DEPLETION AND SKIN CANCER INCIDENCE: AN INTEGRATED MODELING APPROACH AND SCENARIO STUDY

Harry Slaper, Michel G.J. den Elzen, and Hans J. v.d. Woerd

National Institute of Public Health and Environmental Protection (RIVM)
Bilthoven, The Netherlands

**ABSTRACT**

A decrease in stratospheric ozone, probably caused by chlorofluorocarbon (CFC) emissions, has been observed over large parts of the globe. Further depletion, leading to an increase in biologically-effective solar UV at ground level, is expected. An integrated source-risk model has been developed and used to evaluate the increased skin cancer incidence and mortality related to various CFC emission scenarios. The model provides estimates tropospheric concentrations of CFCs, chlorine levels in the stratosphere, ozone depletion, UV irradiance, and effects on the incidence of skin cancer incidence and mortality rates.

The results obtained show that even if full worldwide compliance with the Copenhagen Amendments to the Montreal Protocol is realized, full recovery of the stratospheric ozone balance is not expected prior to 2080. At 40-50° N, full compliance with the Copenhagen Amendments is expected to lead to a maximal depletion of approximately 10 percent by the year 2000 followed by a slow recovery thereafter. In this case, skin cancer risks associated with ozone depletion, expressed in terms of incidence rates and excess mortality, are expected to increase to a maximum around the years 2040-2050. Excess skin cancer risks in the Netherlands are expected to result in an excess mortality rate of three million per year and an excess incidence of 200 per million per year. The calculated mortality risk level is above the Maximum Tolerable Risk Limit adopted in the risk-oriented policy approach for chemical substances and ionizing radiation sources in the Netherlands. Implications of risk modeling for environmental monitoring strategies and sustainable development are discussed.

NATO ASI Series, Vol. I 18
Stratospheric Ozone Depletion/
UV-B Radiation in the Biosphere
Edited by R. H. Biggs and M. E. B. Joyner

# APPENDIX

# UV-B RADIATION AND THE PHOTOSYNTHETIC PROCESS

Janet F. Bornman, Cecilia Sundby-Emanuelsson,*
Yan-Ping Cen, and Camilla Ålenius

Departments of Plant Physiology and *Biochemistry
Lund University, S-220 07 Lund, Sweden

Key words: biochemical changes, *Brassica napus*, leaf morphology, ozone, PAR, physiological changes, Rubisco, UV-B radiation

Continuing decreases in stratospheric ozone have been documented and it is postulated that this may result in a greater proportion of shortwave ultraviolet radiation (UV-B, 290-320 nm) reaching the earth's surface. The on-going enlargement of the hole in the ozone in the Antarctic is a particular cause for concern, since the geographical region over which this ozone-poor air extends continues to increase. In addition, a lesser ozone hole has also been reported over the Arctic region during springtime (Pyle 1991). The amount of UV-B radiation received at the earth's surface will depend on a number of ground factors, not least of which is the level of tropospheric ozone and other pollutants.

UV-B radiation causes changes in biochemistry, plant morphology, and ultrastructure. These changes can in turn result in indirect, subtle effects on cell processes and growth. One of the more subtle consequences is the alteration of photosynthetically active radiation (PAR), that part of the radiation most active in photosynthesis, within plant organs. If there is a change in the internal radiation environment, one might then also expect changes in the photosynthetic process itself, as well as changed light microenvironments for photoreceptors of other plant processes.

With enhanced UV-B, some of the induced changes which can modify subsequent plant response include anatomical alterations, quantitative and qualitative changes in epicuticular wax, and pigment changes, among others. Potential biochemical changes include the xanthophyll cycle. This cycle is a protective mechanism for the dissipation of excess amounts of visible radiation, otherwise known as photo-inhibition. The cycle occurs within the chloroplast membrane, the thylakoids, and involves the conversion of three xanthophylls (oxygenated carotenoids). So far, evidence has shown that this cycle is itself a target of UV radiation (Pfündel 1992), which means that photoinhibitory conditions together with enhanced levels of UV radiation may result in an enhanced and possibly synergistic stress effect.

With regard to anatomical changes, one of many is the increase in leaf thickness; although this is not a universal response, when expressed, it is dependent on the amount of UV-B and visible radiation. The way in which the different tissues of the leaf are affected seems also to depend on visible radiation and species. Figure 1 shows some of the results obtained for *Brassica napus* after plants had been grown for 16 days with the addition of approximately 9 kJ $m^{-2}$ $day^{-1}$ of biologically-effective UV-B radiation, weighted according to Caldwell's generalized plant action spectrum (Caldwell 1971). The distribution of UV-screening pigments (Figure 2) also reflects very well the attenuation gradient of UV radiation within a leaf. One can see a greater penetration of the longer UV-B radiation wavelengths just below the upper epidermis with increasing attenuation of, in particular, the shorter wavelengths with increasing depth (Cen and Bornman 1993).

NATO ASI Series, Vol. I 18
Stratospheric Ozone Depletion/
UV-B Radiation in the Biosphere
Edited by R. H. Biggs and M. E. B. Joyner

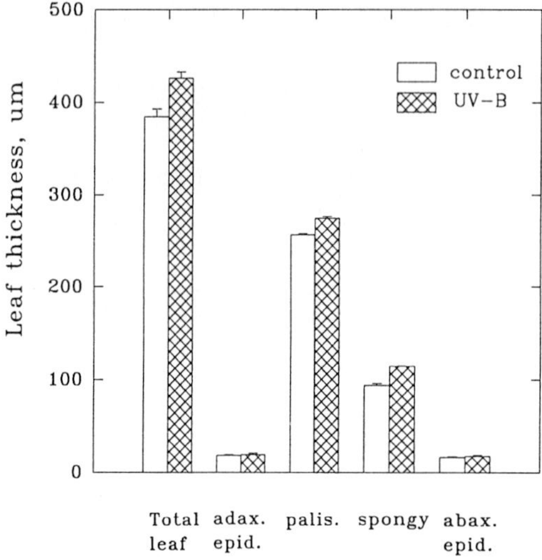


Fig. 1. Changes in leaf structure in *Brassica napus* after plants were grown for 16 days with the addition of approximately 9 kJ $m^{-2}$ $day^{-1}$ biologically effective UV-B radiation.

Quartz fiber optics have been used as microprobes to directly determine the distribution and relative amount of radiation within leaves (Vogelmann and Björn 1984; Bornman and Vogelmann 1988, 1991; Day et al. 1992; Cen and Bornman 1993). Measurements of visible radiation in the 400-700 nm (PAR) range have shown that the backscattered component of the PAR is especially increased in leaves of plants grown under an enhanced level of UV-B radiation (Bornman and Vogelmann 1991). This is probably due mainly to changes in internal structure and could serve as a way in which the PAR is able to be distributed to greater depths within a leaf of increased thickness. The penetration of UV-A radiation has been demonstrated in needles of two conifers, fir and spruce. Internal anatomy rather than flavonoid content appears to have played a role in the differential penetration between the two species, with more radiation being found in spruce needles (Bornman and Vogelmann 1988). A cross-section of spruce and fir needles (Figure 3) shows the radial arrangement of sheet-like palisade cells of spruce. These presumably aid in guiding the light, in contrast to the more sponqy, randomly-arranged cells in fir.

Many target sites of UV radiation have been reported, and in the case of the photosynthetic system, the list is long, so that only a few will be mentioned. The photosynthetic system consists of photosystems I and II, and although these are both composed of pigment-protein complexes, it seems that PS II is much more responsive to UV-B radiation than PS I. Much research has focused on the reaction center (RC) which, among other components, contains the two polypeptides, D1 and D2. In 1981 it was discovered that D1 turns over very rapidly in higher plants and unicellular algae in the presence of light; it is also the protein involved in the process of photoinhibition and recovery. Broad target areas of PS II include both the oxidizing and reducing sides. In addition, UV-B radiation has been seen to affect energy transfer and the lifetime of electron migration within the systems (Bornman 1989; Renger et al. 1991). One may also see increased quenching or dissipation of energy by non-photochemical means with concomitant decreases in photochemical quenching as a result of UV-B radiation.

The two phenomena of photoinhibition by PAR and changes induced by UV-B radiation can be inves-

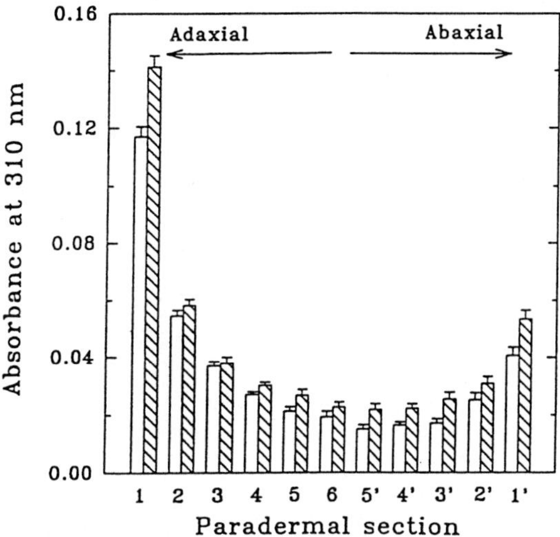


Fig. 2. Distribution of UV-screening pigments throughout the leaf of Brassica napus, analyzed from 40 m thick paradermal sections. (Cen and Bornman 1993; reproduced with permission).

tigated in a number of ways. First, one needs to ask whether they are separate effects. In many cases, the same processes may be affected, but with certain differences. For example, after photoinhibition the kinetics of recovery by visible light of several photosynthetic parameters is much faster than recovery from UV-B radiation (Chow et al. 1992). One of the reasons put forward for this delay in recovery is that

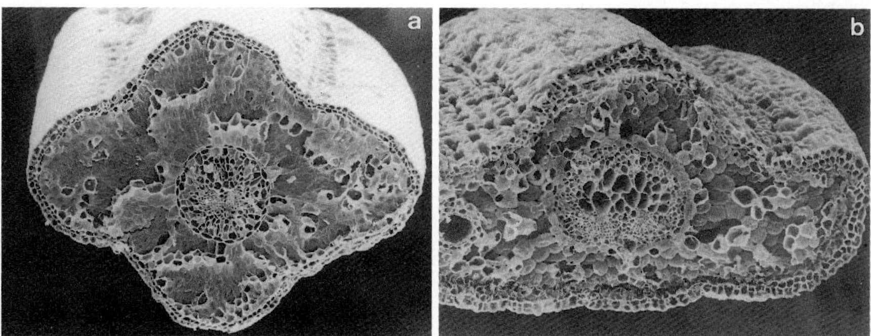


Fig. 3. Transverse sections of naturally grown spruce (a) and fir needles showing the radial arrangement of sheet-like palisade cells of spruce and the spongy, randomly arranged cells of fir (b). (Bornman and Vogelmann 1988; reproduced with permission).

membrane integrity is affected by UV-B radiation, thus decreasing the ATP supply, which in turn would hamper repair processes (Chow et al. 1992).

In a recent study (Sundby-Emanuelsson et al. 1993, unpubl.), parameters which were used to investigate the effect of UV-B radiation under high light conditions included measurement of quantum yield, dissipation or quenching of energy, and the extent of D1 turnover. Cloudy conditions were simulated by keeping plant material under either relatively low PAR (400 mol $m^{-2}$ $s^{-1}$) with low UV-B levels (3.3 kJ $m^{-2}$ $day^{-1}$ $UV\text{-}B_{BE}$) or high PAR (1 600 mol $m^{-2}$ $s^{-1}$) with enhanced levels of UV-B radiation (13 kJ $m^{-2}$ $day^{-1}$ $UV\text{-}B_{BE}$). Quantum yield decreased with time of exposure under high PAR and enhanced levels of UV-B radiation, while it remained unaffected under low radiation levels. Other work by Jordan et al. (1991) has shown that for the D1 protein there were reduced RNA transcript levels, indicating rapid changes of gene expression in response to UV-B radiation. Decreased RNA transcript levels were also found for an important enzyme, ribulose bisphosphate carboxylase (Rubisco), involved in the fixing of $C0_2$ inside the plant, which also showed reduced activity with UV-B radiation (Jordan et al. 1992). Thus although the measured effects on net photosynthesis, or the assimilation of $C0_2$, are often much smaller than the larger changes measured for the electron transport chain and in particular on PS II, there may be more subtle changes in the carbon cycle, which may have a more gradual impact on plant productivity. For example, Takeuchi et al. (1989) found that for carbon metabolism, UV-B radiation decreased the amounts of organic acids and soluble sugars.

With regard to research priorities in the future, while the field covering the effects of UV-B radiation is, of necessity, of an *applied* nature, basic research is very important if we are to answer the questions of *why* and *how*.

## References

Bornman, J.F. 1989. Target sities of UV-B radiation in photosynthesis of higher plants. J. Photochem. Photobiol. 4:145-158.

Bornman, J.F., and T.C. Vogelmann. 1988. Penetration of blue and UV radiation measured by fibre optics in spruce and fir needles. Phys. Plant. 72:699-705.

Bornman, J.F., and T.C. Vogelmann. 1991. The effect of UV-B radiation on leaf opticial propertiies measured with fibre optics. J. Exp. Bot. 42: 547-554.

Caldwell, M.M. 1971. Solar UV irradiation and the growth and development of higher plants. In A.C. Giese (ed.), Photophysiology, 6:131-177. Academic Press, New York.

Cen, Y.-P., and J.F. Bornman. 1993. The effect of exposure to enhanced W-B radiataion on the penetration of monochromatic and polychromatic UV-B radiation in leaves of Brassica napus. Physiol. Plant. 87:249-255.

Chow, W.S., Å. Strid, and J.M. Anderson. 1992. Short-term treatment of pea plants with supplementary ultraviolet-B radiation: Recovery time-courses of some photosynthetic functions and components. In N. Murata (ed.), Research in Photosynthesis, IV:361-364. Kluwer, The Netherlands.

Day, T.A., T.C. Vogelmann, and E.H. DeLucia. 1992. Are some plant life forms more effective than others in screening out ultraviolet-B radiation? Oecologia 92:513-519.

Jordan, B.R., W.S. Chow, Å. Strid, and J.M. Anderson. 1991. Reduction in cab and psb A RNA transcripts in response to supplementary ultraviolet-B radiation. FEBS Lett. 284:5-8.

Jordan, B.R., J. He, W.S. Chow, and J.M. Anderson. 1992. Changes in mRNA levels and polypeptide subunits of ribulose 1,5-bisphosphate carboxylase in response to supplementary ultraviolet-B radiation. Plant Cell Environ. 15:91-98.

Pfündel, E.E., R.S. Pan, and R.A. Dilley. 1992. Inhibition of violaxanthin deepoxidation by ultraviolet-B radiation in isolated chloroplasts and intact leaves. Plant Physiol. 98:1372-1380.

Pyle, J. 1991. Closing in on Arctic ozone. New Scientist 9:49-52.

Renger, G. and W. Rettig. 1991. The effect of UV-B irradiation on the lifetimes of singlet excitons in isolated photosystem II membrane fragments from spinach. J. Photochem. Photobiol. 9:201-210.

Takeuchi, Y., M. Akizuki, H. Shimizu, N. Kondo, and K. Sugahara. 1989. Effect of UV-B (290-320 nm) irradiation on growth and metabolism of cucumber cotyledons. Physiol. Plant. 76:425-430.

Vogelmann, T.C., L.O. Björn. 1984. Measurement of light gradients and spectral regime in plant tissue with a fiber optic probe. Phys. Plant. 60:361-368.

# THE EFFECTS OF UV-B RADIATION ON THE MAMMALIAN IMMUNE SYSTEM

Edward C. DeFabo and Frances P. Noonan

Department of Dermatology
The George Washington University Medical Center
Washington, DC 20037

Key words: antigen-specific T suppressor cells, immunology, infectious disease, mammal, photoimmunology, UV-B radiation

## ABSTRACT

Photoimmunology is the study of UV light and its effects on the mammalian immune system. UV light in this context is defined as radiation with wavelengths between 250 and 320 nm. These wavelengths describe part of the UV-C (200-280 nm) and all of the UV-A (280-320) waveband ranges. The effects of this radiation on immunity include suppression of the contact hypersensitivity response (CHS), an immune response similar to a poison ivy reaction, and suppression of the rejection of transplanted UV-induced skin tumors, a reaction directly involved in the outgrowth of skin cancer. Some evidence also exists that suggests that certain infectious diseases may be affected by UV immune suppression.

To understand the basic mechanism involved, a variety of experiments have been employed including the identification of the active waveband emitted from experimental sunlamps, detailed dose-response analyses, the identification of the initial light absorber, or the photoreceptor to determine the process responsible for initiating the steps leading to systemic immune suppression and the involvement of antigen-specific T suppressor cells. These experiments will be summarized and a working hypothesis presented detailing what we postulate to be some of the major steps leading to immune suppression. The practical as well as the evolutionary advantages of such an unusual sunlight-activated immune-regulating mechanism on mammalian skin will be discussed.

NATO ASI Series, Vol. I 18
Stratospheric Ozone Depletion/
UV-B Radiation in the Biosphere
Edited by R. H. Biggs and M. E. B. Joyner

# UV-B AND THE IRRIGATED RICE ECOSYSTEM: EFFECTS AND MODELING RESEARCH

David M. Olszyk

US EPA
Environmental Research Lab, Corvallis, OR

**ABSTRACT**

This project provides data to characterize the risks to the Asian irrigated rice ecosystem from UV-B radiation and global climate change. In terms of UV-B effects, areas investigated so far are:

1) characterization of UV-B climate for Asian rice-producing areas under current conditions with a doubling of the atmospheric $CO_2$ concentration;
2) evaluation of the relative UV-B response for a wide range of cultivars representing major rice types and production areas;
3) determination of the response of rice to enhanced UV-B under field conditions;
4) documentation of the impact of UV-B on the rice-blast disease produced by *Pyricularia oryzae*.

Comments are made on the potential for modeling effects of UV-B on rice yield based on the responses of different components of the rice system.

NATO ASI Series, Vol. I 18
Stratospheric Ozone Depletion/
UV-B Radiation in the Biosphere
Edited by R. H. Biggs and M. E. B. Joyner

## INFORMAL SURVEY
## TUESDAY MORNING COFFEE BREAK
## ASKING:

## WHAT IS THE QUESTION AND/OR ISSUE THAT WON'T GO AWAY AND YOU'RE SICK TO DEATH OF REHASHING

REPLIES:

Too much talk interferes with the urgency; the problem is happening NOW and we're still eternally talking.

Quit talking about resolution; ONE nm is fine.

There is too much of a continual drive to bring together issues of research assessment and regulatory needs. What we need is CURIOSITY-DRIVEN science.

Before you do ANYTHING, ask yourself: What is the *question* you're trying to answer. Are you characterizing each species just for the hell of it???

Get a realistic perspective regarding UV-B and plants. Changes in percent yield is most likely the most important point. The effect will be different for each species, maybe even each variety--this kind of research is NOT realistic.

Instrumentation: Talk about SCIENCE. Not yet evident that the ozone hole and the whole related ball of wax is *truly* a trend; it may be a normal fluctuation of a natural phenomenon.

Everything is talked about in terms of what *should* be. Just get the measurements DONE. Get some good intercomparisons done. These should be incorporated and integrated, not just side-by-side-by-side. Techniques need to be addressed and, one hopes, agreed upon.

Build on a data base that can be made readily available. Get it defined properly and resolved in a *quanti*tative way instead of yelling at each other *quali*tatively.

Are the parameters/constraints such that it is just not *possible* to do a risk assessment?

Do they just *want* one or do they truly understand *why* (or IF) we need a network . . . and WHO will use it?

NATO ASI Series, Vol. I 18
Stratospheric Ozone Depletion/
UV-B Radiation in the Biosphere
Edited by R. H. Biggs and M. E. B. Joyner

Many people may not be aware of *what* networks and *which* networks *are* available and *how* these networks can be used.

Broadband vs. Spectroradiometer and similar issues:

Resolution of the selection of instruments is absolutely necessary. Is it a question of *an* instrument or of *instruments*? It's not a question of a magic meter. We probably need *two* instruments: a *super* (i.e., huge) to measure trends and one of a more normal size to measure effects.

Each side should recognize that, even within practical cost constraints, neither instrument can answer ALL application requirements.

Tired of the talk that Broadband instruments can't detect trends; we need to get *quantitative* about just what the instrument CAN do.

There is an unrealistic drive to get one instrument to do *everything*. I can't relate *my* data to the next guy's data if there isn't a method to integrate and relate data for different sources, and we wont be able to do this until a *written* concensus is reached.

We shouldn't try to define a wish-list for a multipurpose instrument that will never exist. We need to begin using the data we *have, knowing its uncertainties,* and learn that way what improvements are needed. We must characterize instruments, define calibration uncertainties, cross-calibrate to relate the different instruments to each other. Data may have imperfections, but we can't waste time dreaming up technical impracticalities to answer questions that haven't even been *defined*. IN SHORT: STOP YAKKING AND GET ON WITH USING WHAT WE *HAVE*!

And said in passing, on the subject of endless debate:

> "I can make nonsense with the best of them, but what does it accomplish?"

# INDEX

# INDEX

Printing: Druckhaus Beltz, Hemsbach
Binding: Buchbinderei Schäffer, Grünstadt

## The ASI Series Books Published as a Result of Activities of the Special Programme on Global Environmental Change

---

This book contains the proceedings of a NATO Advanced Research Workshop held within the activities of the NATO Special Programme on Global Environmental Change, which started in 1991 under the auspices of the NATO Science Committee.

The volumes published as a result of the activities of the Special Programme are:

Vol. 1: **Global Environmental Change.** Edited by R. W. Corell and P. A. Anderson. 1991.
Vol. 2: **The Last Deglaciation: Absolute and Radiocarbon Chronologies.** Edited by E. Bard and W. S. Broecker. 1992.
Vol. 3: **Start of a Glacial.** Edited by G. J. Kukla and E. Went. 1992.
Vol. 4: **Interactions of C, N, P and S Biogeochemical Cycles and Global Change.** Edited by R. Wollast, F. T. Mackenzie and L. Chou. 1993.
Vol. 5: **Energy and Water Cycles in the Climate System.** Edited by E. Raschke and D. Jacob. 1993.
Vol. 6: **Prediction of Interannual Climate Variations.** Edited by J. Shukla. 1993.
Vol. 7: **The Tropospheric Chemistry of Ozone in the Polar Regions.** Edited by H. Niki and K. H. Becker. 1993.
Vol. 8: **The Role of the Stratosphere in Global Change.** Edited by M.-L. Chanin. 1993.
Vol. 9: **High Spectral Resolution Infrared Remote Sensing for Earth's Weather and Climate Studies.** Edited by A. Chedin, M.T. Chahine and N.A. Scott. 1993.
Vol. 10: **Towards a Model of Ocean Biogeochemical Processes.** Edited by G. T. Evans and M.J.R. Fasham. 1993.
Vol. 11: **Modelling Oceanic Climate Interactions.** Edited by J. Willebrand and D.L.T. Anderson. 1993.
Vol. 12: **Ice in the Climate System.** Edited by W. Richard Peltier. 1993.
Vol. 13: **Atmospheric Methane: Sources, Sinks, and Role in Global Change.** Edited by M. A. K. Khalil. 1993.
Vol. 14: **The Role of Regional Organizations in the Context of Climate Change.** Edited by M. H. Glantz. 1993.
Vol. 15: **The Global Carbon Cycle.** Edited by M. Heimann. 1993.
Vol. 16: **Interacting Stresses on Plants in a Changing Climate.** Edited by M. B. Jackson and C. R. Black. 1993.
Vol. 17: **Carbon Cycling in the Glacial Ocean: Constraints on the Ocean's Role in Global Change.** Edited by R. Zahn, T. F. Pedersen, M. A. Kaminski and L. Labeyrie. 1994.

Vol. 18: **Stratospheric Ozone Depletion/UV-B Radiation in the Biosphere.** Edited by R. H. Biggs and M. E. B. Joyner. 1994.

Vol. 19: **Data Assimilation: Tools for Modelling the Ocean in a Global Change Perspective.** Edited by P. O. Brasseur and J. Nihoul. 1994.

Vol. 20: **Biodiversity, Temperate Ecosystems, and Global Change.** Edited by T. J. B. Boyle and C. E. B. Boyle. 1994.